Differential Equations

K. A. Stroud
Formerly Principal Lecturer
Department of Mathematics
Coventry University

Dexter J. Booth
Principal Lecturer
School of Computing and Engineering
University of Huddersfield

INDUSTRIAL PRESS, INC.
NEW YORK

Library of Congress Cataloging-in-Publication Data

Stroud, K. A.
  Differential equations / by K.A. Stroud and Dexter Booth.
    p. cm.
  Includes index.
  ISBN 0–8311–3187–X
  1. Differential equations. 2. Differential equations—Problems, exercises, etc. 3. Laplace transformation—Problems, exercises, etc. I. Booth, Dexter J. II. Title.

QA371.S86 2004
515'.35—dc22

2004056731

Published in North America under license from Palgrave Publishers Ltd, Houndmills, Basingstoke, Hants RG21 6XS, United Kingdom

Industrial Press, Inc.

200 Madison Avenue
New York, NY 10016-4078

10 9 8 7 6 5 4 3 2

Contents

Program 6   Introduction to Laplace transforms   179

Program 7   Laplace transforms 2   201

Hints on using this book

This book contains lessons called *Programs*. Each Program has been written in such a way as to make learning more effective and more interesting. It is like having a personal tutor because you proceed at your own rate of learning and any difficulties you may have are cleared before you have the chance to practise incorrect ideas or techniques.

You will find that each Program is divided into numbered sections called *frames*. When you start a Program, begin at Frame 1. Read each frame carefully and carry out any instructions or exercise that you are asked to do. In almost every frame, you are required to make a response of some kind, testing your understanding of the information in the frame, and you can immediately compare your answer with the correct answer given in the next frame. To obtain the greatest benefit, you are strongly advised to cover up the following frame until you have made your response. When a series of dots occurs, you are expected to supply the missing word, phrase, number or mathematical expression. At every stage you will be guided along the right path. There is no need to hurry: read the frames carefully and follow the directions exactly. In this way, you must learn.

Each Program opens with a list of **Learning outcomes** which specify exactly what you will learn by studying the contents of the Program. The Program ends with a matching checklist of **Can You?** questions that enables you to rate your success in having achieved the **Learning outcomes**. If you feel sufficiently confident then tackle the short **Test exercise** which follows. This is set directly on what you have learned in the Program: the questions are straightforward and contain no tricks. To provide you with the necessary practice, a set of **Further problems** is also included: do as many of these problems as you can. Remember, that in mathematics, as in many other situations, practice makes perfect – or more nearly so.

Useful background information

Symbols used in the text

$=$	is equal to	$\rightarrow$	tends to
$\approx$	is approximately equal to	$\neq$	is not equal to
$>$	is greater than	$\equiv$	is identical to
$\geq$	is greater than or equal to	$<$	is less than
$n!$	factorial $n = 1 \times 2 \times 3 \times \ldots \times n$	$\leq$	is less than or equal to
$\lvert k \rvert$	modulus of k, i.e. size of k irrespective of sign	∞	infinity
$\sum$	summation	$\underset{n \to \infty}{Lim}$	limiting value as $n \to \infty$

Useful mathematical information

1 Algebraic identities

$$(a+b)^2 = a^2 + 2ab + b^2 \qquad (a+b)^3 = a^3 + 3a^2b + 3ab^2 + b^3$$

$$(a-b)^2 = a^2 - 2ab + b^2 \qquad (a-b)^3 = a^3 - 3a^2b + 3ab^2 - b^3$$

$$(a+b)^4 = a^4 + 4a^3b + 6a^2b^2 + 4ab^3 + b^4$$

$$(a-b)^4 = a^4 - 4a^3b + 6a^2b^2 - 4ab^3 + b^4$$

$$a^2 - b^2 = (a-b)(a+b) \qquad a^3 - b^3 = (a-b)(a^2 + ab + b^2)$$

$$a^3 + b^3 = (a+b)(a^2 - ab + b^2)$$

2 Trigonometrical identities

(a) $\sin^2\theta + \cos^2\theta = 1$; $\sec^2\theta = 1 + \tan^2\theta$; $\operatorname{cosec}^2\theta = 1 + \cot^2\theta$

(b) $\sin(A+B) = \sin A \cos B + \cos A \sin B$

$\sin(A-B) = \sin A \cos B - \cos A \sin B$

$\cos(A+B) = \cos A \cos B - \sin A \sin B$

$\cos(A-B) = \cos A \cos B + \sin A \sin B$

$$\tan(A+B) = \frac{\tan A + \tan B}{1 - \tan A \tan B}$$

$$\tan(A-B) = \frac{\tan A - \tan B}{1 + \tan A \tan B}$$

(c) Let $A = B = \theta$ $\therefore$ $\sin 2\theta = 2\sin\theta\cos\theta$

$$\cos 2\theta = \cos^2\theta - \sin^2\theta = 1 - 2\sin^2\theta = 2\cos^2\theta - 1$$

$$\tan 2\theta = \frac{2\tan\theta}{1 - \tan^2\theta}$$

(d) Let $\theta = \dfrac{\phi}{2}$ $\therefore$ $\sin\phi = 2\sin\dfrac{\phi}{2}\cos\dfrac{\phi}{2}$

$$\cos\phi = \cos^2\dfrac{\phi}{2} - \sin^2\dfrac{\phi}{2} = 1 - 2\sin^2\dfrac{\phi}{2} = 2\cos^2\dfrac{\phi}{2} - 1$$

$$\tan\phi = \dfrac{2\tan\dfrac{\phi}{2}}{1 - 2\tan^2\dfrac{\phi}{2}}$$

(e) $\sin C + \sin D = 2\sin\dfrac{C+D}{2}\cos\dfrac{C-D}{2}$

$\sin C - \sin D = 2\cos\dfrac{C+D}{2}\sin\dfrac{C-D}{2}$

$\cos C + \cos D = 2\cos\dfrac{C+D}{2}\cos\dfrac{C-D}{2}$

$\cos D - \cos C = 2\sin\dfrac{C+D}{2}\sin\dfrac{C-D}{2}$

(f) $2\sin A\cos B = \sin(A+B) + \sin(A-B)$
$2\cos A\sin B = \sin(A+B) - \sin(A-B)$
$2\cos A\cos B = \cos(A+B) + \cos(A-B)$
$2\sin A\sin B = \cos(A-B) - \cos(A+B)$

(g) Negative angles: $\sin(-\theta) = -\sin\theta$
$\cos(-\theta) = \cos\theta$
$\tan(-\theta) = -\tan\theta$

(h) Angles having the same trigonometrical ratios:
  (i)   Same sine: θ and $(180° - \theta)$
  (ii) Same cosine: θ and $(360° - \theta)$, i.e. $(-\theta)$
  (iii) Same tangent: θ and $(180° + \theta)$

(i) $a\sin\theta + b\cos\theta = A\sin(\theta + \alpha)$
$a\sin\theta - b\cos\theta = A\sin(\theta - \alpha)$
$a\cos\theta + b\sin\theta = A\cos(\theta - \alpha)$
$a\cos\theta - b\sin\theta = A\cos(\theta + \alpha)$

where $\begin{cases} A = \sqrt{a^2 + b^2} \\ \alpha = \tan^{-1}\dfrac{b}{a} \quad (0° < \alpha < 90°) \end{cases}$

3 Standard curves
(a) *Straight line*

Slope, $m = \dfrac{dy}{dx} = \dfrac{y_2 - y_1}{x_2 - x_1}$

Angle between two lines, $\tan\theta = \dfrac{m_2 - m_1}{1 + m_1 m_2}$

For parallel lines, $m_2 = m_1$
For perpendicular lines, $m_1 m_2 = -1$

Equation of a straight line (slope $= m$)

(i) Intercept c on real y-axis: $y = mx + c$

(ii) Passing through (x_1, y_1): $y - y_1 = m(x - x_1)$

(iii) Joining (x_1, y_1) and (x_2, y_2): $\dfrac{y - y_1}{y_2 - y_1} = \dfrac{x - x_1}{x_2 - x_1}$

(b) *Circle*

Centre at origin, radius r: $x^2 + y^2 = r^2$

Centre (h, k), radius r: $(x - h)^2 + (y - k)^2 = r^2$

General equation: $x^2 + y^2 + 2gx + 2fy + c = 0$

with centre $(-g, -f)$: radius $= \sqrt{g^2 + f^2 - c}$

Parametric equations: $x = r\cos\theta,\ y = r\sin\theta$

(c) *Parabola*

Vertex at origin, focus $(a, 0)$: $y^2 = 4ax$

Parametric equations : $x = at^2,\ y = 2at$

(d) *Ellipse*

Centre at origin, foci $\left(\pm\sqrt{a^2 + b^2},\ 0 \right)$: $\dfrac{x^2}{a^2} + \dfrac{y^2}{b^2} = 1$

where $a =$ semi-major axis, $b =$ semi-minor axis

Parametric equations: $x = a\cos\theta,\ y = b\sin\theta$

(e) *Hyperbola*

Centre at origin, foci $\left(\pm\sqrt{a^2 + b^2},\ 0 \right)$: $\dfrac{x^2}{a^2} - \dfrac{y^2}{b^2} = 1$

Parametric equations: $x = a\sec\theta,\ y = b\tan\theta$

Rectangular hyperbola:

Centre at origin, vertex $\pm\left(\dfrac{a}{\sqrt{2}}, \dfrac{a}{\sqrt{2}} \right)$: $xy = \dfrac{a^2}{2} = c^2$

i.e. $xy = c^2$ where $c = \dfrac{a}{\sqrt{2}}$

Parametric equations: $x = ct,\ y = c/t$

4 Laws of mathematics

(a) *Associative laws* – for addition and multiplication

$a + (b + c) = (a + b) + c$

$a(bc) = (ab)c$

(b) *Commutative laws* – for addition and multiplication

$a + b = b + a$

$ab = ba$

(c) *Distributive laws* – for multiplication and division

$a(b + c) = ab + ac$

$\dfrac{b + c}{a} = \dfrac{b}{a} + \dfrac{c}{a}$ (provided $a \neq 0$)

Preface

It is now over 30 years since Ken Stroud first developed his approach to personalized learning with his classic text *Engineering Mathematics*, now in its fifth edition. That unique and hugely successful programmed learning style is exemplified in this text and I am delighted to have been asked to contribute to it. I have endeavored to retain the very essence of his style that has contributed to so many students' mathematical abilities over the years, particularly the time-tested Stroud format with its close attention to technique development throughout. In my task I have been greatly assisted by a first-rate team of academics who have worked alongside me in the development of this book. To them I should like to express my sincere gratitude for all the detailed care and consideration they have given to all my contributions.

The first two Programs deal with first and second-order ordinary differential equations and present a standard approach but with a thorough coverage. The next two Programs deal with sequences and series in readiness for the fifth Program that considers power series solutions of ordinary differential equations, providing a firm underpinning for Bessel's equation, Legendre polynomials and Sturm-Liouville systems. The following three Programs consider the Laplace transform and its use in the solution of systems subjected to continuous, piecewise continuous and impulsive inputs. These are followed by an introduction to the Z-transform as applicable to discrete systems. The penultimate Program covers matrix algebra and its use in the solutions of systems of first and second order ordinary differential equations. The final Program deals with the numerical methods of solving ordinary differential equations,, covering the Euler, Euler-Cauchy, Runge-Kutta and predictor-corrector methods.

To give the student as much assistance as possible in organizing the students' study there are specific **Learning outcomes** at the beginning and **Can You?** checklists at the end of each Program. In this way the learning experience is made more explicit and the student is given greater confidence in what has been learnt. Test exercises and Further problems follow, in which the student can consolidate their newly found knowledge.

This is the third opportunity that I have had to work on the Stroud books, having made additions to both the *Engineering Mathematics* and *Advanced Engineering Mathematics* texts. It is as ever a challenge and an honour to be able to work with Ken Stroud's material. Ken had an understanding of his students and their learning and thinking processes which was second to none, and this is reflected in every page of this book. As always my thanks go to the Stroud family for their continuing support for and encouragement of new projects and ideas which are allowing Ken's work an ever wider public.

It is only in working on this book that the enormity of Ken Stroud's achievement can be really understood. The vast amount of work involved, the care and attention to detail and above all the complete understanding of his students and their learning processes are apparent in every page. It has been both a challenge and an honour to be able to work on such a book. I should like to thank the Stroud family again for their support in my work for this book.

Huddersfield   Dexter J Booth
July 2004

First-order differential equations

Frames
1 to 83

Learning outcomes

When you have completed this Program you will be able to:

- Recognize the order of a differential equation
- Appreciate that a differential equation of order n can be derived from a function containing n arbitrary constants
- Solve certain first-order differential equations by direct integration
- Solve certain first-order differential equations by separating the variables
- Solve certain first-order homogeneous differential equations by an appropriate substitution
- Solve certain first-order differential equations by using an integrating factor
- Solve Bernoulli's equation

Introduction

1

A *differential equation* is a relationship between an independent variable, x, a dependent variable y, and one or more derivatives of y with respect to x.

e.g. $x^2 \dfrac{dy}{dx} = y \sin x = 0$

$xy \dfrac{d^2y}{dx^2} + y \dfrac{dy}{dx} + e^{3x} = 0$

Differential equations represent dynamic relationships, i.e. quantities that change, and are thus frequently occurring in scientific and engineering problems.

The *order* of a differential equation is given by the highest derivative involved in the equation.

$x \dfrac{dy}{dx} - y^2 = 0$ is an equation of the 1st order

$xy \dfrac{d^2y}{dx^2} - y^2 \sin x = 0$ is an equation of the 2nd order

$\dfrac{d^3y}{dx^3} - y \dfrac{dy}{dx} + e^{4x} = 0$ is an equation of the 3rd order

So that $\dfrac{d^2y}{dx^2} + 2 \dfrac{dy}{dx} + 10y = \sin 2x$ is an equation of the *Second* ...order.

2

$$\boxed{\text{second}}$$

Because

In the equation $\dfrac{d^2y}{dx^2} + 2 \dfrac{dy}{dx} + 10y = \sin 2x$, the highest derivative involved is $\dfrac{d^2y}{dx^2}$.

Similarly:

(a) $x \dfrac{dy}{dx} = y^2 + 1$ is a ....order equation

(b) $\cos^2 x \dfrac{dy}{dx} + y = 1$ is a ....order equation

(c) $\dfrac{d^2y}{dx^2} - 3 \dfrac{dy}{dx} + 2y = x^2$ is a ....order equation

(d) $(y^3 + 1) \dfrac{dy}{dx} - xy^2 = x$ is a ....order equation

On to Frame 3

(a) first	(b) first	(c) second	(d) first

3

Next frame

Formation of differential equations

4

Differential equations may be formed in practice from a consideration of the physical problems to which they refer. Mathematically, they can occur when arbitrary constants are eliminated from a given function. Here are a few examples.

Example 1

Consider $y = A \sin x + B \cos x$, where A and B are two arbitrary constants.
  If we differentiate, we get:

$$\frac{dy}{dx} = A \cos x - B \sin x$$

and $\quad \dfrac{d^2y}{dx^2} = -A \sin x - B \cos x$

which is identical to the original equation, but with the sign changed.

i.e. $\dfrac{d^2y}{dx^2} = -y \qquad \therefore \dfrac{d^2y}{dx^2} + y = 0$

This is a differential equation of the ....2...... order.

5

second

Example 2

Form a differential equation from the function $y = x + \dfrac{A}{x}$.

  We have $y = x + \dfrac{A}{x} = x + Ax^{-1}$

$\therefore \dfrac{dy}{dx} = 1 - Ax^{-2} = 1 - \dfrac{A}{x^2}$

From the given equation, $\dfrac{A}{x} = y - x \ \therefore \ A = x(y - x)$

$\therefore \dfrac{dy}{dx} = 1 - \dfrac{x(y-x)}{x^2}$

$= 1 - \dfrac{y-x}{x} = \dfrac{x-y+x}{x} = \dfrac{2x-y}{x}$

$\therefore x\dfrac{dy}{dx} = 2x - y$

This is an equation of the ....1....... order.

6

| first |

Now one more.

Example 3

Form the differential equation for $y = Ax^2 + Bx$.

We have $\qquad y = Ax^2 + Bx \qquad\qquad (1)$

$$\therefore\ \frac{dy}{dx} = 2Ax + B \qquad\qquad (2)$$

$$\therefore\ \frac{d^2y}{dx^2} = 2A \qquad\qquad (3)\quad A = \frac{1}{2}\frac{d^2y}{dx^2}$$

Substitute for $2A$ in (2): $\dfrac{dy}{dx} = x\dfrac{d^2y}{dx^2} + B$

$$\therefore\ B = \frac{dy}{dx} - x\frac{d^2y}{dx^2}$$

Substituting for A and B in (1), we have:

$$y = x^2 \cdot \frac{1}{2}\frac{d^2y}{dx^2} + x\left(\frac{dy}{dx} - x\frac{d^2y}{dx^2}\right)$$

$$= \frac{x^2}{2}\cdot\frac{d^2y}{dx^2} + x\cdot\frac{dy}{dx} - x^2\cdot\frac{d^2y}{dx^2}$$

$$\therefore\ y = x\frac{dy}{dx} - \frac{x^2}{2}\cdot\frac{d^2y}{dx^2}$$

and this is an equation of the .....Second..... order.

7

| second |

If we collect our last few results together, we have:

$y = A\sin x + B\cos x$ gives the equation $\dfrac{d^2y}{dx^2} + y = 0$ (2nd order)

$y = Ax^2 + Bx$ gives the equation $y = x\dfrac{dy}{dx} - \dfrac{x^2}{2}\dfrac{d^2y}{dx^2}$ (2nd order)

$y = x + \dfrac{A}{x}$ gives the equation $x\dfrac{dy}{dx} = 2x - y$ (1st order)

If we were to investigate the following, we should also find that:

$y = Axe^x$ gives the differential equation $x\dfrac{dy}{dx} - y(1 + x) = 0$ (1st order)

$y = Ae^{-4x} + Be^{-6x}$ gives the differential equation $\dfrac{d^2y}{dx^2} + 10\dfrac{dy}{dx} + 24y = 0$
(2nd order)

Some of the functions give 1st-order equations: some give 2nd-order equations. Now look at the five results above and see if you can find any distinguishing features in the functions which decide whether we obtain a 1st-order equation or a 2nd-order equation in any particular case.

When you have come to a conclusion, move on to Frame 8

> A function with 1 arbitrary constant gives a 1st-order equation.
> A function with 2 arbitrary constants gives a 2nd-order equation.

8

Correct, and in the same way:

A function with 3 arbitrary constants would give a 3rd order equation.

So, without working each out in detail, we can say that:

(a) $y = e^{-2x}(A + Bx)$ would give a differential equation of ...2...... order.

(b) $y = A\dfrac{x-1}{x+1}$ would give a differential equation of ...1...... order.

(c) $y = e^{3x}(A\cos 3x + B\sin 3x)$ would give a differential equation of ..2........ order.

9

(a) 2nd   (b) 1st   (c) 2nd

Because

(a) and (c) each have 2 arbitrary constants,

while (b) has only 1 arbitrary constant.

Similarly:

(a) $x^2\dfrac{dy}{dx} + y = 1$ is derived from a function having ...1...... arbitrary constants.

(b) $\cos^2 x \dfrac{dy}{dx} = 1 - y$ is derived from a function having ...1...... arbitrary constants.

(c) $\dfrac{d^2y}{dx^2} + 4\dfrac{dy}{dx} + y = e^{2x}$ is derived from a function having ..2........ arbitrary constants.

10

(a) 1   (b) 1   (c) 2

So, from all this, the following rule emerges:

A 1st-order differential equation is derived from a function having 1 arbitrary constant.

A 2nd-order differential equation is derived from a function having 2 arbitrary constants.

An *n*th-order differential equation is derived from a function having *n* arbitrary constants.

Copy this last statement into your record book. It is important to remember this rule and we shall make use of it at various times in the future.

Then on to Frame 11

Solution of differential equations

11 To solve a differential equation, we have to find the function for which the equation is true. This means that we have to manipulate the equation so as to eliminate all the derivatives and leave a relationship between y and x. The rest of this particular Program is devoted to the various methods of solving *first-order differential equations*. Second-order equations will be dealt with in the next Program.

So, for the first method, move on to Frame 12

12 ## Method 1: *By direct integration*

If the equation can be arranged in the form $\dfrac{dy}{dx} = f(x)$, then the equation can be solved by simple integration.

Example 1

$$\frac{dy}{dx} = 3x^2 - 6x + 5$$

Then $y = \int (3x^2 - 6x + 5)dx = x^3 - 3x^2 + 5x + C$

i.e. $y = x^3 - 3x^2 + 5x + C$

As always, of course, the constant of integration must be included. Here it provides the one arbitrary constant which we always get when solving a first-order differential equation.

Example 2

Solve $x\dfrac{dy}{dx} = 5x^3 + 4$

In this case, $\dfrac{dy}{dx} = 5x^2 + \dfrac{4}{x}$ So, $y = \int \left(5x^2 + \frac{4}{x}\right)dx$

$y = 5/3 x^3 - 4 \ln x + c$

13

$$\boxed{y = \frac{5x^3}{3} + 4\ln x + C}$$

As you already know from your work on integration, the value of C cannot be determined unless further information about the function is given. In this present form, the function is called the *general solution* (or *primitive*) of the given equation.

  If we are told the value of y for a given value of x, C can be evaluated and the result is then a *particular solution* of the equation.

Example 3

Find the particular solution of the equation $e^x \dfrac{dy}{dx} = 4$, given that $y = 3$ when $x = 0$.

First rewrite the equation in the form $\dfrac{dy}{dx} = \dfrac{4}{e^x} = 4e^{-x}$.

Then $\quad y = \displaystyle\int 4e^{-x}\, dx = -4e^{-x} + C$ $3 + 4e^{-0} = c = 7$

Knowing that when $x = 0$, $y = 3$, we can evaluate C in this case, so that the required particular solution is $y = \ldots\ldots\ldots\ldots$

$$\boxed{y = -4e^{-x} + 7}$$

14

Method 2: *By separating the variables*

If the given equation is of the form $\dfrac{dy}{dx} = f(x, y)$, the variable y on the right-hand side prevents solving by direct integration. We therefore have to devise some other method of solution.

Let us consider equations of the form $\dfrac{dy}{dx} = f(x) \cdot F(y)$ and of the form $\dfrac{dy}{dx} = \dfrac{f(x)}{F(y)}$, i.e. equations in which the right-hand side can be expressed as products or quotients of functions of x or of y.

A few examples will show how we proceed.

Example 1

Solve $\dfrac{dy}{dx} = \dfrac{2x}{y+1}$

We can rewrite this as $(y + 1)\dfrac{dy}{dx} = 2x$

Now integrate both sides with respect to x:

$$\int (y+1)\frac{dy}{dx}\, dx = \int 2x\, dx \quad \text{i.e.}$$

$$\int (y+1)\, dy = \int 2x\, dx$$

$$\text{and this gives } \frac{y^2}{2} + y = x^2 + C$$

15

Example 2

Solve $\dfrac{dy}{dx} = (1 + x)(1 + y)$

$\dfrac{1}{1 + y}\dfrac{dy}{dx} = 1 + x$

Integrate both sides with respect to x:

$\displaystyle\int \dfrac{1}{1 + y}\dfrac{dy}{dx}\,dx = \int (1 + x)\,dx \qquad \therefore \quad \int \dfrac{1}{1 + y}\,dy = \int (1 + x)\,dx$

$\ln(1 + y) = x + \dfrac{x^2}{2} + C$

The method depends on our being able to express the given equation in the form $F(y) \cdot \dfrac{dy}{dx} = f(x)$. If this can be done, the rest is then easy, for we have

$\displaystyle\int F(y) \cdot \dfrac{dy}{dx}\,dx = \int f(x)\,dx \qquad \therefore \quad \int F(y)\,dy = \int f(x)\,dx$

and we then continue as in the examples.

Let us see another example, so move on to Frame 16

16

Example 3

Solve $\dfrac{dy}{dx} = \dfrac{1 + y}{2 + x}$ $\qquad\qquad$ (1)

This can be written as $\dfrac{1}{1 + y}\dfrac{dy}{dx} = \dfrac{1}{2 + x}$

Integrate both sides with respect to x:

$\displaystyle\int \dfrac{1}{1 + y}\dfrac{dy}{dx}\,dx = \int \dfrac{1}{2 + x}\,dx$

$\therefore \quad \displaystyle\int \dfrac{1}{1 + y}\,dy = \int \dfrac{1}{2 + x}\,dx$ $\qquad\qquad$ (2)

$\therefore \quad \ln(1 + y) = \ln(2 + x) + C$

It is convenient to write the constant C as the logarithm of some other constant A:

$\ln(1 + y) = \ln(2 + x) + \ln A = \ln A(2 + x)$

$\therefore \quad 1 + y = A(2 + x)$

Note: We can, in practice, get from the given equation (1) to the form of the equation in (2) by a simple routine, thus:

$\dfrac{dy}{dx} = \dfrac{1 + y}{2 + x}$

First multiply across by dx:

$dy = \dfrac{1 + y}{2 + x}dx$

▶

Now collect the 'y-factor' with the dy on the left, i.e. divide by $(1+y)$:

$$\frac{1}{1+y}\,dy = \frac{1}{2+x}\,dx$$

Finally, add the integral signs:

$$\int \frac{1}{1+y}\,dy = \int \frac{1}{2+x}\,dx$$

and then continue as before.

This is purely a routine which enables us to sort out the equation algebraically, the whole of the work being done in one line. Notice, however, that the RHS of the given equation must be expressed as 'x-factors' and 'y-factors'.

Now for another example, using this routine.

Example 4

Solve $\quad \dfrac{dy}{dx} = \dfrac{y^2 + xy^2}{x^2 y - x^2}$

First express the RHS in 'x-factors' and 'y-factors':

$$\frac{dy}{dx} = \frac{y^2(1+x)}{x^2(y-1)}$$

Now rearrange the equation so that we have the 'y-factors' and dy on the LHS and the 'x-factors' and dx on the RHS.

So we get .... $\dfrac{(y-1)}{y^2}\,dy = \dfrac{1+x}{x^2}\,dx$

17

$$\boxed{\dfrac{y-1}{y^2}\,dy = \dfrac{1+x}{x^2}\,dx}$$

We now add the integral signs:

$$\int \frac{y-1}{y^2}\,dy = \int \frac{1+x}{x^2}\,dx$$

and complete the solution:

$$\int \left\{ \frac{1}{y} - y^{-2} \right\} dy = \int \left\{ x^{-2} + \frac{1}{x} \right\} dx$$

$$\therefore \quad \ln y + y^{-1} = \ln x - x^{-1} + C$$

$$\therefore \quad \ln y + \frac{1}{y} = \ln x - \frac{1}{x} + C$$

Here is another.

Example 5

Solve
$$\frac{dy}{dx} = \frac{y^2 - 1}{x}$$

Rearranging, we have
$$dy = \frac{y^2 - 1}{x} dx$$

$$\frac{1}{y^2 - 1} dy = \frac{1}{x} dx$$

$$\int \frac{1}{y^2 - 1} dy = \int \frac{1}{x} dx$$

Which gives .... $= \ln x + c$

18

$$\frac{1}{2} \ln \frac{y-1}{y+1} = \ln x + C$$

$$\therefore \quad \ln \frac{y-1}{y+1} = 2 \ln x + \ln A$$

$$\therefore \quad \frac{y-1}{y+1} = Ax^2$$

$$y - 1 = Ax^2(y+1)$$

You see they are all done in the same way. Now here is one for you to do:

Example 6

Solve $xy \frac{dy}{dx} = \frac{x^2 + 1}{y + 1}$ $xy\,dy = \frac{x^2+1}{y+1}\,dx$ $(y+1)y\,dy = \frac{x^2+1}{x}dx$

First of all, rearrange the equation into the form:

$F(y)dy = f(x)dx$

i.e. arrange the 'y-factors' and dy on the LHS and the 'x-factors' and dx on the RHS.

What do you get?

19

$$y(y+1)dy = \frac{x^2 + 1}{x} dx$$

Because

$$xy \frac{dy}{dx} = \frac{x^2 + 1}{y + 1} \qquad \therefore \quad xy\,dy = \frac{x^2+1}{y+1} dx \qquad \therefore \quad y(y+1)\,dy = \frac{x^2+1}{x} dx$$

So we now have $\int (y^2 + y) dy = \int \left(x + \frac{1}{x}\right) dx$

Now finish it off, then move on to the next frame

$$\frac{y^3}{3} + \frac{y^2}{2} - \frac{x^2}{2} + \ln x + c$$

$$\frac{y^3}{3} + \frac{y^2}{2} = \frac{x^2}{2} + \ln x + C$$

Provided that the RHS of the equation $\frac{dy}{dx} = f(x, y)$ can be separated into '*x*-factors' and '*y*-factors', the equation can be solved by the method of *separating the variables*.

Now do this one entirely on your own.

Example 7

Solve $x\frac{dy}{dx} = y + xy$

When you have finished it completely, move on to Frame 21 and check your solution

Here is the result. Follow it through carefully, even if your own answer is correct.

$$x\frac{dy}{dx} = y + xy \qquad \therefore \ x\frac{dy}{dx} = y(1 + x)$$

$$x\,dy = y(1 + x)\,dx \qquad \therefore \ \frac{dy}{y} = \frac{1 + x}{x}\,dx$$

$$\therefore \ \int \frac{1}{y}\,dy = \int \left(\frac{1}{x} + 1\right)dx$$

$$\therefore \ \ln y = \ln x + x + C$$

At this stage we have eliminated the derivatives and so we have solved the equation. However, we can express the result in a neater form, thus:

$$\ln y - \ln x = x + C$$

$$\therefore \ \ln\left\{\frac{y}{x}\right\} = x + C$$

$$\therefore \ \frac{y}{x} = e^{x+c} = e^x \cdot e^c \quad \text{Now } e^c \text{ is a constant; call it } A.$$

$$\therefore \ \frac{y}{x} = Ae^x \qquad \therefore \ y = Axe^x$$

Next frame

This final example looks more complicated, but it is solved in just the same way. We go through the same steps as before. Here it is.

Example 8

Solve $y\tan x\frac{dy}{dx} = (4 + y^2)\sec^2 x$

First separate the variables, i.e. arrange the '*y*-factors' and d*y* on one side and the '*x*-factors' and d*x* on the other.
 So we get . . . .

23

$$\frac{y}{4+y^2}\,dy = \frac{\sec^2 x}{\tan x}\,dx$$

Adding the integral signs, we get:

$$\int\frac{y}{4+y^2}\,dy = \int\frac{\sec^2 x}{\tan x}\,dx$$

Now determine the integrals, so that we have ....

24

$$\frac{1}{2}\ln(4+y^2) = \ln\tan x + C$$

This result can now be simplified into:

$\ln(4+y^2) = 2\ln\tan x + \ln A$ (expressing the constant $2C$ as $\ln A$)

$\therefore\ 4+y^2 = A\tan^2 x$

$\therefore\ y^2 = A\tan^2 x - 4$

So there we are. Provided we can factorize the equation in the way we have indicated, solution by separating the variables is not at all difficult. So now for a short review exercise to wind up this part of the Program.

Move on now to Frame 25

25   **Review exercise**

Work all the exercise before checking your results.

Find the general solutions of the following equations:

1 $\dfrac{dy}{dx} = \dfrac{y}{x}$
          2 $\dfrac{dy}{dx} = (y+2)(x+1)$

3 $\cos^2 x\dfrac{dy}{dx} = y+3$
          4 $\dfrac{dy}{dx} = xy - y$

5 $\dfrac{\sin x}{1+y}\cdot\dfrac{dy}{dx} = \cos x$

When you have finished them all, move to Frame 26
and check your answers with the solutions given there

26   **1** $\dfrac{dy}{dx} = \dfrac{y}{x}\quad\therefore\quad \int\frac{1}{y}\,dy = \int\frac{1}{x}\,dx$

$\therefore\ \ln y = \ln x + C$

$= \ln x + \ln A$

$\therefore\ y = Ax$

2 $\dfrac{dy}{dx} = (y+2)(x+1)$

$\therefore \quad \displaystyle\int \dfrac{1}{y+2}\,dy = \int (x+1)\,dx$

$\therefore \quad \ln(y+2) = \dfrac{x^2}{2} + x + C$

3 $\cos^2 x \dfrac{dy}{dx} = y + 3$

$\therefore \quad \displaystyle\int \dfrac{1}{y+3}\,dy = \int \dfrac{1}{\cos^2 x}\,dx$

$\qquad\qquad\qquad = \displaystyle\int \sec^2 x\,dx$

$\qquad \ln(y+3) = \tan x + C$

4 $\dfrac{dy}{dx} = xy - y \quad \therefore \quad \dfrac{dy}{dx} = y(x-1)$

$\therefore \quad \displaystyle\int \dfrac{1}{y}\,dy = \int (x-1)\,dx$

$\therefore \quad \ln y = \dfrac{x^2}{2} - x + C$

5 $\dfrac{\sin x}{1+y} \cdot \dfrac{dy}{dx} = \cos x$

$\displaystyle\int \dfrac{1}{1+y}\,dy = \int \dfrac{\cos x}{\sin x}\,dx$

$\therefore \quad \ln(1+y) = \ln \sin x + C$

$\qquad\qquad\qquad = \ln \sin x + \ln A$

$\qquad\qquad 1+y = A \sin x$

$\qquad\qquad \therefore \quad y = A \sin x - 1$

If you are quite happy about those, we can start
the next part of the Program, so move on now to Frame 27

Method 3: *Homogeneous equations – by substituting y = vx* 　27

Here is an equation:

$$\dfrac{dy}{dx} = \dfrac{x+3y}{2x}$$

This looks simple enough, but we find that we cannot express the RHS in the form of 'x-factors' and 'y-factors', so we cannot solve by the method of separating the variables.

▶

In this case we make the substitution $y = vx$, where v is a function of x. So $y = vx$. Differentiate with respect to x (using the product rule):

$$\therefore \quad \frac{dy}{dx} = v \cdot 1 + x\frac{dv}{dx} = v + x\frac{dv}{dx}$$

Also $\quad \dfrac{x+3y}{2x} = \dfrac{x+3vx}{2x} = \dfrac{1+3v}{2}$

The equation now becomes $v + x\dfrac{dv}{dx} = \dfrac{1+3v}{2}$

$$\therefore \quad x\frac{dv}{dx} = \frac{1+3v}{2} - v$$
$$= \frac{1+3v-2v}{2} = \frac{1+v}{2}$$
$$\therefore \quad x\frac{dv}{dx} = \frac{1+v}{2}$$

The given equation is now expressed in terms of v and x, and in this form we find that we can solve by separating the variables. Here goes:

$$\int \frac{2}{1+v}\,dv = \int \frac{1}{x}\,dx$$
$$\therefore \quad 2\ln(1+v) = \ln x + C = \ln x + \ln A$$
$$(1+v)^2 = Ax$$

But $y = vx \quad \therefore \quad v = \left\{\dfrac{y}{x}\right\} \quad \therefore \quad \left(1 + \dfrac{y}{x}\right)^2 = Ax$

which gives $(x+y)^2 = Ax^3$

Note: $\dfrac{dy}{dx} = \dfrac{x+3y}{2x}$ is an example of a *homogeneous differential equation*.

This is determined by the fact that the total degree in x and y for each of the terms involved is the same (in this case, of degree 1). The key to solving every homogeneous equation is to substitute $y = vx$ where v is a function of x. This converts the equation into a form which we can solve by separating the variables.

Let us work some examples, so move on to Frame 28

If $y \equiv vx$

then

$$\frac{dy}{dx} = \frac{d(vx)}{dx} = v\frac{dx}{dx} + x\frac{dv}{dx}$$
$$= v + x\frac{dv}{dx}$$

Example 1

28

Solve $\dfrac{dy}{dx} = \dfrac{x^2 + y^2}{xy}$

Here, all terms of the RHS are of degree 2, i.e. the equation is homogeneous.
∴ We substitute $y = vx$ (where v is a function of x)

$$\therefore \ \frac{dy}{dx} = v + x\frac{dv}{dx}$$

and $\dfrac{x^2 + y^2}{xy} = \dfrac{x^2 + v^2 x^2}{vx^2} = \dfrac{1 + v^2}{v}$

The equation now becomes:

$$v + x\frac{dv}{dx} = \frac{1 + v^2}{v}$$

$$\therefore \ x\frac{dv}{dx} = \frac{1 + v^2}{v} - v$$

$$= \frac{1 + v^2 - v^2}{v} = \frac{1}{v}$$

$$\therefore \ x\frac{dv}{dx} = \frac{1}{v}$$

Now you can separate the variables and get the result in terms of v and x.

Off you go: when you have finished, move on to Frame 29

29

$$\boxed{\frac{v^2}{2} = \ln x + C}$$

Because

$$\int v\,dv = \int \frac{1}{x}\,dx$$

$$\therefore \ \frac{v^2}{2} = \ln x + C$$

All that remains now is to express v back in terms of x and y. The substitution
we used was $y = vx$ ∴ $v = \dfrac{y}{x}$

$$\therefore \ \frac{1}{2}\left(\frac{y}{x}\right)^2 = \ln x + C$$
$$y^2 = 2x^2(\ln x + C)$$

Now, what about this one?

Example 2

Solve $\dfrac{dy}{dx} = \dfrac{2xy + 3y^2}{x^2 + 2xy}$

Is this a homogeneous equation? If you think so, what are your reasons? Yes

When you have decided, move on to Frame 30 $x^2 y^2$

30

> Yes, because the degree of each term is the same

Correct. They are all, of course, of degree 2.

So we now make the substitution, $y = \dots\dots\dots$

31

> $y = vx$, where v is a function of x

Right. That is the key to the whole process.

$$\frac{dy}{dx} = \frac{2xy + 3y^2}{x^2 + 2xy}$$

So express each side of the equation in terms of v and x:

$$\frac{dy}{dx} = \dots\dots\dots$$

and $$\frac{2xy + 3y^2}{x^2 + 2xy} = \dots\dots\dots$$

When you have finished, move on to the next frame

32

> $$\frac{dy}{dx} = v + x\frac{dv}{dx}$$
> $$\frac{2xy + 3y^2}{x^2 + 2xy} = \frac{2vx^2 + 3v^2x^2}{x^2 + 2vx^2} = \frac{2v + 3v^2}{1 + 2v}$$

So that $v + x\dfrac{dv}{dx} = \dfrac{2v + 3v^2}{1 + 2v}$

Now take the single v over to the RHS and simplify, giving:

$$x\frac{dv}{dx} = \dots\dots\dots$$

33

> $$x\frac{dv}{dx} = \frac{2v + 3v^2}{1 + 2v} - v$$
> $$= \frac{2v + 3v^2 - v - 2v^2}{1 + 2v}$$
> $$x\frac{dv}{dx} = \frac{v + v^2}{1 + 2v}$$

Now you can separate the variables, giving $\dots\dots\dots$

34

$$\int \frac{1+2v}{v+v^2}\,\mathrm{d}v = \int \frac{1}{x}\,\mathrm{d}x$$

Integrating both sides, we can now obtain the solution in terms of v and x. What do you get?

35

$$\ln(v+v^2) = \ln x + C = \ln x + \ln A$$
$$\therefore\ v+v^2 = Ax$$

We have almost finished the solution. All that remains is to express v back in terms of x and y.

Remember the substitution was $y=vx$, so that $v = \dfrac{y}{x}$

So finish it off.

Then move on

36

$$xy+y^2 = Ax^3$$

Because

$$v+v^2 = Ax \text{ and } v = \frac{y}{x} \qquad \therefore\ \frac{y}{x}+\frac{y^2}{x^2} = Ax$$
$$xy+y^2 = Ax^3$$

And that is all there is to it.

Move to Frame 37

37

Here is the solution of the previous equation, all in one piece. Follow it through again.

To solve $\quad \dfrac{\mathrm{d}y}{\mathrm{d}x} = \dfrac{2xy+3y^2}{x^2+2xy}$

This is homogeneous, all terms of degree 2. Put $y=vx$

$$\therefore\ \frac{\mathrm{d}y}{\mathrm{d}x} = v + x\frac{\mathrm{d}v}{\mathrm{d}x}$$

$$\frac{2xy+3y^2}{x^2+2xy} = \frac{2vx^2+3v^2x^2}{x^2+2vx^2} = \frac{2v+3v^2}{1+2v} \qquad \therefore\ v+x\frac{\mathrm{d}v}{\mathrm{d}x} = \frac{2v+3v^2}{1+2v}$$

$$x\frac{\mathrm{d}v}{\mathrm{d}x} = \frac{2v+3v^2}{1+2v} - v = \frac{2v+3v^2-v-2v^2}{1+2v}$$

$$\therefore\ x\frac{\mathrm{d}v}{\mathrm{d}x} = \frac{v+v^2}{1+2v} \qquad \therefore\ \int\frac{1+2v}{v+v^2}\,\mathrm{d}v = \int\frac{1}{x}\,\mathrm{d}x$$

$$\therefore\ \ln(v+v^2) = \ln x + C = \ln x + \ln A$$

$$v+v^2 = Ax$$

But $y=vx \qquad \therefore\ v = \dfrac{y}{x} \qquad \therefore\ \dfrac{y}{x}+\dfrac{y^2}{x^2} = Ax \qquad \therefore\ xy+y^2 = Ax^3$

▶

Now, in the same way, you do this one. Take your time and be sure that you understand each step.

Example 3

Solve $(x^2 + y^2)\dfrac{dy}{dx} = xy$

When you have completely finished it, move to Frame 38 and check your solution

---

38

Here is the solution in full.

$$(x^2 + y^2)\frac{dy}{dx} = xy \qquad \therefore \quad \frac{dy}{dx} = \frac{xy}{x^2 + y^2}$$

$$\text{Put} \quad y = vx \quad \therefore \quad \frac{dy}{dx} = v + x\frac{dv}{dx}$$

$$\text{and} \quad \frac{xy}{x^2 + y^2} = \frac{vx^2}{x^2 + v^2 x^2} = \frac{v}{1 + v^2}$$

$$\therefore \quad v + x\frac{dv}{dx} = \frac{v}{1 + v^2}$$

$$x\frac{dv}{dx} = \frac{v}{1 + v^2} - v$$

$$x\frac{dv}{dx} = \frac{v - v - v^3}{1 + v^2} = \frac{-v^3}{1 + v^2}$$

$$\therefore \quad \int \frac{1 + v^2}{v^3}\, dv = -\int \frac{1}{x}\, dx$$

$$\therefore \quad \int \left(v^{-3} + \frac{1}{v} \right) dv = -\ln x + C$$

$$\therefore \quad \frac{-v^{-2}}{2} + \ln v = -\ln x + \ln A$$

$$\ln v + \ln x + \ln K = \frac{1}{2v^2} \qquad (\ln K = -\ln A)$$

$$\ln Kvx = \frac{1}{2v^2}$$

$$\text{But} \quad v = \frac{y}{x} \quad \therefore \quad \ln Ky = \frac{x^2}{2y^2}$$

$$2y^2 \ln Ky = x^2$$

This is one form of the solution: there are of course other ways of expressing it.

Now for a short review exercise on this part of the work, move on to Frame 39

Review exercise

Solve the following:

1 $(x - y)\dfrac{dy}{dx} = x + y$

2 $2x^2\dfrac{dy}{dx} = x^2 + y^2$

3 $(x^2 + xy)\dfrac{dy}{dx} = xy - y^2$

When you have finished all three, move on and check your results

The solution of equation **1** can be written as:

$$\tan^{-1}\left\{\frac{y}{x}\right\} = \ln A + \ln x + \frac{1}{2}\ln\left\{1 + \frac{y^2}{x^2}\right\}$$

Did you get that? If so, move straight on to Frame 41. If not, check your working with the following.

1 $(x - y)\dfrac{dy}{dx} = x + y$ $\therefore$ $\dfrac{dy}{dx} = \dfrac{x + y}{x - y}$

Put $y = vx$ $\therefore$ $\dfrac{dy}{dx} = v + x\dfrac{dv}{dx}$ $\dfrac{x + y}{x - y} = \dfrac{1 + v}{1 - v}$

$\therefore$ $v + x\dfrac{dv}{dx} = \dfrac{1 + v}{1 - v}$ $\therefore$ $x\dfrac{dv}{dx} = \dfrac{1 + v}{1 - v} - v = \dfrac{1 + v - v + v^2}{1 - v} = \dfrac{1 + v^2}{1 - v}$

$\therefore$ $\displaystyle\int\dfrac{1 - v}{1 + v^2}\,dv = \int\dfrac{1}{x}\,dx$ $\therefore$ $\displaystyle\int\left\{\dfrac{1}{1 + v^2} - \dfrac{v}{1 + v^2}\right\}dv = \ln x + C$

$\therefore$ $\tan^{-1}v - \dfrac{1}{2}\ln(1 + v^2) = \ln x + \ln A$

But $v = \dfrac{y}{x}$ $\therefore$ $\tan^{-1}\left\{\dfrac{y}{x}\right\} = \ln A + \ln x + \dfrac{1}{2}\ln\left(1 + \dfrac{y^2}{x^2}\right)$

This result can, in fact, be simplified further.

Now on to Frame 41

Equation **2** gives the solution: $\boxed{\dfrac{2x}{x - y} = \ln x + C}$

If you agree, move straight on to Frame 42. Otherwise, follow through the working. Here it is.

▶

2 $2x^2 \dfrac{dy}{dx} = x^2 + y^2$ $\therefore$ $\dfrac{dy}{dx} = \dfrac{x^2 + y^2}{2x^2}$

Put $y = vx$ $\therefore$ $\dfrac{dy}{dx} = v + x\dfrac{dv}{dx}$; $\dfrac{x^2 + y^2}{2x^2} = \dfrac{x^2 + v^2 x^2}{2x^2} = \dfrac{1 + v^2}{2}$

$\therefore$ $v + x\dfrac{dv}{dx} = \dfrac{1 + v^2}{2}$ $\therefore$ $x\dfrac{dv}{dx} = \dfrac{1 + v^2}{2} - v = \dfrac{1 - 2v + v^2}{2} = \dfrac{(v-1)^2}{2}$

$\therefore$ $\displaystyle\int \dfrac{2}{(v-1)^2}\, dv = \int \dfrac{1}{x}\, dx$ $\therefore$ $-2\dfrac{1}{v-1} = \ln x + C$

But $v = \dfrac{y}{x}$ and $\dfrac{2}{1-v} = \ln x + C$ $\therefore$ $\dfrac{2x}{x-y} = \ln x + C$

On to Frame 42

42

One form of the result for equation **3** is: $\boxed{xy = Ae^{x/y}}$. Follow through the working and check yours.

3 $(x^2 + xy)\dfrac{dy}{dx} = xy - y^2$ $\therefore$ $\dfrac{dy}{dx} = \dfrac{xy - y^2}{x^2 + xy}$

Put $y = vx$ $\therefore$ $\dfrac{dy}{dx} = v + x\dfrac{dv}{dx}$; $\dfrac{xy - y^2}{x^2 + xy} = \dfrac{vx^2 - v^2 x^2}{x^2 + vx^2} = \dfrac{v - v^2}{1 + v}$

$\therefore$ $v + x\dfrac{dv}{dx} = \dfrac{v - v^2}{1 + v}$

$x\dfrac{dv}{dx} = \dfrac{v - v^2}{1 + v} - v = \dfrac{v - v^2 - v - v^2}{1 + v} = \dfrac{-2v^2}{1 + v}$

$\therefore$ $\displaystyle\int \dfrac{1 + v}{v^2}\, dv = \int \dfrac{-2}{x}\, dx$

$\displaystyle\int \left(v^{-2} + \dfrac{1}{v} \right) dv = -\int \dfrac{2}{x}\, dx$

$\therefore$ $\ln v - \dfrac{1}{v} = -2\ln x + C$ Let $C = \ln A$

$\ln v + 2\ln x = \ln A + \dfrac{1}{v}$

$\ln \left\{ \dfrac{y}{x} \cdot x^2 \right\} = \ln A + \dfrac{x}{y}$ $\therefore$ $xy = Ae^{x/y}$

Now move to the next frame

---

43   ## Method 4: *Linear equations – use of integrating factor*

Consider the equation $\dfrac{dy}{dx} + 5y = e^{2x}$

This is clearly an equation of the first order, but different from those we have dealt with so far. In fact, none of our previous methods could be used to solve this one, so we have to find a further method of attack.

▶

In this case, we begin by multiplying both sides by e^{5x}. This gives

$$e^{5x}\frac{dy}{dx} + y5e^{5x} = e^{2x} \cdot e^{5x} = e^{7x}$$

We now find that the LHS is, in fact, the derivative of $y \cdot e^{5x}$.

$$\therefore \frac{d}{dx}\left\{y \cdot e^{5x}\right\} = e^{7x}$$

$$\frac{d(y \cdot e^{5x})}{dx} = e^{5x}\frac{dy}{dx} + y5e^{5x}\frac{dy}{dx}$$

Now, of course, the rest is easy. Integrate both sides with respect to x:

$$\therefore y \cdot e^{5x} = \int e^{7x}\,dx = \frac{e^{7x}}{7} + C \quad \therefore y = \frac{e^{2x}}{7} + Ce^{-5x}$$

44

$$y = \frac{e^{2x}}{7} + Ce^{-5x}$$

Did you forget to divide the C by the e^{5x}? It is a common error so watch out for it.

The equation we have just solved is an example of a set of equations of the form $\frac{dy}{dx} + Py = Q$, where P and Q are functions of x (or constants). This equation is called a *linear equation of the first order* and to solve any such equation, we multiply both sides by an *integrating factor* which is always $e^{\int P dx}$. This converts the LHS into the derivative of a product.

In our previous example, $\frac{dy}{dx} + 5y = e^{2x}$, $P = 5$

$$\therefore \int P dx = 5x \text{ and the integrating factor was therefore } e^{5x}.$$

Note: In determining $\int P\,dx$, we do not include a constant of integration. This omission is purely for convenience, for a constant of integration here would in practice give a constant factor on both sides of the equation, which would subsequently cancel. This is one of the rare occasions when we do not write down the constant of integration.

So: *To solve a differential equation of the form*

$$\frac{dy}{dx} + Py = Q$$

where P and Q are constants or functions of x, multiply both sides by the integrating factor $e^{\int P dx}$

This is important, so copy this rule down into your record book.

Then move on to Frame 45

45

Example 1

To solve $\dfrac{dy}{dx} - y = x$

If we compare this with $\dfrac{dy}{dx} + Py = Q$, we see that in this case
$$P = -1 \text{ and } Q = x.$$

The integrating factor is always $e^{\int P\,dx}$ and here $P = -1$.

$\therefore \displaystyle\int P\,dx = -x$ and the integrating factor is therefore ...$\underset{e^{-x}}{\underline{e}}$.....

46

$$\boxed{e^{-x}}$$

We therefore multiply both sides by e^{-x}.

$$\therefore\ e^{-x}\frac{dy}{dx} - ye^{-x} = xe^{-x}$$

$$\frac{d}{dx}\left\{e^{-x}y\right\} = xe^{-x} \quad \therefore\ ye^{-x} = \int xe^{-x}\,dx$$

The RHS integral can now be determined by integrating by parts:

$$ye^{-x} = x(-e^{-x}) + \int e^{-x}\,dx = -xe^{-x} - e^{-x} + C$$

$$\therefore\ y = -x - 1 + Ce^x \quad \therefore\ y = Ce^x - x - 1$$

The whole method really depends on:

(a) being able to find the integrating factor

(b) being able to deal with the integral that emerges on the RHS.

Let us consider the general case.

47

Consider $\dfrac{dy}{dx} + Py = Q$ where P and Q are functions of x. Integrating factor,

IF $= e^{\int P\,dx}$ $\therefore\ \dfrac{dy}{dx}\cdot e^{\int P\,dx} + Pye^{\int P\,dx} = Qe^{\int P\,dx}$ *multiply all terms by* $e^{\int P\,dx}$

You will now see that the LHS is the derivative of $ye^{\int P\,dx}$

$$\therefore\ \frac{d}{dx}\left\{ye^{\int P\,dx}\right\} = Qe^{\int P\,dx}$$

Integrate both sides with respect to x:

$$ye^{\int P\,dx} = \int Qe^{\int P\,dx}\cdot dx$$

This result looks far more complicated than it really is. If we indicate the integrating factor by IF, this result becomes:

$$y\cdot\text{IF} = \int Q\cdot\text{IF}\,dx$$

and, in fact, we remember it in that way.

▶

So, the solution of an equation of the form

$$\frac{dy}{dx} + Py = Q \text{ (where } P \text{ and } Q \text{ are functions of } x)$$

is given by $\quad y \cdot \text{IF} = \displaystyle\int Q \cdot \text{IF} \, dx \qquad$ where $\text{IF} = e^{\int P dx}$

Copy this into your record book.

Then move to Frame 48

48

So if we have the equation:

$$\frac{dy}{dx} + 3y = \sin x$$

$$\left[\frac{dy}{dx} + Py = Q \right]$$

then in this case:

(a) $P = \overset{3}{\ldots\ldots}$

(b) $\displaystyle\int P \, dx = \overset{\displaystyle\int 3 \, dx}{\ldots} \overset{\downarrow 3x}{\ldots}$

(c) $\text{IF} = e^{\ldots} \overset{= e^{3x}}{}$

49

$$\boxed{\text{(a) } P = 3 \qquad \text{(b) } \int P \, dx = 3x \qquad \text{(c) } \text{IF} = e^{3x}}$$

Before we work through any further examples, let us establish a very useful piece of simplification, which we can make good use of when we are finding integrating factors. We want to simplify $e^{\ln F}$, where F is a function of x.

Let $y = e^{\ln F}$ $\qquad\qquad\qquad\qquad \therefore \ln y = \ln e^{\ln F} = \ln F$

Then, by the very definition of a logarithm, $\ln y = \ln F$

$\therefore y = F \quad \therefore F = e^{\ln F} \quad$ i.e. $e^{\ln F} = F$

This means that $\quad e^{\ln(\text{function})} = \text{function}$. Always!

$$e^{\ln x} = x$$

$$e^{\ln \sin x} = \sin x$$

$$e^{\ln \tanh x} = \tanh x$$

$$e^{\ln(x^2)} = \overset{2}{X} \ldots\ldots$$

50

$$\boxed{x^2}$$

Similarly, what about $e^{k \ln F}$? If the log in the index is multiplied by any external coefficient, this coefficient must be taken inside the log as a power.

e.g. $\qquad\qquad e^{2 \ln x} = e^{\ln(x^2)} = x^2$

$$e^{3 \ln \sin x} = e^{\ln(\sin^3 x)} = \sin^3 x$$

$$e^{-\ln x} = e^{\ln(x^{-1})} = x^{-1} = \frac{1}{x}$$

and $\quad e^{-2 \ln x} = \dots \frac{1}{x^2} \dots$

51

$$\boxed{\frac{1}{x^2}} \quad \text{because } e^{-2 \ln x} = e^{\ln(x^{-2})} = x^{-2} = \frac{1}{x^2}$$

So here is the rule once again: $e^{\ln F} = F$

Make a note of this rule in your record book.

Then on to Frame 52

52 Now let us see how we can apply this result to our working.

Example 2

Solve $\quad x\dfrac{dy}{dx} + y = x^3$

First we divide through by x to reduce the first term to a single $\dfrac{dy}{dx}$

i.e. $\dfrac{dy}{dx} + \dfrac{1}{x} \cdot y = x^2$

Compare with $\left[\dfrac{dy}{dx} + Py = Q\right] \quad \therefore \ P = \dfrac{1}{x}$ and $Q = x^2$

$\text{IF} = e^{\int P dx} \qquad \int P \, dx = \int \dfrac{1}{x} \, dx = \ln x$

$\therefore \ \text{IF} = e^{\ln x} = x \quad \therefore \ \text{IF} = x$

The solution is $\quad y \cdot \text{IF} = \displaystyle\int Q \cdot \text{IF} \, dx$

so $yx = \displaystyle\int x^2 \cdot x \, dx = \int x^3 \, dx = \dfrac{x^4}{4} + C \quad \therefore \ xy = \dfrac{x^4}{4} + C$

Move to Frame 53

Example 3

Solve $\dfrac{dy}{dx} + y\cot x = \cos x$

Compare with $\left[\dfrac{dy}{dx} + Py = Q\right]$ $\quad \therefore \begin{cases} P = \cot x \\ Q = \cos x \end{cases}$

$\text{IF} = e^{\int P\,dx}$ $\quad \int P\,dx = \int \cot x\,dx = \int \dfrac{\cos x}{\sin x}\,dx = \ln\sin x$

$\quad \therefore \ \text{IF} = e^{\ln\sin x} = \sin x$

$y \cdot \text{IF} = \int Q \cdot \text{IF}\,dx \quad \therefore \ y\sin x = \int \sin x\cos x\,dx = \dfrac{\sin^2 x}{2} + C$

$\quad \therefore \ y = \dfrac{\sin x}{2} + C\operatorname{cosec} x$

— In general

$\left[\dfrac{f'(x)}{f(x)} = \ln[f(x)]c \right.$

Now here is another.

Example 4

Solve $(x+1)\dfrac{dy}{dx} + y = (x+1)^2$

The first thing is to .... *divide by (x+1)*

$\boxed{\text{Divide through by } (x+1)}$

Correct, since we must reduce the coefficient of $\dfrac{dy}{dx}$ to 1.

$\quad \therefore \ \dfrac{dy}{dx} + \dfrac{1}{x+1} \cdot y = x + 1$

Compare with $\dfrac{dy}{dx} + Py = Q$

In this case $P = \dfrac{1}{x+1}$ and $Q = x + 1$

Now determine the integrating factor, which simplifies to

$\text{IF} = ...x+1.....$

55

$$\boxed{\text{IF} = x + 1}$$

Because

$$\int P\,dx = \int \frac{1}{x+1}\,dx = \ln(x+1)$$

$$\therefore\ \text{IF} = e^{\ln(x+1)} = (x+1)$$

The solution is always $\quad y \cdot \text{IF} = \int Q \cdot \text{IF}\,dx$

and we know that, in this case, IF $= x + 1$ and $Q = x + 1$.

So finish off the solution and then move on to Frame 56

56

$$\boxed{y = \frac{(x+1)^2}{3} + \frac{C}{x+1}}$$

Here is the solution in detail:

$$y \cdot (x+1) = \int (x+1)(x+1)\,dx$$

$$= \int (x+1)^2\,dx$$

$$= \frac{(x+1)^3}{3} + C$$

$$\therefore\ y = \frac{(x+1)^2}{3} + \frac{C}{x+1}$$

Now let us do another one.

Example 5

Solve $\quad x\dfrac{dy}{dx} - 5y = x^7$

In this case, $P = \dots\dfrac{-5}{x}\dots\quad Q = \dots x^6\dots$

57

$$\boxed{P = -\frac{5}{x} \quad Q = x^6}$$

Because if

$$x\frac{dy}{dx} - 5y = x^7$$

$$\therefore\ \frac{dy}{dx} - \frac{5}{x} \cdot y = x^6$$

Compare with $\left[\dfrac{dy}{dx} + Py = Q\right]\quad \therefore\ P = -\dfrac{5}{x};\quad Q = x^6$

So integrating factor, IF $= e^{\int P\,dx} = e^{\int -5/x\,dx} = e^{-\int \frac{5}{x}\,dx} = e^{-5\ln x}$

$$= e^{+\ln x^{-5}} = \frac{1}{x^5}$$

$$\boxed{\text{IF} = x^{-5} = \frac{1}{x^5}}$$

Because

$$\text{IF} = e^{\int P \, dx} \qquad \int P \, dx = -\int \frac{5}{x} \, dx = -5 \ln x$$

$$\therefore \text{IF} = e^{-5 \ln x} = e^{\ln(x^{-5})} = x^{-5} = \frac{1}{x^5}$$

So the solution is:

$$y \cdot \frac{1}{x^5} = \int x^6 \cdot \frac{1}{x^5} \, dx$$

$$\frac{y}{x^5} = \int x \, dx = \frac{x^2}{2} + C \qquad \therefore \; y = \frac{x^7 + x^5 C}{2}$$

$$\boxed{y = \frac{x^7}{2} + Cx^5}$$

Did you remember to multiply the C by x^5?

Fine. Now you do this one entirely on your own.

Example 6

Solve $\quad (1 - x^2)\dfrac{dy}{dx} - xy = 1$.

When you have finished it, move to Frame 60

$$\boxed{y\sqrt{1 - x^2} = \sin^{-1} x + C}$$

Here is the working in detail. Follow it through.

$$(1 - x^2)\frac{dy}{dx} - xy = 1$$

$$\therefore \frac{dy}{dx} - \frac{x}{1 - x^2} \cdot y = \frac{1}{1 - x^2}$$

$$\text{IF} = e^{\int P \, dx} \qquad \int P \, dx = \int \frac{-x}{1 - x^2} \, dx = \frac{1}{2}\ln(1 - x^2)$$

$$\therefore \text{IF} = e^{\frac{1}{2}\ln(1-x^2)} = e^{\ln\{(1-x^2)^{\frac{1}{2}}\}} = (1 - x^2)^{\frac{1}{2}}$$

$$\text{Now } y \cdot \text{IF} = \int Q \cdot \text{IF} \, dx$$

$$\therefore \; y\sqrt{1 - x^2} = \int \frac{1}{1 - x^2}\sqrt{1 - x^2} \cdot dx$$

$$= \int \frac{1}{\sqrt{1 - x^2}} \, dx = \sin^{-1} x + C$$

$$y\sqrt{1 - x^2} = \sin^{-1} x + C$$

Now on to Frame 61

61

In practically all the examples so far, we have been concerned with finding the general solutions. If further information is available, of course, particular solutions can be obtained. Here is one final example for you to do.

Example 7

Solve the equation

$$(x - 2)\frac{dy}{dx} - y = (x - 2)^3$$

given that $y = 10$ when $x = 4$.

Off you go then. It is quite straightforward.

When you have finished, move on to Frame 62 and check your solution

62

$$\boxed{2y = (x - 2)^3 + 6(x - 2)}$$

Here it is:

$$(x - 2)\frac{dy}{dx} - y = (x - 2)^3$$

$$\frac{dy}{dx} - \frac{1}{x - 2} \cdot y = (x - 2)^2$$

$$\int P\,dx = \int \frac{-1}{x - 2}\,dx = -\ln(x - 2)$$

$$\therefore \ \text{IF} = e^{-\ln(x-2)} = e^{\{\ln\{(x-2)^{-1}\}\}} = (x - 2)^{-1}$$

$$= \frac{1}{x - 2}$$

$$\therefore \ y \cdot \frac{1}{x - 2} = \int (x - 2)^2 \cdot \frac{1}{(x - 2)}\,dx$$

$$= \int (x - 2)\,dx$$

$$= \frac{(x - 2)^2}{2} + C$$

$$\therefore \ y = \frac{(x - 2)^3}{2} + C(x - 2) \qquad \text{General solution}$$

When $x = 4$, $y = 10$:

$$10 = \frac{8}{2} + C \cdot 2 \quad \therefore \ 2C = 6 \quad \therefore \ C = 3$$

$$\therefore \ 2y = (x - 2)^3 + 6(x - 2)$$

Review exercise

Finally, for this part of the Program, here is a short review exercise.

Solve the following:

1 $\dfrac{dy}{dx} + 3y = e^{4x}$

2 $x\dfrac{dy}{dx} + y = x\sin x$

3 $\tan x\dfrac{dy}{dx} + y = \sec x$

Work through them all: then check your results with those given in Frame 64

1 $\quad y = \dfrac{e^{4x}}{7} + Ce^{-3x}$ $\qquad$ (IF $= e^{3x}$)

2 $\quad xy = \sin x - x\cos x + C$ $\qquad$ (IF $= x$)

3 $\quad y\sin x = x + C$ $\qquad$ (IF $= \sin x$)

There is just one other type of equation that we must consider. Here is an example: let us see how it differs from those we have already dealt with.

To solve $\dfrac{dy}{dx} + \dfrac{1}{x}\cdot y = xy^2$

Note that if it were not for the factor y^2 on the right-hand side, this equation would be of the form $\dfrac{dy}{dx} + Py = Q$ that we know of old.

To see how we deal with this new kind of equation, we will consider the general form, so move on to Frame 65.

Bernoulli's equation

These are equations of the form:

$$\frac{dy}{dx} + Py = Qy^n$$

where, as before, P and Q are functions of x (or constants).

The trick is the same every time:

(a) Divide both sides by y^n. This gives:

$$y^{-n}\frac{dy}{dx} + Py^{1-n} = Q$$

(b) Now put $z = y^{1-n}$

so that, differentiating, $\dfrac{dz}{dx} = \dfrac{(1-n)y^{-n}dy}{dx}$

66

$$\frac{dz}{dx} = (1-n)y^{-n}\frac{dy}{dx}$$

So we have:

$$\frac{dy}{dx} + Py = Qy^n \qquad\qquad (1)$$

$$\therefore\; y^{-n}\frac{dy}{dx} + Py^{1-n} = Q \qquad\qquad (2)$$

Put $z = y^{1-n}$ so that $\dfrac{dz}{dx} = (1-n)y^{-n}\dfrac{dy}{dx}$

If we now multiply (2) by $(1-n)$ we shall convert the first term into $\dfrac{dz}{dx}$.

$$(1-n)y^{-n}\frac{dy}{dx} + (1-n)Py^{1-n} = (1-n)Q$$

Remembering that $z = y^{1-n}$ and that $\dfrac{dz}{dx} = (1-n)y^{-n}\dfrac{dy}{dx}$, this last line can now be written $\dfrac{dz}{dx} + P_1 z = Q_1$ with P_1 and Q_1 functions of x.

This we can solve by use of an integrating factor in the normal way. Finally, having found z, we convert back to y using $z = y^{1-n}$.

Let us see this routine in operation – so on to Frame 67

67

Example 1

Solve $\quad \dfrac{dy}{dx} + \dfrac{1}{x}y = xy^2$

(a) Divide through by y^2, giving $\dots\dots\dots$

68

$$y^{-2}\frac{dy}{dx} + \frac{1}{x}\cdot y^{-1} = x$$

(b) Now put $z = y^{1-n}$, i.e. in this case $z = y^{1-2} = y^{-1}$

$$z = y^{-1} \quad \therefore\; \frac{dz}{dx} = -y^{-2}\frac{dy}{dx}$$

(c) Multiply through the equation by (-1), to make the first term $\dfrac{dz}{dx}$.

$$-y^{-2}\frac{dy}{dx} - \frac{1}{x}y^{-1} = -x$$

so that $\dfrac{dz}{dx} - \dfrac{1}{x}z = -x$ which is of the form $\dfrac{dz}{dx} + Pz = Q$ so that you can now solve the equation by the normal integrating factor method. What do you get?

When you have done it, move on to the next frame

69

$$y = (Cx - x^2)^{-1}$$

Check the working:

$$\frac{dz}{dx} - \frac{1}{x}z = -x$$

$$\text{IF} = e^{\int P dx} \qquad \int P\,dx = \int -\frac{1}{x}\,dx = -\ln x$$

$$\therefore \ \text{IF} = e^{-\ln x} = e^{\ln(x^{-1})} = x^{-1} = \frac{1}{x}$$

$$z \cdot \text{IF} = \int Q \cdot \text{IF}\,dx \quad \therefore \ z\frac{1}{x} = \int -x \cdot \frac{1}{x}\,dx$$

$$\therefore \ \frac{z}{x} = \int -1\,dx = -x + C$$

$$\therefore \ z = Cx - x^2$$

But $z = y^{-1}$ $\quad \therefore \ \dfrac{1}{y} = Cx - x^2 \quad \therefore \ y = (Cx - x^2)^{-1}$

Right! Here is another.

Example 2

Solve $\quad x^2 y - x^3 \dfrac{dy}{dx} = y^4 \cos x$

First of all, we must rewrite this in the form $\dfrac{dy}{dx} + Py = Qy^n$

So, what do we do?

70

$$\text{Divide both sides by } (-x^3)$$

giving $\dfrac{dy}{dx} - \dfrac{1}{x} \cdot y = -\dfrac{y^4 \cos x}{x^3}$

Now divide by the power of y on the RHS giving ....

71

$$y^{-4}\frac{dy}{dx} - \frac{1}{x}y^{-3} = -\frac{\cos x}{x^3}$$

Next we make the substitution $z = y^{1-n}$ which, in this example, is $z = y^{1-4} = y^{-3}$

$$\therefore \ z = y^{-3} \quad \text{and} \quad \therefore \ \frac{dz}{dx} = \underset{\overline{dx}}{-3y^{-4}\frac{dy}{dx}}$$

72

$$\frac{dz}{dx} = -3y^{-4}\frac{dy}{dx}$$

If we now multiply the equation by (-3) to make the first term into $\dfrac{dz}{dx}$, we have:

$$-3y^{-4}\frac{dy}{dx} + 3\frac{1}{x}\cdot y^{-3} = \frac{3\cos x}{x^3}$$

$$\text{i.e.}\quad \frac{dz}{dx} + \frac{3}{x}z = \frac{3\cos x}{x^3}$$

This you can now solve to find z and so back to y.

Finish it off and then check with the next frame

73

$$y^3 = \frac{x^3}{3\sin x + C}$$

Because

$$\frac{dz}{dx} + \frac{3}{x}\cdot z = \frac{3\cos x}{x^3}$$

$$\text{IF} = e^{\int P dx} \qquad \int P\, dx = \int \frac{3}{x}\, dx = 3\ln x$$

$$\therefore\ \text{IF} = e^{3\ln x} = e^{\ln(x^3)} = x^3$$

$$z.\text{IF} = \int Q.\text{IF}\, dx$$

$$\therefore\ zx^3 = \int \frac{3\cos x}{x^3} x^3\, dx$$

$$= \int 3\cos x\, dx$$

$$\therefore\ zx^3 = 3\sin x + C$$

But, in this example, $z = y^{-3}$

$$\therefore\ \frac{x^3}{y^3} = 3\sin x + C$$

$$\therefore\ y^3 = \frac{x^3}{3\sin x + C}$$

Let us look at the complete solution as a whole, so on to Frame 74

Here it is:

To solve
$$x^2 y - x^3 \frac{dy}{dx} = y^4 \cos x$$

$$\therefore \quad \frac{dy}{dx} - \frac{1}{x} y = -\frac{y^4 \cos x}{x^3}$$

$$\therefore \quad y^{-4} \frac{dy}{dx} - \frac{1}{x} y^{-3} = -\frac{\cos x}{x^3}$$

Put $\quad z = y^{1-n} = y^{1-4} = y^{-3} \qquad \therefore \quad \frac{dz}{dx} = -3y^{-4} \frac{dy}{dx}$

Equation becomes

$$-3y^{-4} \frac{dy}{dx} + \frac{3}{x} \cdot y^{-3} = \frac{3 \cos x}{x^3}$$

i.e. $\quad \dfrac{dz}{dx} + \dfrac{3}{x} \cdot z = \dfrac{3 \cos x}{x^3}$

IF $= e^{\int P \, dx} \qquad \displaystyle\int P \, dx = \int \frac{3}{x} \, dx = 3 \ln x$

$$\therefore \quad \text{IF} = e^{3 \ln x} = e^{\ln(x^3)} = x^3$$

$$\therefore \quad zx^3 = \int \frac{3 \cos x}{x^3} x^3 \, dx$$

$$= \int 3 \cos x \, dx$$

$$\therefore \quad zx^3 = 3 \sin x + C$$

But $\quad z = y^{-3}$

$$\therefore \quad \frac{x^3}{y^3} = 3 \sin x + C$$

$$\therefore \quad y^3 = \frac{x^3}{3 \sin x + C}$$

They are all done in the same way. Once you know the trick, the rest is very straightforward.

On to the next frame

Here is one for you to do entirely on your own.

Example 3

Solve $\quad 2y - 3 \dfrac{dy}{dx} = y^4 e^{3x}$

Work through the same steps as before. When you have finished, check your working with the solution in Frame 76.

76

$$y^3 = \frac{5e^{2x}}{e^{5x} + A}$$

Solution in detail:

$$2y - 3\frac{dy}{dx} = y^4 e^{3x}$$

$$\therefore \frac{dy}{dx} - \frac{2}{3}y = -\frac{y^4 e^{3x}}{3}$$

$$\therefore y^{-4}\frac{dy}{dx} - \frac{2}{3}y^{-3} = -\frac{e^{3x}}{3}$$

Put $z = y^{1-4} = y^{-3}$ $\therefore \frac{dz}{dx} = -3y^{-4}\frac{dy}{dx}$

Multiplying through by (-3), the equation becomes:

$$-3y^{-4}\frac{dy}{dx} + 2y^{-3} = e^{3x}$$

i.e. $\dfrac{dz}{dx} + 2z = e^{3x}$

IF $= e^{\int P dx}$ $\int P\,dx = \int 2\,dx = 2x$ $\therefore$ IF $= e^{2x}$

$$\therefore ze^{2x} = \int e^{3x}e^{2x}dx = \int e^{5x}dx$$

$$= \frac{e^{5x}}{5} + C$$

But $z = y^{-3}$ $\therefore \dfrac{e^{2x}}{y^3} = \dfrac{e^{5x} + A}{5}$

$$\therefore y^3 = \frac{5e^{2x}}{e^{5x} + A}$$

On to Frame 77

77

Finally, one example for you, just to be sure.

Example 4

Solve $y - 2x\dfrac{dy}{dx} = x(x+1)y^3$

First rewrite the equation in standard form $\dfrac{dy}{dx} + Py = Qy^n$

This gives ....

78

$$\frac{dy}{dx} - \frac{1}{2x} \cdot y = -\frac{(x+1)y^3}{2}$$

Now off you go and complete the solution. When you have finished, check with the working in Frame 79.

$$y^2 = \frac{6x}{2x^3 + 3x^2 + A}$$

Working:

$$\frac{dy}{dx} - \frac{1}{2x} \cdot y = -\frac{(x+1)y^3}{2}$$

$$\therefore \; y^{-3} \frac{dy}{dx} - \frac{1}{2x} \cdot y^{-2} = -\frac{(x+1)}{2}$$

Put $z = y^{1-3} = y^{-2}$ $\therefore \; \dfrac{dz}{dx} = -2y^{-3} \dfrac{dy}{dx}$

Equation becomes:

$$-2y^{-3} \frac{dy}{dx} + \frac{1}{x} \cdot y^{-2} = x + 1$$

i.e. $\dfrac{dz}{dx} + \dfrac{1}{x} \cdot z = x + 1$

IF $= e^{\int P dx}$ $\displaystyle\int P\,dx = \int \frac{1}{x}\,dx = \ln x$

$$\therefore \; \text{IF} = e^{\ln x} = x$$

$z.\text{IF} = \displaystyle\int Q.\text{IF}\,dx$ $\therefore \; zx = \displaystyle\int (x+1)x\,dx$

$$= \int (x^2 + x)\,dx$$

$$\therefore \; zx = \frac{x^3}{3} + \frac{x^2}{2} + C$$

But $z = y^{-2}$ $\therefore \; \dfrac{x}{y^2} = \dfrac{2x^3 + 3x^2 + A}{6}$ $(A = 6C)$

$$\therefore \; y^2 = \frac{6x}{2x^3 + 3x^2 + A}$$

There we are. You have now reached the end of this Program, except for the **Can You?** checklist and the **Test exercise** which follow. Before you tackle them, however, read down the **Review summary** presented in the next frame. It will remind you of the main points that we have covered in this Program on first-order differential equations.

Move on then to Frame 80

80 Review summary

1 The *order* of a differential equation is given by the highest derivative present.

An equation of *order n* is derived from a function containing *n arbitrary constants*.

2 *Solution of first-order differential equations*

(a) By direct integration: $\dfrac{dy}{dx} = f(x)$

gives $y = \displaystyle\int f(x)\,dx$

(b) By separating the variables: $F(y) \cdot \dfrac{dy}{dx} = f(x)$

gives $\displaystyle\int F(y)\,dy = \int f(x)\,dx$

(c) Homogeneous equations: Substituting $y = vx$

gives $v + x\dfrac{dv}{dx} = F(v)$

(d) Linear equations: $\dfrac{dy}{dx} + Py = Q$

Integrating factor, $\mathrm{IF} = e^{\int P\,dx}$

and remember that $e^{\ln F} = F$

gives $y\mathrm{IF} = \displaystyle\int Q \cdot \mathrm{IF}\,dx$

(e) Bernoulli's equation: $\dfrac{dy}{dx} + Py = Qy^n$

Divide by y^n: then put $z = y^{1-n}$

Reduces to type (d) above.

✅ Can You?

81 Checklist 1

Check this list before and after you try the end of Program test.

On a scale of 1 to 5 how confident are you that you can: Frames

- Recognize the order of a differential equation?

  Yes ☐ ☐ ☐ ☐ ☐ *No*   1 to 3

- Appreciate that a differential equation of order *n* can be derived from a function containing *n* arbitrary constants?

  Yes ☐ ☐ ☐ ☐ ☐ *No*   4 to 10

- Solve certain first-order differential equations by direct integration?

  Yes ☐ ☐ ☐ ☐ ☐ *No*   11 to 13

▶

Frames

- Solve certain first-order differential equations by separating the variables?

  Yes ☐ ☐ ☐ ☐ ☐ *No*

  14 to **26**

- Solve certain first-order homogeneous differential equations by an appropriate substitution?

  Yes ☐ ☐ ☐ ☐ ☐ *No*

  27 to **42**

- Solve certain first-order differential equations by using an integrating factor?

  Yes ☐ ☐ ☐ ☐ ☐ *No*

  43 to **64**

- Solve Bernoulli's equation

  Yes ☐ ☐ ☐ ☐ ☐ *No*

  65 to **79**

 Test exercise 1

The questions are similar to the equations you have been solving in the Program. They cover all the methods, but are quite straightforward. Do not hurry: take your time and work carefully and you will find no difficulty with them.

82

Solve the following differential equations:

1 $x\dfrac{dy}{dx} = x^2 + 2x - 3$ Direct

2 $(1+x)^2\dfrac{dy}{dx} = 1 + y^2$ Direct

3 $\dfrac{dy}{dx} + 2y = e^{3x}$ Linear

4 $x\dfrac{dy}{dx} - y = x^2$ Linear

5 $x^2\dfrac{dy}{dx} = x^3\sin 3x + 4$ Direct + u.v − v∫u

6 $x\cos y\dfrac{dy}{dx} - \sin y = 0$ Direct

7 $(x^3 + xy^2)\dfrac{dy}{dx} = 2y^3$ Substitution

8 $(x^2 - 1)\dfrac{dy}{dx} + 2xy = x$

9 $\dfrac{dy}{dx} + y\tanh x = 2\sinh x$

10 $x\dfrac{dy}{dx} - 2y = x^3\cos x$

11 $\dfrac{dy}{dx} + \dfrac{y}{x} = y^3$

12 $x\dfrac{dy}{dx} + 3y = x^2y^2$

🚲 **Further problems 1**

83 Solve the following equations.

I. *Separating the variables*

1 $x(y-3)\dfrac{dy}{dx} = 4y$

2 $(1+x^3)\dfrac{dy}{dx} = x^2 y$, given that $x = 1$ when $y = 2$.

3 $x^3 + (y+1)^2 \dfrac{dy}{dx} = 0$

4 $\cos y + (1 + e^{-x})\sin y \dfrac{dy}{dx} = 0$, given that $y = \pi/4$ when $x = 0$.

5 $x^2(y+1) + y^2(x-1)\dfrac{dy}{dx} = 0$

II. *Homogeneous equations*

6 $(2y-x)\dfrac{dy}{dx} = 2x + y$, given that $y = 3$ when $x = 2$.

7 $(xy + y^2) + (x^2 - xy)\dfrac{dy}{dx} = 0$

8 $(x^3 + y^3) = 3xy^2 \dfrac{dy}{dx}$

9 $y - 3x + (4y + 3x)\dfrac{dy}{dx} = 0$

10 $(x^3 + 3xy^2)\dfrac{dy}{dx} = y^3 + 3x^2 y$

III. *Integrating factor*

11 $x\dfrac{dy}{dx} - y = x^3 + 3x^2 - 2x$

12 $\dfrac{dy}{dx} + y\tan x = \sin x$

13 $x\dfrac{dy}{dx} - y = x^3 \cos x$, given that $y = 0$ when $x = \pi$.

14 $(1+x^2)\dfrac{dy}{dx} + 3xy = 5x$, given that $y = 2$ when $x = 1$.

15 $\dfrac{dy}{dx} + y\cot x = 5e^{\cos x}$, given that $y = -4$ when $x = \pi/2$.

▶

IV. *Transformations.* Make the given substitutions and work in much the same way as for first-order homogeneous equations.

16 $(3x + 3y - 4)\dfrac{dy}{dx} = -(x + y)$ Put $x + y = v$

17 $(y - xy^2) = (x + x^2 y)\dfrac{dy}{dx}$ Put $y = \dfrac{v}{x}$

18 $(x - y - 1) + (4y + x - 1)\dfrac{dy}{dx} = 0$ Put $v = x - 1$

19 $(3y - 7x + 7) + (7y - 3x + 3)\dfrac{dy}{dx} = 0$ Put $v = x - 1$

20 $y(xy + 1) + x(1 + xy + x^2 y^2)\dfrac{dy}{dx} = 0$ Put $y = \dfrac{v}{x}$

V. *Bernoulli's equation*

21 $\dfrac{dy}{dx} + y = xy^3$

22 $\dfrac{dy}{dx} + y = y^4 e^x$

23 $2\dfrac{dy}{dx} + y = y^3(x - 1)$

24 $\dfrac{dy}{dx} - 2y \tan x = y^2 \tan^2 x$

25 $\dfrac{dy}{dx} + y \tan x = y^3 \sec^4 x$

VI. *Miscellaneous.* Choose the appropriate method in each case.

26 $(1 - x^2)\dfrac{dy}{dx} = 1 + xy$

27 $xy\dfrac{dy}{dx} - (1 + x)\sqrt{y^2 - 1} = 0$

28 $(x^2 - 2xy + 5y^2) = (x^2 + 2xy + y^2)\dfrac{dy}{dx}$

29 $\dfrac{dy}{dx} - y \cot x = y^2 \sec^2 x$, given $y = -1$ when $x = \pi/4$.

30 $y + (x^2 - 4x)\dfrac{dy}{dx} = 0$

VII. *Further examples*

31 Solve the equation $\dfrac{dy}{dx} - y \tan x = \cos x - 2x \sin x$, given that $y = 0$ when $x = \pi/6$.

32 Find the general solution of the equation

$$\frac{dy}{dx} = \frac{2xy + y^2}{x^2 + 2xy}.$$

▶

33 Find the general solution of $(1 + x^2)\dfrac{dy}{dx} = x(1 + y^2)$.

34 Solve the equation $x\dfrac{dy}{dx} + 2y = 3x - 1$, given that $y = 1$ when $x = 2$.

35 Solve $x^2\dfrac{dy}{dx} = y^2 - xy\dfrac{dy}{dx}$, given that $y = 1$ when $x = 1$.

36 Solve $\dfrac{dy}{dx} = e^{3x-2y}$, given that $y = 0$ when $x = 0$.

37 Find the particular solution of $\dfrac{dy}{dx} + \dfrac{1}{x} \cdot y = \sin 2x$, such that $y = 2$ when $x = \pi/4$.

38 Find the general solution of $y^2 + x^2\dfrac{dy}{dx} = xy\dfrac{dy}{dx}$.

39 Obtain the general solution of the equation $2xy\dfrac{dy}{dx} = x^2 - y^2$.

40 By substituting $z = x - 2y$, solve the equation $\dfrac{dy}{dx} = \dfrac{x - 2y + 1}{2x - 4y}$, given that $y = 1$ when $x = 1$.

41 Find the general solution of $(1 - x^3)\dfrac{dy}{dx} + x^2y = x^2(1 - x^3)$.

42 Solve $\dfrac{dy}{dx} + \dfrac{y}{x} = \sin x$, given $y = 0$ when $x = \pi/2$.

43 Solve $\dfrac{dy}{dx} + x + xy^2 = 0$, given $y = 0$ when $x = 1$.

44 Determine the general solution of the equation $\dfrac{dy}{dx} + \left\{\dfrac{1}{x} - \dfrac{2x}{1 - x^2}\right\}y = \dfrac{1}{1 - x^2}$

45 Solve $(1 + x^2)\dfrac{dy}{dx} + xy = (1 + x^2)^{3/2}$.

46 Solve $x(1 + y^2) - y(1 + x^2)\dfrac{dy}{dx} = 0$, given $y = 2$ when $x = 0$.

47 Solve $\dfrac{r\tan\theta}{a^2 - r^2} \cdot \dfrac{dr}{d\theta} = 1$, given $r = 0$ when $\theta = \pi/4$.

48 Solve $\dfrac{dy}{dx} + y\cot x = \cos x$, given $y = 0$ when $x = 0$.

49 Use the substitution $y = \dfrac{v}{x}$, where v is a function of x only, to transform the equation $\dfrac{dy}{dx} + \dfrac{y}{x} = xy^2$ into a differential equation in v and x. Hence find y in terms of x.

50 The rate of decay of a radioactive substance is proportional to the amount A remaining at any instant. If $A = A_0$ at $t = 0$, prove that, if the time taken for the amount of the substance to become $\dfrac{1}{2}A_0$ is T, then $A = A_0e^{-(t\ln 2)/T}$. Prove also that the time taken for the amount remaining to be reduced to $\dfrac{1}{20}A_0$ is $4.32T$.

Second-order differential equations

Frames
1 to 52

Learning outcomes

When you have completed this Program you will be able to:

- Use the auxiliary equation to solve certain second-order homogeneous equations
- Use the complementary function and the particular integral to solve certain second-order inhomogeneous equations

1 Homogeneous equations

Many practical problems in engineering give rise to second-order differential equations of the form

$$a\frac{d^2y}{dx^2} + b\frac{dy}{dx} + cy = f(x)$$

where a, b and c are constant coefficients and $f(x)$ is a given function of x. By the end of this Program you will have no difficulty with equations of this type.

Let us first take the case where $f(x) = 0$, so that the equation becomes

$$a\frac{d^2y}{dx^2} + b\frac{dy}{dx} + cy = 0$$

Let $y = u$ and $y = v$ (where u and v are functions of x) be two solutions of the equation:

$$\therefore \quad a\frac{d^2u}{dx^2} + b\frac{du}{dx} + cu = 0 \quad \text{and} \quad a\frac{d^2v}{dx^2} + b\frac{dv}{dx} + cv = 0$$

Adding these two lines together, we get:

$$a\left(\frac{d^2u}{dx^2} + \frac{d^2v}{dx^2}\right) + b\left(\frac{du}{dx} + \frac{dv}{dx}\right) + c(u + v) = 0$$

Now $\dfrac{d}{dx}(u + v) = \dfrac{du}{dx} + \dfrac{dv}{dx}$ and $\dfrac{d^2}{dx^2}(u + v) = \dfrac{d^2u}{dx^2} + \dfrac{d^2v}{dx^2}$, therefore the equation can be written

$$a\frac{d^2}{dx^2}(u + v) + b\frac{d}{dx}(u + v) + c(u + v) = 0$$

which is our original equation with y replaced by $(u + v)$.

i.e. If $y = u$ and $y = v$ are solutions of the equation

$$a\frac{d^2y}{dx^2} + b\frac{dy}{dx} + cy = 0, \text{ so also is } y = u + v.$$

This is an important result and we shall be referring to it later, so make a note of it in your record book.

Move on to Frame 2

2

Our equation was $a\dfrac{d^2y}{dx^2} + b\dfrac{dy}{dx} + cy = 0$. If $a = 0$, we get the first-order equation of the same family:

$$b\frac{dy}{dx} + cy = 0 \quad \text{i.e.} \quad \frac{dy}{dx} + ky = 0 \quad \text{where } k = \frac{c}{b}$$

Solving this by the method of separating variables, we have

$$\frac{dy}{dx} = -ky \quad \therefore \quad \int\frac{dy}{y} = -\int k\,dx$$

which gives $\ln y = -kx + C$

3

$$\ln y = -kx + c$$

$\therefore\ y = e^{-kx+c} = e^{-kx} \cdot e^{c} = A e^{-kx}$ (since e^c is a constant)

i.e. $y = A e^{-kx}$

If we write the symbol m for $-k$, the solution is $y = A e^{mx}$

In the same way, $y = A e^{mx}$ will be a solution of the second-order equation

$a\dfrac{d^2y}{dx^2} + b\dfrac{dy}{dx} + cy = 0$, if it satisfies this equation.

Now, if $\quad y = A e^{mx}$

$$\frac{dy}{dx} = Am e^{mx}$$

$$\frac{d^2y}{dx^2} = Am^2 e^{mx}$$

and substituting these expressions for the differential coefficients in the left-hand side of the equation, we get $aAm^2e^{mx} + bAme^{mx} + cAe^{mx} = 0$

On to Frame 4

4

$$aAm^2 e^{mx} + bAm e^{mx} + cA e^{mx} = 0$$

Right. So dividing both sides by $A e^{mx}$ we obtain

$am^2 + bm + c = 0$

which is a quadratic equation giving two values for m. Let us call these

$m = m_1$ and $m = m_2$

i.e. $y = A e^{m_1 x}$ and $y = B e^{m_2 x}$ are two solutions of the given equation.

Now we have already seen that if $y = u$ and $y = v$ are two solutions so also is $y = u + v$.

$\therefore$ If $y = A e^{m_1 x}$ and $y = B e^{m_2 x}$ are solutions so also is

$y = A e^{m_1 x} + B e^{m_2 x}$

Note: This contains the necessary two arbitrary constants for a second-order differential equation, so there can be no further solution.

Move to Frame 5

5

The solution, then, of $a\dfrac{d^2y}{dx^2} + b\dfrac{dy}{dx} + cy = 0$ is seen to be

$y = A e^{m_1 x} + B e^{m_2 x}$

where A and B are two arbitrary constants and m_1 and m_2 are the roots of the quadratic equation $am^2 + bm + c = 0$.

▶

This quadratic equation is called the *auxiliary equation* and is obtained directly from the equation $a\dfrac{d^2y}{dx^2} + b\dfrac{dy}{dx} + cy = 0$, by writing m^2 for $\dfrac{d^2y}{dx^2}$, m for $\dfrac{dy}{dx}$, 1 for y.

Example

For the equation $2\dfrac{d^2y}{dx^2} + 5\dfrac{dy}{dx} + 6y = 0$, the auxiliary equation is $2m^2 + 5m + 6 = 0$.

   In the same way, for the equation $\dfrac{d^2y}{dx^2} + 3\dfrac{dy}{dx} + 2y = 0$, the auxiliary equation is $m^2 + 3m + 2 = 0$

Then on to Frame 6

6

$$m^2 + 3m + 2 = 0$$

Since the auxiliary equation is always a quadratic equation, the values of m can be determined in the usual way.

   i.e. if $m^2 + 3m + 2 = 0$

$$(m+1)(m+2) = 0 \quad \therefore \ m = -1 \text{ and } m = -2$$

$\therefore$ the solution of $\dfrac{d^2y}{dx^2} + 3\dfrac{dy}{dx} + 2y = 0$ is

$$y = Ae^{-x} + Be^{-2x}$$

In the same way, if the auxiliary equation were $m^2 + 4m - 5 = 0$, this factorizes into $(m+5)(m-1) = 0$ giving $m = 1$ or -5, and in this case the solution would be $y = Ae^{x} + Be^{-5x}$

7

$$y = Ae^{x} + Be^{-5x}$$

The type of solution we get depends on the roots of the auxiliary equation.

1 *Real and different roots* to the auxiliary equation

Example 1

$$\dfrac{d^2y}{dx^2} + 5\dfrac{dy}{dx} + 6y = 0$$

Auxiliary equation: $m^2 + 5m + 6 = 0$

$\therefore \ (m+2)(m+3) = 0 \quad \therefore \ m = -2 \text{ or } m = -3$

$\therefore$ Solution is $y = Ae^{-2x} + Be^{-3x}$

Example 2

$$\frac{d^2y}{dx^2} - 7\frac{dy}{dx} + 12y = 0$$

Auxiliary equation: $m^2 - 7m + 12 = 0$

$(m-3)(m-4) = 0$ $\therefore$ $m = 3$ or $m = 4$

So the solution is $y = \ldots$ $Ae^{3x} + Be^{4x}$

Move to Frame 8

8

$$y = Ae^{3x} + Be^{4x}$$

Here you are. Do this one.

Solve the equation $\dfrac{d^2y}{dx^2} + 3\dfrac{dy}{dx} - 10y = 0$

Auxiliary eq = $m^2 + 3x - 10 = 0$ or $m+5$ $m+2$ $m = -5$
$m = 2$

When you have finished, move on to Frame 9

9

$$y = Ae^{2x} + Be^{-5x}$$

Now consider the next case.

2 *Real and equal roots* **to the auxiliary equation**

Let us take $\dfrac{d^2y}{dx^2} + 6\dfrac{dy}{dx} + 9y = 0.$

The auxiliary equation is: $m^2 + 6m + 9 = 0$

$\therefore$ $(m+3)(m+3) = 0$ $\therefore$ $m = -3$ (twice)

If $m_1 = -3$ and $m_2 = -3$ then these would give the solution $y = Ae^{-3x} + Be^{-3x}$ and their two terms would combine to give $y = Ce^{-3x}$. But every second-order differential equation has two arbitrary constants, so there must be another term containing a second constant. In fact, it can be shown that $y = Kxe^{-3x}$ also satisfies the equation, so that the complete general solution is of the form $y = Ae^{-3x} + Bxe^{-3x}$

i.e. $y = e^{-3x}(A + Bx)$

In general, if the auxiliary equation has real and equal roots, giving $m = m_1$ twice, the solution of the differential equation is

$y = e^{m_1 x}(A + Bx)$

Make a note of this general statement and then move on to Frame 10

10

Here is an example:

Example 1

Solve $\dfrac{d^2y}{dx^2} + 4\dfrac{dy}{dx} + 4y = 0$

Auxiliary equation: $m^2 + 4m + 4 = 0$

$(m+2)(m+2) = 0$ $\therefore$ $m = -2$ (twice)

The solution is: $y = e^{-2x}(A + Bx)$

Here is another:

Example 2

Solve $\dfrac{d^2y}{dx^2} + 10\dfrac{dy}{dx} + 25y = 0$

Auxiliary equation: $m^2 + 10m + 25 = 0$

$(m+5)^2 = 0$ $\therefore$ $m = -5$ (twice)

$y = e^{-5x}(A + Bx)$

Example 3

Now here is one for you to do:

Solve $\dfrac{d^2y}{dx^2} + 8\dfrac{dy}{dx} + 16y = 0$

When you have done it, move on to Frame 11

11

$$\boxed{y = e^{-4x}(A + Bx)}$$

Because if $\dfrac{d^2y}{dx^2} + 8\dfrac{dy}{dx} + 16y = 0$

the auxiliary equation is

$m^2 + 8m + 16 = 0$

$\therefore (m+4)^2 = 0$ $\therefore$ $m = -4$ (twice)

$\therefore y = e^{-4x}(A + Bx)$

So, for *real and different roots* $m = m_1$ and $m = m_2$ the solution is

$y = Ae^{m_1 x} + Be^{m_2 x}$

and for *real and equal roots* $m = m_1$ (twice) the solution is

$y = e^{m_1 x}(A + Bx)$

Just find the values of m from the auxiliary equation and then substitute these values in the appropriate form of the result.

Move to Frame 12

3 *Complex roots* to the auxiliary equation

12

Now let us see what we get when the roots of the auxiliary equation are complex.

Suppose $m = \alpha \pm i\beta$, i.e. $m_1 = \alpha + i\beta$ and $m_2 = \alpha - i\beta$. Then the solution would be of the form:

$$y = Ce^{(\alpha+i\beta)x} + De^{(\alpha-i\beta)x} = Ce^{\alpha x} \cdot e^{i\beta x} + De^{\alpha x} \cdot e^{-i\beta x}$$
$$= e^{\alpha x}\{Ce^{i\beta x} + De^{-i\beta x}\}$$

Now from our previous work on complex numbers, we know that:

$$e^{ix} = \cos x + i\sin x$$
$$e^{-ix} = \cos x - i\sin x$$

and that $\begin{cases} e^{i\beta x} = \cos \beta x + i\sin \beta x \\ e^{-i\beta x} = \cos \beta x - i\sin \beta x \end{cases}$

Our solution above can therefore be written:

$$y = e^{\alpha x}\{C(\cos \beta x + i\sin \beta x) + D(\cos \beta x - i\sin \beta x)\}$$
$$= e^{\alpha x}\{(C + D)\cos \beta x + i(C - D)\sin \beta x\}$$
$$y = e^{\alpha x}\{A\cos \beta x + B\sin \beta x\}$$
$$\text{where} \quad A = C + D \quad \text{and} \quad B = i(C - D)$$

$\therefore$ If $m = \alpha \pm i\beta$, the solution can be written in the form:

$$y = e^{\alpha x}\Big\{A\cos \beta x + B\sin \beta x\Big\}$$

Here is an example: If $m = -2 \pm 3i$

then $y = e^{-2x}\Big\{A\cos 3x + B\sin 3x\Big\}$

Similarly, if $m = 5 \pm 2i$ then $y = e^{5x}\Big(A\cos 2x + B\sin 2x\Big)$

$$y = e^{5x}[A\cos 2x + B\sin 2x]$$

Here is one of the same kind:

Solve $\dfrac{d^2y}{dx^2} + 4\dfrac{dy}{dx} + 9y = 0$

Auxiliary equation: $m^2 + 4m + 9 = 0$

$$\therefore m = \frac{-4 \pm \sqrt{16 - 36}}{2} = \frac{-4 \pm \sqrt{-20}}{2} = \frac{-4 \pm 2i\sqrt{5}}{2} = -2 \pm i\sqrt{5}$$

In this case $\alpha = -2$ and $\beta = \sqrt{5}$

Solution is: $y = e^{-2x}(A\cos \sqrt{5}x + B\sin \sqrt{5}x)$

Now you can solve this one: $\dfrac{d^2y}{dx^2} - 2\dfrac{dy}{dx} + 10y = 0$

When you have finished it, move on to Frame 14

14

$$y = e^x(A\cos 3x + B\sin 3x)$$

Just check your working:

$$\frac{d^2y}{dx^2} - 2\frac{dy}{dx} + 10y = 0$$

Auxiliary equation: $m^2 - 2m + 10 = 0$

$$m = \frac{2 \pm \sqrt{4-40}}{2} = \frac{2 \pm \sqrt{-36}}{2} = 1 \pm 3i$$

$$y = e^x(A\cos 3x + B\sin 3x)$$

Move to Frame 15

15   Here is a *summary* of the work so far.

Equations of the form $a\dfrac{d^2y}{dx^2} + b\dfrac{dy}{dx} + cy = 0$

Auxiliary equation: $am^2 + bm + c = 0$

1 *Roots real and different* $m = m_1$ and $m = m_2$

Solution is $y = Ae^{m_1x} + Be^{m_2x}$

2 *Real and equal roots* $m = m_1$ (twice)

Solution is $y = e^{m_1x}(A + Bx)$

3 *Complex roots* $m = \alpha \pm i\beta$

Solution is $y = e^{\alpha x}(A\cos\beta x + B\sin\beta x)$

In each case, we simply solve the auxiliary equation to establish the values of m and substitute in the appropriate form of the result.

On to Frame 16

16   We shall now consider equations of the form $\dfrac{d^2y}{dx^2} \pm n^2y = 0$

This is a special case of the equation

$$a\frac{d^2y}{dx^2} + b\frac{dy}{dx} + cy = 0 \text{ when } b = 0$$

i.e. $a\dfrac{d^2y}{dx^2} + cy = 0$ i.e. $\dfrac{d^2y}{dx^2} + \dfrac{c}{a}y = 0$

which can be written as $\dfrac{d^2y}{dx^2} \pm n^2y = 0$ to cover the two cases when the coefficient of y is positive or negative.

(a) If $\dfrac{d^2y}{dx^2} + n^2y = 0$, $m^2 + n^2 = 0$ $\therefore$ $m^2 = -n^2$ $\therefore$ $m = \pm in$

(This is like $m = \alpha \pm i\beta$, when $\alpha = 0$ and $\beta = n$)

$\therefore$ $y = A\cos nx + B\sin nx$

▶

(b) If $\dfrac{d^2y}{dx^2} - n^2y = 0$, $m^2 - n^2 = 0$ ∴ $m^2 = n^2$ ∴ $m = \pm n$

∴ $y = Ce^{nx} + De^{-nx}$

This last result can be written in another form which is sometimes more convenient, so move on to the next frame and we will see what it is.

17

You will remember from your work on hyperbolic functions that

$$\cosh nx = \frac{e^{nx} + e^{-nx}}{2} \quad \therefore \ e^{nx} + e^{-nx} = 2\cosh nx$$

$$\sinh nx = \frac{e^{nx} - e^{-nx}}{2} \quad \therefore \ e^{nx} - e^{-nx} = 2\sinh nx$$

Adding these two results: $\qquad 2e^{nx} = 2\cosh nx + 2\sinh nx$

$$\therefore \ e^{nx} = \cosh nx + \sinh nx$$

Similarly, by subtracting: $\qquad e^{-nx} = \cosh nx - \sinh nx$

Therefore, the solution of our equation $y = Ce^{nx}$ can be written:

$$y = C(\cosh nx + \sinh nx) + D(\cosh nx - \sinh nx)$$
$$= (C + D)\cosh nx + (C - D)\sinh nx$$

i.e. $\ y = A\cosh nx + B\sinh nx$

Note: In this form the two results are very much alike:

(a) $\dfrac{d^2y}{dx^2} + n^2y = 0 \qquad y = A\cos nx + B\sin nx$

(b) $\dfrac{d^2y}{dx^2} - n^2y = 0 \qquad y = A\cosh nx + B\sinh nx$

Make a note of these results in your record book.

Then – next frame

18

Here are some examples:

Example 1

$\dfrac{d^2y}{dx^2} + 16y = 0 \quad \therefore \ m^2 = -16 \quad \therefore \ m = \pm 4i$

$\qquad \therefore \ y = A\cos 4x + B\sin 4x$

Example 2

$\dfrac{d^2y}{dx^2} - 3y = 0 \quad \therefore \ m^2 = 3 \quad \therefore \ m = \pm\sqrt{3}$

$\qquad y = A\cosh\sqrt{3}x + B\sinh\sqrt{3}x$

Similarly

Example 3

$\dfrac{d^2y}{dx^2} + 5y = 0 \qquad m^2 + 5 = 0 \quad m^2 = -5 \quad m = \pm i\sqrt{5}$

$\qquad y = A\cos\sqrt{5}x + B\sin\sqrt{5}x$

Then move on to Frame 19

19

$$y = A \cos \sqrt{5}x + B \sin \sqrt{5}x$$

And now this one:

Example 4

$M^2 - 4 = 0 \quad m^2 = 4 \quad m = \pm 2$

$\dfrac{d^2y}{dx^2} - 4y = 0 \quad \therefore \ m^2 = 4 \quad \therefore \ m = \pm 2$

$y = A\cosh 2x + B\sinh 2x$

20

$$y = A \cosh 2x + B \sinh 2x$$

Now before we go on to the next section of the Program, here is a Review exercise on what we have covered so far. The questions are set out in the next frame. Work them all before checking your results.

So on you go to Frame 21

21   **Review exercise**

Solve the following:

1 $\dfrac{d^2y}{dx^2} - 12\dfrac{dy}{dx} + 36y = 0$

2 $\dfrac{d^2y}{dx^2} + 7y = 0$

3 $\dfrac{d^2y}{dx^2} + 2\dfrac{dy}{dx} - 3y = 0$

4 $2\dfrac{d^2y}{dx^2} + 4\dfrac{dy}{dx} + 3y = 0$

5 $\dfrac{d^2y}{dx^2} - 9y = 0$

For the answers, move to Frame 22

22   Here are the answers:

1 $y = e^{6x}(A + Bx)$

2 $y = A \cos \sqrt{7}x + B \sin \sqrt{7}x$

3 $y = A e^x + B e^{-3x}$

4 $y = e^{-x}\left(A \cos \dfrac{x}{\sqrt{2}} + B \sin \dfrac{x}{\sqrt{2}}\right)$

5 $y = A \cosh 3x + B \sinh 3x$

By now we are ready for the next section of the Program, so move on to Frame 23

Inhomogeneous equations

Complementary function and particular integral

23

So far we have considered equations of the form:

$$a\frac{d^2y}{dx^2} + b\frac{dy}{dx} + cy = f(x) \text{ for the case where } f(x) = 0$$

If $f(x) = 0$, then $am^2 + bm + c = 0$ giving $m = m_1$ and $m = m_2$ and the solution is in general $y = Ae^{m_1x} + Be^{m_2x}$.

In the equation $a\dfrac{d^2y}{dx^2} + b\dfrac{dy}{dx} + cy = f(x)$, the substitution

$y = Ae^{m_1x} + Be^{m_2x}$ would make the left-hand side zero. Therefore, there must be a further term in the solution which will make the LHS equal to $f(x)$ and not zero. The complete solution will therefore be of the form

$y = Ae^{m_1x} + Be^{m_2x} + X$, where X is the extra function yet to be found.

$y = Ae^{m_1x} + Be^{m_2x}$ is called the *complementary function* (CF)

$y = X$ (a function of x) is called the *particular integral* (PI)

Note: The complete general solution is given by:

general solution = complementary function + particular integral

Our main problem at this stage is how are we to find the particular integral for any given equation? This is what we are now going to deal with.

So on then to Frame 24

24

To solve an equation $a\dfrac{d^2y}{dx^2} + n\dfrac{dy}{dx} + cy = f(x)$

(I) The *complementary function* is obtained by solving the equation with $f(x) = 0$, as in the previous part of this Program. This will give one of the following types of solution:

(a) $y = Ae^{m_1x} + Be^{m_2x}$ (b) $y = e^{m_1x}(A + Bx)$

(c) $y = e^{\alpha x}(A \cos \beta x + B \sin \beta x)$ (d) $y = A \cos nx + B \sin nx$

(e) $y = A \cosh nx + B \sinh nx$

(II) The *particular integral* is found by assuming the general form of the function on the right-hand side of the given equation, substituting this in the equation, and equating coefficients. An example will make this clear:

Solve $\dfrac{d^2y}{dx^2} - 5\dfrac{dy}{dx} + 6y = x^2$

(1) *To find the CF* solve LHS $= 0$, i.e. $m^2 - 5m + 6 = 0$

$\therefore (m-2)(m-3) = 0$ $\therefore m = 2$ or $m = 3$

$\therefore$ Complementary function is $y = Ae^{2x} + Be^{3x}$ (1)

▶

(2) *To find the PI* we assume the general form of the RHS which is a second-degree function. Let $y = Cx^2 + Dx + E$.

Then $\dfrac{\mathrm{d}y}{\mathrm{d}x} = 2Cx + D$ and $\dfrac{\mathrm{d}^2 y}{\mathrm{d}x^2} = 2C$

Substituting these in the given equation, we get:

$$2C - 5(2Cx + D) + 6(Cx^2 + Dx + E) = x^2$$

$$2C - 10Cx - 5D + 6Cx^2 + 6Dx + 6E = x^2$$

$$6Cx^2 + (6D - 10C)x + (2C - 5D + 6E) = x^2$$

Equating coefficients of powers of x, we have:

$[x^2]$ $6C = 1$ $\therefore C = \dfrac{1}{6}$

$[x]$ $6D - 10C = 0$ $\therefore 6D = \dfrac{10}{6} = \dfrac{5}{3}$ $\therefore D = \dfrac{5}{18}$

$[CT]$ $2C - 5D + 6E = 0$ $\therefore 6E = \dfrac{25}{18} - \dfrac{2}{6} = \dfrac{19}{18}$ $\therefore E = \dfrac{19}{108}$

$$\therefore \text{ Particular integral is } y = \dfrac{x^2}{6} + \dfrac{5x}{18} + \dfrac{19}{108} \tag{2}$$

Complete general solution = CF + PI

$$\text{General solution is } y = A e^{2x} + B e^{3x} + \dfrac{x^2}{6} + \dfrac{5x}{18} + \dfrac{19}{108}$$

This frame is quite important, since all equations of this type are solved in this way.

On to Frame 25

25

We have seen that to find the particular integral, we assume the general form of the function on the RHS of the equation and determine the values of the constants by substitution in the whole equation and equating coefficients. These will be useful:

If $f(x) = k\ldots$ $\ldots$	Assume	$y = C$
$f(x) = kx\ldots$ $\ldots$		$y = Cx + D$
$f(x) = kx^2 \ldots$ $\ldots$		$y = Cx^2 + Dx + E$
$f(x) = k\sin x$ or $k\cos x$		$y = C\cos x + D\sin x$
$f(x) = k\sinh x$ or $k\cosh x$		$y = C\cosh x + D\sinh x$
$f(x) = e^{kx} \ldots$ $\ldots$		$y = C e^{kx}$

This list covers all the cases you are likely to meet at this stage.

So if the function on the RHS of the equation is $f(x) = 2x^2 + 5$, you would take as the assumed PI:

$y = Cx^2 + Dx + E$

$$\boxed{y = Cx^2 + Dx + E}$$

26

Correct, since the assumed PI will be the general form of the second-degree function.

What would you take as the assumed PI in each of the following cases:

1 $f(x) = 2x - 3$ $Cx+D$
2 $f(x) = e^{5x}$ Ce^{5x}
3 $f(x) = \sin 4x$ $C\cos 4x + D \sin 4x$
4 $f(x) = 3 - 5x^2$ $Cx^2 + Dx + E$
5 $f(x) = 27$ $y = C$
6 $f(x) = 5 \cosh 4x$ $y = C \cosh 4x + D \sinh 4x$

When you have decided all six, check you answers with those in Frame 27

Here are the answers:

27

1 $f(x) = 2x - 3$ PI is of the form $y = Cx + D$
2 $f(x) = e^{5x}$ $f(x) = 5e^{5x}$ still $\Rightarrow y = Ce^{5x}$
3 $f(x) = \sin 4x$ $y = C \cos 4x + D \sin 4x$
4 $f(x) = 3 - 5x^2$ $y = Cx^2 + Dx + E$
5 $f(x) = 27$ $y = C$
6 $f(x) = 5 \cosh 4x$ $y = C \cosh 4x + D \sinh 4x$

All correct? If you have made a slip with any one of them, be sure that you understand where and why your result was incorrect before moving on.

Next frame

Let us work through a few examples. Here is the first:

28

Example 1

Solve $\dfrac{d^2y}{dx^2} - 5\dfrac{dy}{dx} + 6y = 24$

(1)   *CF*   Solve LHS $= 0$ $\therefore m^2 - 5m + 6 = 0$

         $\therefore (m-2)(m-3) = 0$ $\therefore m = 2$ and $m = 3$

         $\therefore y = Ae^{2x} + Be^{3x}$ (1)

(2)   *PI* $f(x) = 24$, i.e. a constant. Assume $y = C$

        Then $\dfrac{dy}{dx} = 0$ and $\dfrac{d^2y}{dx^2} = 0$

Substituting in the given equation:

   $0 - 5(0) + 6C = 24$ $C = 24/6 = 4$

   $\therefore$ PI is $y = 4$ (2)

   General solution is $y = \text{CF} + \text{PI}$,   i.e. $y = \underbrace{Ae^{2x} + Be^{3x}}_{\text{CF}} \underbrace{+4}_{\text{PI}}$

▶

Now another:

Example 2

Solve $\dfrac{d^2y}{dx^2} - 5\dfrac{dy}{dx} + 6y = 2\sin 4x$

(1) *CF* This will be the same as in the previous example, since the LHS of this equation is the same

   i.e. $y = Ae^{2x} + Be^{3x}$

(2) *PI* The general form of the PI in this case will be ....

29

$$y = C\cos 4x + D\sin 4x$$

Note: Although the RHS is $f(x) = 2\sin 4x$, it is necessary to include the full general function $y = C\cos 4x + D\sin 4x$ since, in finding the derivatives, the cosine term will also give rise to $\sin 4x$.

So we have:

$$y = C\cos 4x + D\sin 4x$$
$$\frac{dy}{dx} = -4C\sin 4x + 4D\cos 4x$$
$$\frac{d^2y}{dx^2} = -16C\cos 4x - 16D\sin 4x$$

We now substitute these expressions in the LHS of the equation and by equating coefficients, find the values of C and D.

Away you go then.

Complete the job and then move on to Frame 30

30

$$C = \frac{2}{25} \quad D = -\frac{1}{25} \quad y = \frac{1}{25}(2\cos 4x - \sin 4x)$$

Here is the working:

$$-16C\cos 4x - 16D\sin 4x + 20C\sin 4x - 20D\cos 4x$$
$$+ 6C\cos 4x + 6D\sin 4x = 2\sin 4x$$
$$(20C - 10D)\sin 4x - (10C + 20D)\cos 4x = 2\sin 4x$$

$$\left.\begin{array}{ll} 20C - 10D = 2 & 40C - 20D = 4 \\ 10C + 20D = 0 & 10C + 20D = 0 \end{array}\right\} 50C = 4 \quad \therefore\ C = \frac{2}{25}$$

Solve simultaneous

$$D = -\frac{1}{25}$$

In each case the PI is $y = \dfrac{1}{25}(2\cos 4x - \sin 4x)$

The CF was $y = Ae^{2x} + Be^{3x}$

The general solution is: $y = Ae^{2x} + Be^{3x} + \dfrac{1}{25}(2\cos 4x - \sin 4x)$

Here is an example we can work through together:

31

Example 3

Solve $\dfrac{d^2y}{dx^2} + 14\dfrac{dy}{dx} + 49y = 4e^{5x}$

First we have to find the CF. To do this we solve the equation ....

32

$$\boxed{\dfrac{d^2y}{dx^2} + 14\dfrac{dy}{dx} + 49y = 0}$$

Correct. So start off by writing down the auxiliary equation, which is
....

33

$$\boxed{m^2 + 14m + 49 = 0}$$

This gives $(m+7)(m+7) = 0$, i.e. $m = -7$ (twice)

$\therefore$ The CF is $y = e^{-7x}(A + Bx)$ 　　　　　　　　　(1)

Now for the PI. To find this we take the general form of the RHS of the given equation, i.e. we assume $y = $ ....

34

$$\boxed{y = Ce^{5x}}$$

Right. So we now differentiate twice which gives us:

$$\dfrac{dy}{dx} = \text{............ and } \dfrac{d^2y}{dx^2} = \text{............}$$

35

$$\boxed{\dfrac{dy}{dx} = 5Ce^{5x}, \quad \dfrac{d^2y}{dx^2} = 25\,Ce^{5x}}$$

14×5

The equation now becomes:

$25Ce^{5x} + 14.5Ce^{5x} + 49Ce^{5x} = 4e^{5x}$

Dividing through by e^{5x}: 　$25C + 70C + 49C = 4$

$$144C = 4 \quad \therefore \quad C = \dfrac{1}{36}$$

$$\text{The PI is } \quad y = \dfrac{e^{5x}}{36} \qquad (2)$$

So there we are. The CF is $y = e^{-7x}(A + Bx)$

and the PI is $y = \dfrac{e^{5x}}{36}$

and the complete general solution is therefore $y = e^{-7x}(A+Bx) + \dfrac{e^{5x}}{36}$

$$y = e^{-7x}(A + Bx) + \frac{e^{5x}}{36}$$

Correct, because in every case, the general solution is the sum of the complementary function and the particular integral.

Here is another example.

Example 4

Solve $\dfrac{d^2y}{dx^2} + 6\dfrac{dy}{dx} + 10y = 2\sin 2x$

(1) *To find CF*   Solve LHS $= 0$ $\therefore$ $m^2 + 6m + 10 = 0$

$$\therefore m = \frac{-6 \pm \sqrt{36 - 40}}{2} = \frac{-6 \pm \sqrt{-4}}{2} = -3 \pm i$$

$$y = e^{-3x}(A\cos x + B\sin x) \qquad\qquad (1)$$

(2) *To find PI*   Assume the general form of the RHS

i.e. $y = \ldots\ldots\ldots$

On to Frame 37

$$y = C\cos 2x + D\sin 2x$$

Do not forget that we have to include the cosine term as well as the sine term, since that will also give $\sin 2x$ when the derivatives are found.

As usual, we now differentiate twice and substitute in the given equation $\dfrac{d^2y}{dx^2} + 6\dfrac{dy}{dx} + 10y = 2\sin 2x$ and equate coefficients of $\sin 2x$ and of $\cos 2x$.

Off you go then. Find the PI on your own.

When you have finished, check your result with that in Frame 38

$$y = \frac{1}{15}(\sin 2x - 2\cos 2x)$$

Because if:

$$y = C\cos 2x + D\sin 2x$$

$$\therefore \frac{dy}{dx} = -2C\sin 2x + 2D\cos 2x$$

$$\therefore \frac{d^2y}{dx^2} = -4C\cos 2x - 4D\sin 2x$$

▶

Substituting in the equation gives:

$$-4C\cos 2x - 4D\sin 2x - 12C\sin 2x + 12D\cos 2x$$
$$+ 10C\cos 2x + 10D\sin 2x = 2\sin 2x$$
$$(6C + 12D)\cos 2x + (6D - 12C)\sin 2x = 2\sin 2x$$
$$6C + 12D = 0 \quad \therefore \ C = -2D$$

$$6D - 12C = 2 \quad \therefore \ 6D + 24D = 2 \quad \therefore \ 30D = 2 \quad \therefore \ D = \frac{1}{15}$$

$$\therefore \ C = -\frac{2}{15}$$

PI is $y = \dfrac{1}{15}(\sin 2x - 2\cos 2x)$ (2)

So the CF is $y = e^{-3x}(A\cos x + B\sin x)$

and the PI is $y = \dfrac{1}{15}(\sin 2x - 2\cos 2x)$

The complete general solution is therefore

$y = \ldots\ldots\ldots\ldots$

39

$$\boxed{y = e^{-3x}(A\cos x + B\sin x) + \frac{1}{15}(\sin 2x - 2\cos 2x)}$$

Before we do another example, list what you would assume for the PI in an equation when the RHS function was:

1 $f(x) = 3\cos 4x$ $y = C\cos 4x + D\sin 4x$
2 $f(x) = 2e^{7x}$ $y = Ce^{7x}$
3 $f(x) = 3\sinh x$ $y = C\cosh x + D\sinh x$
4 $f(x) = 2x^2 - 7$ $y\ \ Cx^2 + Dx - E$
5 $f(x) = x + 2e^x$ $y = Cx + D + Ee^x$

Jot down all five results before turning to Frame 40 to check your answers

---

40

1	$y = C\cos 4x + D\sin 4x$
2	$y = Ce^{7x}$
3	$y = C\cosh x + D\sinh x$
4	$y = Cx^2 + Dx + E$
5	$y = Cx + D + Ee^x$

Note that in **5** we use the general form of both the terms.

General form for x is $Cx + D$
$\quad\quad\quad\quad\quad\quad e^x$ is Ee^x

$\therefore$ The general form of $x + e^x$ is $y = Cx + D + Ee^x$

▶

Now do this one all on your own:

Example 5

Solve $\dfrac{d^2y}{dx^2} - 3\dfrac{dy}{dx} + 2y = x^2$

Do not forget: find (1) the CF and (2) the PI. Then the general solution is $y = \text{CF} + \text{PI}$.
Off you go.

When you have finished completely, move to Frame 41

41

$$y = A\,e^x + B\,e^{2x} + \frac{1}{4}(2x^2 + 6x + 7)$$

Here is the solution in detail:

$$\frac{d^2y}{dx^2} - 3\frac{dy}{dx} + 2y = x^2$$

(1)   CF $m^2 - 3m + 2 = 0$ $\therefore$ $(m-1)(m-2) = 0$ $\therefore$ $m = 1$ or 2

$\therefore$ $y = A\,e^x + B\,e^{2x}$ (1)

(2)   PI $y = Cx^2 + Dx + E$ $\therefore$ $\dfrac{dy}{dx} = 2Cx + D$ $\therefore$ $\dfrac{d^2y}{dx^2} = 2C$

$2C - 3(2Cx + D) + 2(Cx^2 + Dx + E) = x^2$

$2Cx^2 + (2D - 6C)x + (2C - 3D + 2E) = x^2$

$2C = 1$ $\therefore$ $C = \dfrac{1}{2}$

$2D - 6C = 0$ $\therefore$ $D = 3C$ $\therefore$ $D = \dfrac{3}{2}$

$2C - 3D + 2E = 0$ $\therefore$ $2E = 3D - 2C = \dfrac{9}{2} - 1 = \dfrac{7}{2}$ $\therefore$ $E = \dfrac{7}{4}$

$\therefore$ PI is $y = \dfrac{x^2}{2} + \dfrac{3x}{2} + \dfrac{7}{4} = \dfrac{1}{4}(2x^2 + 6x + 7)$ (2)

General solution: $y = A\,e^x + B\,e^{2x} + \dfrac{1}{4}(2x^2 + 6x + 7)$

Next frame

42

Particular solutions.   The previous result was $y = A\,e^x + B\,e^{2x} + \dfrac{1}{4}(2x^2 + 6x + 7)$

and as with all second-order differential equations, this contains two arbitrary constants A and B. These can be evaluated when the appropriate extra information is provided.

For example, we might have been told that at $x = 0$, $y = \dfrac{3}{4}$ and $\dfrac{dy}{dx} = \dfrac{5}{2}$.

▶

It is *important* to note that the values of A and B can be found only from the complete general solution and not from the CF as soon as you obtain it. This is a common error so do not be caught out by it. Get the complete general solution before substituting to find A and B.

In this case, we are told that when $x = 0$, $y = \dfrac{3}{4}$, so inserting these values gives ....

Move on to Frame 43

$$\boxed{A + B = -1}$$

Because

$$\frac{3}{4} = A + B + \frac{7}{4} \quad \therefore \ A + B = -1$$

We are also told that when $x = 0$, $\dfrac{dy}{dx} = \dfrac{5}{2}$, so we must first differentiate the general solution

$$y = A e^x + B e^{2x} + \frac{1}{4}(2x^2 + 6x + 7)$$

to obtain an expression for $\dfrac{dy}{dx}$. So, $\dfrac{dy}{dx} = $ ....

$$\boxed{\frac{dy}{dx} = A e^x + 2B e^{2x} + \frac{1}{2}(2x + 3)}$$

Now we are given that when $x = 0$, $\dfrac{dy}{dx} = \dfrac{5}{2}$

$$\therefore \ \frac{5}{2} = A + 2B + \frac{3}{2} \quad \therefore \ A + 2B = 1$$

So we have
$$\left.\begin{array}{l} A + B = -1 \\ \text{and } A + 2B = 1 \end{array}\right\}$$

and these simultaneous equations give:

$A = ..\overset{-}{.}3......$ $B = ...\overset{2}{.}........$

Then on to Frame 45

$$\boxed{A = -3 \qquad B = 2}$$

Substituting these values in the general solution

$$y = A e^x + B e^{2x} + \frac{1}{4}(2x^2 + 6x + 7)$$

gives the *particular solution*:

$$y = 2 e^{2x} - 3 e^x + \frac{1}{4}(2x^2 + 6x + 7)$$

▶

And here is one for you, all on your own:

Solve the equation $\dfrac{d^2y}{dx^2} + 4\dfrac{dy}{dx} + 5y = 13e^{3x}$, given that when $x = 0$, $y = \dfrac{5}{2}$ and $\dfrac{dy}{dx} = \dfrac{1}{2}$. Remember:

(1) Find the CF.
(2) Find the PI.
(3) The general solution is $y = \text{CF} + \text{PI}$.
(4) Finally insert the given conditions to obtain the particular solution.

When you have finished, check with the solution in Frame 46

46

$$y = e^{-2x}(2\cos x + 3\sin x) + \frac{e^{3x}}{2}$$

Because

$$\frac{d^2y}{dx^2} + 4\frac{dy}{dx} + 5y = 13e^{3x}$$

(1) CF $m^2 + 4m + 5 = 0$ $\therefore$ $m = \dfrac{-4 \pm \sqrt{16 - 20}}{2} = \dfrac{-4 \pm 2i}{2}$

$\therefore$ $m = -2 \pm i$ $\therefore$ $y = e^{-2x}(A\cos x + B\sin x)$ (1)

(2) PI $y = Ce^{3x}$ $\therefore$ $\dfrac{dy}{dx} = 3Ce^{3x}$, $\dfrac{d^2y}{dx^2} = 9Ce^{3x}$

$\therefore$ $9Ce^{3x} + 12Ce^{3x} + 5Ce^{3x} = 13e^{3x}$

$26C = 13$ $\therefore$ $C = \dfrac{1}{2}$ $\therefore$ PI is $y = \dfrac{e^{3x}}{2}$ (2)

General solution $y = e^{-2x}(A\cos x + B\sin x) + \dfrac{e^{3x}}{2}$; $x = 0$, $y = \dfrac{5}{2}$

$\therefore$ $\dfrac{5}{2} = A + \dfrac{1}{2}$ $\therefore$ $A = 2$ $y = e^{-2x}(2\cos x + B\sin x) + \dfrac{e^{3x}}{2}$

$\dfrac{dy}{dx} = e^{-2x}(-2\sin x + B\cos x) - 2e^{-2x}(2\cos x + B\sin x) + \dfrac{3e^{3x}}{2}$

$x = 0$, $\dfrac{dy}{dx} = \dfrac{1}{2}$ $\therefore$ $\dfrac{1}{2} = B - 4 + \dfrac{3}{2}$ $\therefore$ $B = 3$

$\therefore$ Particular solution is $y = e^{-2x}(2\cos x + 3\sin x) + \dfrac{e^{3x}}{2}$

47

Since the CF makes the LHS$=0$, it is pointless to use as a PI a term already contained in the CF. If this occurs, multiply the assumed PI by x and proceed as before. If this too is already included in the CF, multiply by a further x and proceed as usual. Here is an example:

Solve $\dfrac{d^2y}{dx^2} - 2\dfrac{dy}{dx} - 8y = 3e^{-2x}$

(1)   CF $m^2 - 2m - 8 = 0$ $\therefore$ $(m+2)(m-4) = 0$ $\therefore$ $m = -2$ or 4

      $y = Ae^{4x} + Be^{-2x}$ (1)

(2)   PI   The general form of the RHS is Ce^{-2x}, but this term in e^{-2x} is already contained in the CF. Assume $y = Cxe^{-2x}$, and continue as usual:

$$y = Cxe^{-2x}$$

$$\frac{dy}{dx} = Cx(-2e^{-2x}) + Ce^{-2x} = Ce^{-2x}(1 - 2x)$$

$$\frac{d^2y}{dx^2} = Ce^{-2x}(-2) - 2Ce^{-2x}(1 - 2x) = Ce^{-2x}(4x - 4)$$

Substituting in the given equation, we get:

$$Ce^{-2x}(4x - 4) - 2 \cdot Ce^{-2x}(1 - 2x) - 8Cxe^{-2x} = 3e^{-2x}$$

$$(4C + 4C - 8C)x - 4C - 2C = 3$$

$$-6C = 3 \quad \therefore \quad C = -\frac{1}{2}$$

PI is $y = -\dfrac{1}{2}xe^{-2x}$ (2)

General solution: $y = Ae^{4x} + Be^{-2x} - \dfrac{xe^{-2x}}{2}$

So remember, if the general form of the RHS is already included in the CF, multiply the assumed general form of the PI by x and continue as before.

    Here is one final example for you to work:

Solve $\dfrac{d^2y}{dx^2} + \dfrac{dy}{dx} - 2y = e^x$

Finish it off and then move to Frame 48

48

$$\boxed{y = Ae^x + Be^{-2x} + \frac{xe^x}{3}}$$

Here is the working:

     To solve $\dfrac{d^2y}{dx^2} + \dfrac{dy}{dx} - 2y = e^x$

(1)   CF $m^2 + m - 2 = 0$

      $(m - 1)(m + 2) = 0$ $\therefore$ $m = 1$ or -2

      $\therefore$ $y = Ae^x + Be^{-2x}$ (1)

▶

(2) *PI* Take $y = Ce^x$. But this is already included in the CF. Therefore, assume $y = Cxe^x$.

Then $\dfrac{dy}{dx} = Cxe^x + Ce^x = Ce^x(x+1)$

$\dfrac{d^2y}{dx^2} = Ce^x + Cxe^x + Ce^x = Ce^x(x+2)$

$\therefore\ Ce^x(x+2) + Ce^x(x+1) - 2Cxe^x = e^x$

$C(x+2) + C(x+1) - 2Cx = 1$

$3C = 1\ \ \therefore\ C = \dfrac{1}{3}$

PI is $y = \dfrac{xe^x}{3}$ (2)

and so the general solution is

$$y = Ae^x + Be^{-2x} + \frac{xe^x}{3}$$

You are now almost at the end of the Program. Before you work through the **Can You?** checklist and the **Test exercise**, however, look down the **Review summary** given in Frame 49. It lists the main points that we have established during this Program, and you may find it very useful.

So on now to Frame 49

49   ## Review summary

1 Solution of equations of the form $a\dfrac{d^2y}{dx^2} + b\dfrac{dy}{dx} + cy = f(x)$

(1) Auxiliary equation: $am^2 + bm + c = 0$

(2) Types of solutions:

(a) Real and different roots $m = m_1$ and $m = m_2$

   $y = Ae^{m_1x} + Be^{m_2x}$

(b) Real and equal roots $m = m_1$ (twice)

   $y = e^{m_1x}(A + Bx)$

(c) Complex roots $m = \alpha \pm i\beta$

   $y = e^{\alpha x}(A\cos\beta x + B\sin\beta x)$

2 Equations of the form $\dfrac{d^2y}{dx^2} + n^2y = 0$

   $y = A\cos nx + B\sin nx$

3 Equations of the form $\dfrac{d^2y}{dx^2} - n^2y = 0$

   $y = A\cosh nx + B\sinh nx$

4 General solution

   $y = \text{complementary function} + \text{particular integral}$

▶

5 (1) To find CF solve $a\dfrac{\mathrm{d}^2y}{\mathrm{d}x^2} + b\dfrac{\mathrm{d}y}{\mathrm{d}x} + cy = 0$

(2) To find PI assume the general form of the RHS.
Note: If the general form of the RHS is already included in the CF, multiply by x and proceed as before, etc. Determine the complete general solution before substituting to find the values of the arbitrary constants A and B.

Now all that remains is the **Can You?** *checklist*
and the **Test exercise**, *so on to Frame 50*

☑ Can You?

Checklist 2   50

Check this list before and after you try the end of Program test.

On a scale of 1 to 5 how confident are you that you can: Frames

• Use the auxiliary equation to solve certain second-order homogeneous equations?   1 to 22
  Yes ☐ ☐ ☐ ☐ ☐   *No*

• Use the complementary function and the particular integral to solve certain second-order inhomogeneous equations?   23 to 48
  Yes ☐ ☐ ☐ ☐ ☐   *No*

Test exercise 2

Here are eight differential equations for you to solve, similar to those we have dealt   51
with in the Program. They are quite straightforward, so you should have no difficulty with them. Set your work out neatly and take your time: this will help you to avoid making unnecessary slips.

Solve the following:

1 $\dfrac{\mathrm{d}^2y}{\mathrm{d}x^2} - \dfrac{\mathrm{d}y}{\mathrm{d}x} - 2y = 8$

2 $\dfrac{\mathrm{d}^2y}{\mathrm{d}x^2} - 4y = 10e^{3x}$

3 $\dfrac{\mathrm{d}^2y}{\mathrm{d}x^2} + 2\dfrac{\mathrm{d}y}{\mathrm{d}x} + y = e^{-2x}$

▶

4 $\dfrac{d^2y}{dx^2} + 25y = 5x^2 + x$

5 $\dfrac{d^2y}{dx^2} - 2\dfrac{dy}{dx} + y = 4\sin x$

6 $\dfrac{d^2y}{dx^2} + 4\dfrac{dy}{dx} + 5y = 2e^{-2x}$, given that $x = 0$, $y = 1$ and $\dfrac{dy}{dx} = -2$.

7 $3\dfrac{d^2y}{dx^2} - 2\dfrac{dy}{dx} - y = 2x - 3$

8 $\dfrac{d^2y}{dx^2} - 6\dfrac{dy}{dx} + 8y = 8e^{4x}$

🚲 Further problems 2

52 Solve the following equations:

1 $2\dfrac{d^2y}{dx^2} - 7\dfrac{dy}{dx} - 4y = e^{3x}$

2 $\dfrac{d^2y}{dx^2} - 6\dfrac{dy}{dx} + 9y = 54x + 18$

3 $\dfrac{d^2y}{dx^2} - 5\dfrac{dy}{dx} + 6y = 100\sin 4x$

4 $\dfrac{d^2y}{dx^2} + 2\dfrac{dy}{dx} + y = 4\sinh x$

5 $\dfrac{d^2y}{dx^2} + \dfrac{dy}{dx} - 2y = 2\cosh 2x$

6 $\dfrac{d^2y}{dx^2} - 6\dfrac{dy}{dx} + 10y = 20 - e^{2x}$

7 $\dfrac{d^2y}{dx^2} + 4\dfrac{dy}{dx} + 4y = 2\cos^2 x$

8 $\dfrac{d^2y}{dx^2} - 4\dfrac{dy}{dx} + 3y = x + e^{2x}$

9 $\dfrac{d^2y}{dx^2} - 2\dfrac{dy}{dx} + 3y = x^2 - 1$

10 $\dfrac{d^2y}{dx^2} - 9y = e^{3x} + \sin 3x$

11 For a horizontal cantilever of length l, with load w per unit length, the equation of bending is

$$EI\dfrac{d^2y}{dx^2} = \dfrac{w}{2}(l - x)^2$$

where E, I, w and l are constants. If $y = 0$ and $\dfrac{dy}{dx} = 0$ at $x = 0$, find y in terms of x. Hence find the value of y when $x = l$.

12 Solve the equation

$$\dfrac{d^2x}{dt^2} + 4\dfrac{dx}{dt} + 3x = e^{-3t}$$

given that at $t = 0$, $x = \frac{1}{2}$ and $\dfrac{dx}{dt} = -2$.

▶

13 Obtain the general solution of the equation

$$\frac{d^2y}{dt^2} + 4\frac{dy}{dt} + 5y = 6\sin t$$

and determine the amplitude and frequency of the steady-state function.

14 Solve the equation

$$\frac{d^2x}{dt^2} - 3\frac{dx}{dt} + 2x = \sin t$$

given that at $t = 0$, $x = 0$ and $\frac{dx}{dt} = 0$.

15 Solve $\frac{d^2y}{dx^2} + 3\frac{dy}{dx} + 2y = 3\sin x$, given that when $x = 0$, $y = -0.9$ and $\frac{dy}{dx} = -0.7$.

16 Obtain the general solution of the equation

$$\frac{d^2y}{dx^2} + 6\frac{dy}{dx} + 10y = 50x$$

17 Solve the equation

$$\frac{d^2x}{dt^2} + 2\frac{dx}{dt} + 2x = 85\sin 3t$$

given that when $t = 0$, $x = 0$ and $\frac{dx}{dt} = -20$. Show that the values of t for stationary values of the steady-state solution are the roots of $6\tan 3t = 7$.

18 Solve the equation $\frac{d^2y}{dx^2} = 3\sin x - 4y$, given that $y = 0$, at $x = 0$ and that $\frac{dy}{dx} = 1$ at $x = \pi/2$. Find the maximum value of y in the interval $0 < x < \pi$.

19 A mass suspended from a spring performs vertical oscillations and the displacement x (cm) of the mass at time t (s) is given by $\frac{1}{2}\frac{d^2x}{dt^2} = -48x$. If $x = \frac{1}{6}$ and $\frac{dx}{dt} = 0$ when $t = 0$, determine the period and amplitude of the oscillations.

20 The equation of motion of a body performing damped forced vibrations is $\frac{d^2x}{dt^2} + 5\frac{dx}{dt} + 6x = \cos t$. Solve this equation, given that $x = 0.1$ and $\frac{dx}{dt} = 0$ when $t = 0$. Write the steady-state solution in the form $K\sin(t + a)$.

Series 1

Frames
[1] to [56]

Learning outcomes

When you have completed this Program you will be able to:

- Manipulate arithmetic and geometric series
- Manipulate series of powers of the natural numbers
- Determine the limiting values of arithmetic and geometric series
- Determine the limiting values of simple indeterminate forms
- Apply various convergence tests to infinite series
- Distinguish between absolute and conditional convergence

Sequences

1

A sequence is a set of quantities, $u_1, u_2, u_3, \ldots$, stated in a definite order and each term formed according to a fixed pattern, i.e. $u_r = f(r)$.

e.g. 1, 3, 5, 7,... is a sequence (the next term would be 9).

2, 6, 18, 54,... is a sequence (the next term would be 162).

$1^2, -2^2, 3^2, -4^2, \ldots$ is a sequence (the next term would be 5^2).

Also 1, -5, 37, 6,... is a sequence, but its pattern is more involved and the next term cannot readily be anticipated.

A *finite* sequence contains only a finite number of terms.

An *infinite* sequence is unending.

So which of the following constitutes a finite sequence:

(a) all the natural numbers, i.e. 1, 2, 3, etc.
(b) the page numbers of a book
(c) the ascending sequence of odd multiples of 7 that are less than one billion.

2

> The page numbers of a book
> The odd multiples of 7 less than one billion

Clearly, the page numbers are in fixed order and terminate at the last page. The odd multiples of 7 that are less than one billion are clearly large in number, but are finite nonetheless. The natural numbers form an infinite sequence, since they never come to an end.

Series

3

A *series* is formed by the sum of the terms of a sequence.

e.g. 1, 3, 5, 7, is a sequence

but $1 + 3 + 5 + 7 + \ldots$ is a series.

We shall indicate the terms of a series as follows:

u_1 will represent the first term, u_2 the second term, u_3 the third term, etc., so that u_r will represent the rth term and u_{r+1} the $(r+1)$th term, etc.

Also, the sum of the first 5 terms will be indicated by S_5.

So the sum of the first n terms will be stated as . . . .

4

$$S_n$$

You will already be familiar with two special kinds of series which have many applications. These are (a) *arithmetic series* and (b) *geometric series*. Just by way of review, however, we will first consider the important results relating to these two series.

Arithmetic series (or arithmetic progression), denoted by AP

An example of an AP is the series:

$$2 + 5 + 8 + 11 + 14 + \ldots$$

You will note that each term can be written from the previous term by simply adding on a constant value 3. This regular increment is called the *common difference* and is found by selecting any term and subtracting from it the previous term

e.g. $11 - 8 = 3$; $5 - 2 = 3$; etc.

Move on to the next frame

6

The *general arithmetic series* can therefore be written:

$$a + (a + d) + (a + 2d) + (a + 3d) + \ldots \tag{a}$$

where a = first term and d = common difference.
  You will remember that:

(a) the nth term $= a + (n - 1)d$ (b)

(b) the sum of the first n terms is given by

$$S_n = \frac{n}{2}[2a + (n - 1)d] \tag{c}$$

Make a note of these three items in your record book.
  By way of warming up, find the sum of the first 20 terms of the series:

$$10 + 6 + 2 - 2 - 6 \ldots \text{ etc.}$$

Then move to Frame 7

7

$$S_{20} = -560$$

Because for the series $10 + 6 + 2 - 2 - 6 \ldots$ etc.

$a = 10$ and $d = 2 - 6 = -4$

$$S_n = \frac{n}{2}[2a + (n-1)d]$$

$$\therefore S_{20} = \frac{20}{2}[20 + 19(-4)]$$

$$= 10(20 - 76) = 10(-56) = -560$$

Here is another example:

If the 7th term of an AP is 22 and the 12th term is 37, find the series.

We know 7th term $= 22$ $\quad \therefore a + 6d = 22$ $\left.\right\}$ $5d = 15$ $\quad \therefore d = 3$
and 12th term $= 37$ $\quad \therefore a + 11d = 37$ $\quad \therefore a = 4$

So the series is $4 + 7 + 10 + 13 + 16 + \ldots$ etc.

Here is one for you to do:

The 6th term of an AP is -5 and the 10th term is -21. Find the sum of the first 30 terms.

8

$$S_{30} = -1290$$

Because

$$6\text{th term} = -5 \quad \therefore a + 5d = -5 \left.\right\} \quad 4d = -16 \quad \therefore d = -4$$
$$10\text{th term} = -21 \quad \therefore a + 9d = -21 \quad\quad\quad a = 15$$

$$\therefore a = 15, d = -4, n = 30, S_n = \frac{n}{2}[2a + (n-1)d]$$

$$\therefore S_{30} = \frac{30}{2}[30 + 29(-4)] = 15(30 - 116) = 15(-86) = -1290$$

Arithmetic mean

We are sometimes required to find the arithmetic mean of two numbers, P and Q. This means that we have to insert a number A between P and Q, so that $P + A + Q$ forms an AP.

$$A - P = d \text{ and } Q - A = d$$

$$\therefore A - P = Q - A \quad 2A = P + Q \quad \therefore A = \frac{P+Q}{2}$$

The arithmetic mean of two numbers, then, is simply their average.

Therefore, the arithmetic mean of 23 and 58 is $\ldots\ldots\ldots$

9

The arithmetic mean of 23 and 58 is $\boxed{40.5}$

If we are required to insert 3 arithmetic means between two given numbers, P and Q, it means that we have to supply three numbers, A, B, C between P and Q, so that $P + A + B + C + Q$ forms an AP.

For example: Insert 3 arithmetic means between 8 and 18.

Let the means be denoted by A, B, C.
Then $8 + A + B + C + 18$ forms an AP.

First term, $a = 8$; fifth term $= a + 4d = 18$

$$\left.\begin{array}{l} \therefore\ a = 8 \\ a + 4d = 18 \end{array}\right\} \quad 4d = 10 \qquad \therefore\ d = 2.5$$

$$\left.\begin{array}{l} A = 8 + 2.5 = 10.5 \\ B = 8 + 5 = 13 \\ C = 8 + 7.5 = 15.5 \end{array}\right\} \text{ Required arithmetic means are } 10.5, 13, 15.5$$

Now you find five arithmetic means between 12 and 21.6.

Then move to Frame 10

$\boxed{13.6,\ 15.2,\ 16.8,\ 18.4,\ 20}$

Here is the working:

Let the 5 arithmetic means be A, B, C, D, E.
Then $12 + A + B + C + D + E + 21.6$ forms an AP.

$\therefore\ a = 12; \quad a + 6d = 21.6$
$\therefore\ 6d = 9.6 \qquad \therefore\ d = 1.6$

Then $A = 12 + 1.6 = 13.6$	$A = 13.6$
$B = 12 + 3.2 = 15.2$	$B = 15.2$
$C = 12 + 4.8 = 16.8$	$C = 16.8$
$D = 12 + 6.4 = 18.4$	$D = 18.4$
$E = 12 + 8.0 = 20.0$	$E = 20$

So that is it! Once you have done one, the others are just like it.
Now we will see how much you remember about *geometric series*.

So, on to Frame 11

Geometric series (geometric progression), denoted by GP

11

An example of a GP is the series:

$$1 + 3 + 9 + 27 + 81 + \ldots \text{ etc.}$$

Here you can see that any term can be written from the previous term by multiplying it by a constant factor 3. This constant factor is called the *common ratio* and is found by selecting any term and dividing it by the previous one:

e.g. $27 \div 9 = 3;$ $9 \div 3 = 3;$ etc.

A GP therefore has the form:

$$a + ar + ar^2 + ar^3 + \ldots \text{ etc.}$$

where a = first term, r = common ratio.
So in the geometric series $5 - 10 + 20 - 40 + \ldots$ etc. the common ratio, r, is ....

12

$$r = \frac{20}{-10} = -2$$

The general geometric series is therefore:

$$a + ar + ar^2 + ar^3 + \ldots \qquad \text{etc.} \qquad \text{(d)}$$

and you will remember that:

(1) the nth term = ar^{n-1} (e)

(2) the sum of the first n terms is given by:

$$S_n = \frac{a(1 - r^n)}{1 - r} \qquad \text{(f)}$$

Make a note of these items in your record book.

So, now you can do this one:

For the series $8 + 4 + 2 + 1 + \dfrac{1}{2} + \ldots$ etc., find the sum of the first 8 terms.

Then on to Frame 13

13

$$S_8 = 15\frac{15}{16}$$

Because for the series 8, 4, 2, 1, etc.

$$a = 8; \quad r = \frac{2}{4} = \frac{1}{2}; \quad S_n = \frac{a(1 - r^n)}{1 - r}$$

$$\therefore \; S_8 = \frac{8\left(1 - \left[\frac{1}{2}\right]^8\right)}{1 - \frac{1}{2}}$$

$$= \frac{8\left(1 - \frac{1}{256}\right)}{1 - \frac{1}{2}} = 16 \cdot \frac{255}{256} = \frac{255}{16} = 15\frac{15}{16}$$

Now here is another example.

If the 5th term of a GP is 162 and the 8th term is 4374, find the series.

We have   5th term $= 162$ $\therefore \; ar^4 = 162$

8th term $= 4374$ $\therefore \; ar^7 = 4374$

$$\frac{ar^7}{ar^4} = \frac{4374}{162} \quad \therefore \; r^3 = 27 \quad \therefore \; r = 3$$

$$\therefore \; a = \ldots\ldots\ldots\ldots$$

$$a = 2$$

Because

$$ar^4 = 162; \quad ar^7 = 4374 \text{ and } r = 3$$

$$\therefore \; a \cdot 3^4 = 162 \quad \therefore \; a = \frac{162}{81} \quad \therefore \; a = 2$$

$\therefore$ The series is: $2 + 6 + 18 + 54 + \ldots$ etc.

Of course, now that we know the values of a and r, we could calculate the value of any term or the sum of a given number of terms.

For this same series, find

(a) the 10th term

(b) the sum of the first 10 terms.

When you have finished, move to Frame 15

15

$$\boxed{39{,}366; \ 59{,}048}$$

Because $a = 2$; $r = 3$

(a) 10th term $= ar^9 = 2 \times 3^9 = 2(19{,}683) = 39{,}366$

(b) $S_{10} = \dfrac{a(1 - r^{10})}{1 - r} = \dfrac{2(1 - 3^{10})}{1 - 3}$

$\qquad = \dfrac{2(1 - 59{,}049)}{-2} = 59{,}048$

Geometric mean (GM)

The geometric mean of two given numbers P and Q is a number A such that $P + A + Q$ form a GP.

$$\frac{A}{P} = r \text{ and } \frac{Q}{A} = r$$

$$\therefore \ \frac{A}{P} = \frac{Q}{A} \quad \therefore \ A^2 = PQ \quad A = \sqrt{PQ}$$

So the geomectric mean of 2 numbers is the square root of their product.
    Therefore, the geometric mean of 4 and 25 is ....

16

$$\boxed{A = \sqrt{4 \times 25} = \sqrt{100} = 10}$$

To insert 3 GMs between two given numbers, P and Q, means to insert 3 numbers, A, B, C, such that $P + A + B + C + Q$ form a GP.

For example, insert 4 geometric means between 5 and 1215.

Let the means be A, B, C, D. Then $5 + A + B + C + D + 1215$ form a GP.

i.e. $a = 5$ and $ar^5 = 1215$

$\therefore \ r^5 = \dfrac{1215}{5} = 243 \ \therefore \ r = 3$

$\therefore \ A = 5 \times 3 \ = 15$

$\quad B = 5 \times 9 \ = 45$

$\quad C = 5 \times 27 = 135$ The required geometric means are: 15, 45, 135, 405

$\quad D = 5 \times 81 = 405$

Now here is one for you to do: Insert two geometric means between 5 and 8.64.

Then on to Frame 17

17

> Required geometric means are 6.0, 7.2

Because

Let the means be A and B.

Then $5 + A + B + 8.64$ form a GP.

$\therefore\ a = 5;\quad \therefore\ ar^3 = 8.64;\quad \therefore\ r^3 = 1.728;\quad \therefore\ r = 1.2$

$\left.\begin{array}{l} A = 5 \times 1.2 = 6.0 \\ B = 5 \times 1.44 = 7.2 \end{array}\right\}$ Required means are 6.0 and 7.2

Arithmetic and geometric series are, of course, special kinds of series. There are other special series that are worth knowing. These consist of the series of the powers of the natural numbers. So let us look at these in the next frame.

Series of powers of the natural numbers

18

1 The series $1 + 2 + 3 + 4 + 5 + \ldots + n$ etc. $= \displaystyle\sum_{r=1}^{n} r.$

This series, you will see, is an example of an AP, where $a = 1$ and $d = 1$.
The sum of the first n terms is given by:

$$\sum_{r=1}^{n} r = 1 + 2 + 3 + 4 + 5 = \ldots + n$$

$$= \frac{n}{2}[2a + (n-1)d] = \frac{n(n+1)}{2}$$

$$\sum_{r=1}^{n} r = \frac{n(n+1)}{2}$$

So, the sum of the first 100 natural numbers is ....

Then on to Frame 19

19

> $$\sum_{r=1}^{100} r = 5050$$

Because

$$\sum_{r=1}^{n} r = \frac{100(101)}{2} = 50(101) = 5050$$

▶

2 That was easy enough. Now let us look at this one: To establish the result for the sum of n terms of the series $1^2 + 2^2 + 3^2 + 4^2 + 5^2 + \ldots + n^2$, we make use of the identity:

$$(n+1)^3 = n^3 + 3n^2 + 3n + 1$$

We write this as:

$$(n+1)^3 - n^3 = 3n^2 + 3n + 1$$

Replacing n by $n-1$, we get

$$n^3 - (n-1)^3 = 3(n-1)^2 + 3(n-1) + 1$$

and again $(n-1)^3 - (n-2)^3 = 3(n-2)^2 + 3(n-2) + 1$

and $(n-2)^3 - (n-3)^3 = 3(n-3)^2 + 3(n-3) + 1$

Continuing like this, we should eventually arrive at:

$$3^3 - 2^3 = 3 \times 2^2 + 3 \times 2 + 1$$
$$2^3 - 1^3 = 3 \times 1^2 + 3 \times 1 + 1$$

If we now add all these results together, we find on the left-hand side that all the terms disappear except the first and the last.

$$(n+1)^3 - 1^3 = 3\left\{n^2 + (n-1)^2 + (n-2)^2 + \ldots + 2^2 + 1^2\right\}$$

$$+ 3\left\{n + (n-1) + (n-2) + \ldots + 2 + 1\right\} + n(1)$$

$$= 3 \cdot \sum_{r=1}^{n} r^2 + 3\sum_{r=1}^{n} r + n$$

$$\therefore\ n^3 + 3n^2 + 3n + 1 - 1 = 3\sum_{r=1}^{n} r^2 + 3\sum_{r=1}^{n} r + n = 3\sum_{r=1}^{n} r^2 + 3\frac{n(n+1)}{2} + n$$

$$\therefore\ n^3 + 3n^2 + 2n = 3\sum_{r=1}^{n} r^2 + \frac{3}{2}(n^2 + n)$$

$$\therefore\ 2n^3 + 6n^2 + 4n = 6\sum_{r=1}^{n} r^2 + 3n^2 + 3n$$

$$6\sum_{r=1}^{n} r^2 = 2n^3 + 3n^2 + n$$

$$\therefore\ \sum_{r=1}^{n} r^2 = \frac{n(n+1)(2n+1)}{6}$$

So, the sum of the first 12 terms of the series $1^2 + 2^2 + 3^2 + \ldots$ is . . . .

$$\boxed{650}$$

20

Because $\quad \displaystyle\sum_{r=1}^{n} r^2 = \frac{n(n+1)(2n+1)}{6}$

so $\quad \displaystyle\sum_{r=1}^{12} r^2 = \frac{12(13)(25)}{6} = 26(25) = 650$

3 The sum of the cubes of the natural numbers is found in much the same way. This time, we use the identity:

$$(n+1)^4 = n^4 + 4n^3 + 6n^2 + 4n + 1$$

We rewrite it as before:

$$(n+1)^4 - n^4 = 4n^3 + 6n^2 + 4n + 1$$

If we now do the same trick as before and replace n by $(n-1)$ over and over again, and finally total up the results we get the result:

$$\sum_{r=1}^{n} r^3 = \left\{ \frac{n(n+1)}{2} \right\}^2$$

Note in passing that $\displaystyle\sum_{r=1}^{n} r^3 = \left\{ \sum_{r=1}^{n} r \right\}^2$

Let us collect together these last three results. Here they are:

21

1 $\quad \displaystyle\sum_{r=1}^{n} r = \frac{n(n+1)}{2}$ $\qquad\qquad$ (g)

2 $\quad \displaystyle\sum_{r=1}^{n} r^2 = \frac{n(n+1)(2n+1)}{6}$ $\qquad\qquad$ (h)

3 $\quad \displaystyle\sum_{r=1}^{n} r^3 = \left\{ \frac{n(n+1)}{2} \right\}^2$ $\qquad\qquad$ (i)

These are handy results, so copy them into your record book.

Now move on to Frame 22 and we can see an example of the use of these results

---

Find the sum of the series $\displaystyle\sum_{n=1}^{5} n(3+2n)$

22

$$S_5 = \sum_{n=1}^{5} n(3+2n) = \sum_{n=1}^{5} (3n + 2n^2)$$

$$= \sum_{n=1}^{5} 3n + \sum_{n=1}^{5} 2n^2 = 3\sum_{n=1}^{5} n + 2\sum_{n=1}^{5} n^2$$

$$= 3 \cdot \frac{5 \cdot 6}{2} + 2 \cdot \frac{5 \cdot 6 \cdot 11}{6} = 45 + 110 = 155$$

It is just a question of using the established results. Here is one for you to do in the same manner.

Find the sum of the series $\displaystyle\sum_{n=1}^{4} (2n + n^3)$

Working in Frame 23

23

$$\boxed{120}$$

Because $S_4 = \displaystyle\sum_{n=1}^{4}(2n + n^3)$

$$= 2\sum_{n=1}^{4} n + \sum_{n=1}^{4} n^3$$

$$= \frac{2 \cdot 4 \cdot 5}{2} + \left\{\frac{4 \cdot 5}{2}\right\}^2$$

$$= 20 + 100 = 120$$

Remember:

$$\text{Sum of the first } n \text{ natural numbers} = \frac{n(n+1)}{2}$$

$$\text{Sum of squares of the first } n \text{ natural numbers} = \frac{n(n+1)(2n+1)}{6}$$

$$\text{Sum of cubes of the first } n \text{ natural numbers} = \left\{\frac{n(n+1)}{2}\right\}^2$$

Infinite series

24

So far, we have been concerned with a finite number of terms of a given series. When we are dealing with the sum of an infinite number of terms of a series, we must be careful about the steps we take.

For example, consider the infinite series $1 + \dfrac{1}{2} + \dfrac{1}{4} + \dfrac{1}{8} + \cdots$

This we recognize as a GP in which $a = 1$ and $r = \dfrac{1}{2}$. The sum of the first n terms is therefore given by:

$$S_n = \frac{1\left(1 - \left[\dfrac{1}{2}\right]^n\right)}{1 - \dfrac{1}{2}} = 2\left(1 - \frac{1}{2^n}\right)$$

Now if n is very large, 2^n will be very large and therefore $\dfrac{1}{2^n}$ will be very small.

In fact, as $n \to \infty$, $\dfrac{1}{2^n} \to 0$. The sum of all the terms in this infinite series is therefore given by $S_\infty =$ the limiting value of S_n as $n \to \infty$.

i.e. $S_\infty = \underset{n \to \infty}{Lim} S_n = 2(1 - 0) = 2$

This result means that we can make the sum of the series as near to the value 2 as we please by taking a sufficiently large number of terms.

Next frame

This is not always possible with an infinite series, for in the case of an AP | **25** |
things are very different.

Consider the infinite series $1 + 3 + 5 + 7 + \ldots$

This is an AP in which $a = 1$ and $d = 2$.

Then $S_n = \dfrac{n}{2}[2a + (n-1)d] = \dfrac{n}{2}[2 + (n-1)2]$

$$= \dfrac{n}{2}(2 + 2n - 2)$$

$$S_n = n^2$$

Of course, in this case, if n is large then the value of S_n is very large. In fact, if $n \to \infty$, then $S_n \to \infty$, which is not a definite numerical value and therefore of little use to us.

This always happens with an AP: if we try to find the 'sum to infinity', we always obtain $+\infty$ or $-\infty$ as the result, depending on the actual series.

Move on now to Frame 26

In the previous two frames, we made two important points: | **26** |

(a) We cannot evaluate the sum of an infinite number of terms of an AP because the result is always infinite.

(b) We can sometimes evaluate the sum of an infinite number of terms of a GP since, for such a series, $S_n = \dfrac{a(1 - r^n)}{1 - r}$ and *provided* $|r| < 1$, then as

$n \to \infty$, $r^n \to 0$. In that case $S_\infty = \dfrac{a(1 - 0)}{1 - r} = \dfrac{a}{1 - r}$, i.e. $S_\infty = \dfrac{a}{1 - r}$.

So, find the 'sum to infinity' of the series

$\qquad 20 + 4 + 0.8 + 0.16 + 0.032 + \ldots\ldots\ldots\ldots$

	27
$\boxed{S_\infty = 25}$	

Because, for

$$20 + 4 + 0.8 + 0.16 + 0.032 + \ldots$$

$$a = 20; \quad r = \frac{0.8}{4} = 0.2 = \frac{1}{5}$$

$$\therefore \; S_\infty = \frac{a}{1 - r} = \frac{20}{1 - \dfrac{1}{5}} = \frac{5}{4} \cdot (20) = 25$$

Limiting values

28

In this Program, we have already seen that we have sometimes to determine the limiting value of S_n as $n \to \infty$. Before we leave this topic, let us look a little further into the process of finding limiting values. A few examples will suffice.

So move on to Frame 29

29

Example 1

To find the limiting value of $\dfrac{5n + 3}{2n - 7}$ as $n \to \infty$.

We cannot just substitute $n = \infty$ in the expression and simplify the result, since ∞ is not an ordinary number and does not obey the normal rules. So we do it this way:

$$\frac{5n + 3}{2n - 7} = \frac{5 + 3/n}{2 - 7/n} \quad \text{(dividing top and bottom by } n\text{)}$$

$$\underset{n \to \infty}{Lim} \left\{ \frac{5n + 3}{2n - 7} \right\} = \underset{n \to \infty}{Lim} \frac{5 + 3/n}{2 - 7/n}$$

Now when $n \to \infty$, $3/n \to 0$ and $7/n \to 0$

$$\therefore \underset{n \to \infty}{Lim} \frac{5n + 3}{2n - 7} = \underset{n \to \infty}{Lim} \frac{5 + 3/n}{2 - 7/n} = \frac{5 + 0}{2 - 0} = \frac{5}{2}$$

We can always deal with fractions of the form $\dfrac{c}{n}$, $\dfrac{c}{n^2} 0$, $\dfrac{c}{n^3}$, etc., because when

$n \to \infty$, each of these tends to zero, which is a precise value.

Let us try another example.

On to the next frame then

30

Example 2

To find the limiting value of $\dfrac{2n^2 + 4n - 3}{5n^2 - 6n + 1}$ as $n \to \infty$.

First of all, we divide top and bottom by the highest power of n which is involved, in this case n^2.

$$\frac{2n^2 + 4n - 3}{5n^2 - 6n + 1} = \frac{2 + 4/n - 3/n^2}{5 - 6/n + 1/n^2}$$

$$\therefore \underset{n \to \infty}{Lim} \frac{2n^2 + 4n - 3}{5n^2 - 6n + 1} = \underset{n \to \infty}{Lim} \frac{2 + 4/n - 3/n^2}{5 - 6/n + 1/n^2}$$

$$= \frac{2 + 0 - 0}{5 - 0 + 0} = \frac{2}{5}$$

Example 3

To find $\underset{n \to \infty}{Lim} \dfrac{n^3 - 2}{2n^3 + 3n - 4}$

In this case, the first thing is to . . . .

Move on to Frame 31

31

$$\boxed{\text{Divide top and bottom by } n^3}$$

Right. So we get:

$$\frac{n^3 - 2}{2n^3 + 3n - 4} = \frac{1 - 2/n^3}{2 + 3/n^2 - 4/n^3}$$

$$\therefore \ \lim_{n \to \infty} \frac{n^3 - 2}{2n^3 + 3n - 4} = \cdots\cdots\cdots$$

Finish it off. Then move on to Frame 32

32

$$\boxed{\dfrac{1}{2}}$$

Next frame

Convergent and divergent series

33

A series in which the sum (S_n) of n terms of the series tends to a definite value, as $n \to \infty$, is called a *convergent* series. If S_n does not tend to a definite value as $n \to \infty$, the series is said to be *divergent*.

For example, consider the GP: $1 + \dfrac{1}{3} + \dfrac{1}{9} + \dfrac{1}{27} + \dfrac{1}{81} + \cdots$

We know that for a GP, $S_n = \dfrac{a(1 - r^n)}{1 - r}$, so in this case since $a = 1$ and $r = \dfrac{1}{3}$, we have:

$$S_n = \frac{1\left(1 - \dfrac{1}{3^n}\right)}{1 - \dfrac{1}{3}} = \frac{1 - \dfrac{1}{3^n}}{\dfrac{2}{3}} = \frac{3}{2}\left(1 - \frac{1}{3^n}\right)$$

$$\therefore \ \text{As } n \to \infty, \ \frac{1}{3^n} \to 0 \quad \therefore \ \lim_{n \to \infty} S_n = \frac{3}{2}$$

The sum of n terms of this series tends to the definite value of $\dfrac{3}{2}$ as $n \to \infty$. It is therefore a .... series.

(convergent/divergent)

34

convergent

If S_n tends to a definite value as $n \to \infty$, the series is *convergent*.

If S_n does not tend to a definite value as $n \to \infty$, the series is *divergent*.

Here is another series. Let us investigate this one.

$$1 + 3 + 9 + 27 + 81 + \ldots$$

This is also a GP with $a = 1$ and $r = 3$.

$$\therefore\ S_n = \frac{a(1 - r^n)}{1 - r} = \frac{1(1 - 3^n)}{1 - 3} = \frac{1 - 3^n}{-2}$$
$$= \frac{3^n - 1}{2}$$

Of course, when $n \to \infty$, $3^n \to \infty$ also.

$$\therefore\ \underset{n\to\infty}{Lim}\, S_n = \infty \text{ (which is not a definite numerical value)}$$

So in this case, the series is . . . .

35

divergent

We can make use of infinite series only when they are convergent and it is necessary, therefore, to have some means of testing whether or not a given series is, in fact, convergent.

Of course, we could determine the limiting value of S_n as $n \to \infty$, as we did in the examples a moment ago, and this would tell us directly whether the series in question tended to a definite value (i.e. was convergent) or not.

That is the fundamental test, but unfortunately, it is not always easy to find a formula for S_n and we have therefore to find a test for convergence which uses the terms themselves.

Remember the notation for series in general. We shall denote the terms by $u_1 + u_2 + u_3 + u_4 + \ldots$

So now move on to Frame 36

Test for convergence

36

Test 1. A series cannot be convergent unless its terms ultimately tend to zero, i.e. unless $\underset{n\to\infty}{Lim}\, u_n = 0$

If $\underset{n\to\infty}{Lim}\, u_n \neq 0$, the series is divergent.

This is almost just common sense, for if the sum is to approach some definite value as the value of n increases, the numerical value of the individual terms must diminish.

▶

For example, we have already seen that:

(a) the series $1 + \dfrac{1}{3} + \dfrac{1}{9} + \dfrac{1}{27} + \dfrac{1}{81} + \ldots$ converges

while   (b) the series $1 + 3 + 9 + 27 + 81 + \ldots$ diverges.

So what would you say about the series

$$1 + \dfrac{1}{2} + \dfrac{1}{3} + \dfrac{1}{4} + \dfrac{1}{5} + \dfrac{1}{6} + \ldots ?$$

Just by looking at it, do you think this series converges or diverges?

Most likely you said that the series converges since it was clear that the numerical value of the terms decreases as n increases. If so, I am afraid you were wrong, for we shall show later that, in fact, the series
37

$$1 + \dfrac{1}{2} + \dfrac{1}{3} + \dfrac{1}{4} + \dfrac{1}{5} + \ldots \text{ diverges.}$$

It was rather a trick question, but be very clear about what the rule states. It says:

A series cannot be convergent unless its terms ultimately tend to zero, i.e. $\underset{n \to \infty}{Lim}\, u_n = 0$. It does not say that if the terms tend to zero, then the series is convergent. In fact, it is quite possible for the terms to tend to zero without the series converging – as in the example stated.

In practice, then, we use the rule in the following form:

If $\underset{n \to \infty}{Lim}\, u_n = 0$, the series *may* converge or diverge and we must test further.

If $\underset{n \to \infty}{Lim}\, u_n \neq 0$, we can be sure that the series diverges.

Make a note of these two statements

Before we leave the series
38

$$1 + \dfrac{1}{2} + \dfrac{1}{3} + \dfrac{1}{4} + \dfrac{1}{5} + \dfrac{1}{6} + \ldots + \dfrac{1}{n} + \ldots$$

here is the proof that, although $\underset{n \to \infty}{Lim}\, u_n = 0$, the series does, in fact, diverge.

We can, of course, if we wish, group the terms as follows:

$$1 + \dfrac{1}{2} + \left\{ \dfrac{1}{3} + \dfrac{1}{4} \right\} + \left\{ \dfrac{1}{5} + \dfrac{1}{6} + \dfrac{1}{7} + \dfrac{1}{8} \right\} + \ldots$$

Now $\left\{ \dfrac{1}{3} + \dfrac{1}{4} \right\} > \left\{ \dfrac{1}{4} + \dfrac{1}{4} \right\} = \dfrac{1}{2}$

and $\left\{ \dfrac{1}{5} + \dfrac{1}{6} + \dfrac{1}{7} + \dfrac{1}{8} \right\} > \left\{ \dfrac{1}{8} + \dfrac{1}{8} + \dfrac{1}{8} + \dfrac{1}{8} \right\} = \dfrac{1}{2}$ etc.

So that $S_n > 1 + \dfrac{1}{2} + \dfrac{1}{2} + \dfrac{1}{2} + \dfrac{1}{2} + \dfrac{1}{2} + \ldots$

$\therefore\ S_\infty = \infty$

This is not a definite numerical value, so the series is $\ldots\ldots\ldots\ldots$

39

divergent

The best we can get from Test 1 is that a series *may* converge. We must therefore apply a further test.

Test 2. The comparison test

A series of positive terms is convergent if its terms are less than the corresponding terms of a positive series which is known to be convergent. Similarly, the series is divergent if its terms are greater than the corresponding terms of a series which is known to be divergent.

A couple of examples will show how we apply this particular test.

So move on to the next frame

40   To test the series

$$1 + \frac{1}{2^2} + \frac{1}{3^3} + \frac{1}{4^4} + \frac{1}{5^5} + \frac{1}{6^6} + \ldots + \frac{1}{n^n} + \ldots$$

we can compare it with the series

$$1 + \frac{1}{2^2} + \frac{1}{2^3} + \frac{1}{2^4} + \frac{1}{2^5} + \frac{1}{2^6} + \ldots + \ldots$$

which is known to converge.

If we compare corresponding terms after the first two terms, we see that $\frac{1}{3^3} < \frac{1}{2^3}$; $\frac{1}{4^4} < \frac{1}{2^4}$; and so on for all the further terms, so that, after the first two terms, the terms of the first series are each less than the corresponding terms of the series known to converge.

The first series also, therefore, ...

41

converges

The difficulty with the comparison test is knowing which convergent series to use as a standard. A useful series for this purpose is this one:

$$\frac{1}{1^p} + \frac{1}{2^p} + \frac{1}{3^p} + \frac{1}{4^p} + \frac{1}{5^p} + \ldots + \frac{1}{n^p} + \ldots = \sum_{n=1}^{\infty} \frac{1}{n^p}$$

It can be shown that:

(a) if $p > 1$, the series converges

(b) if $p \leq 1$, the series diverges.

So what about the series $\sum_{n=1}^{\infty} \frac{1}{n^2}$?

Does it converge or diverge?

$$\boxed{\text{Converge}}$$

Because the series $\sum \dfrac{1}{n^2}$ is the series $\sum \dfrac{1}{n^p}$ with $p > 1$.

Let us look at another example:

To test the series $\dfrac{1}{1 \times 2} + \dfrac{1}{2 \times 3} + \dfrac{1}{3 \times 4} + \dfrac{1}{4 \times 5} + \cdots$

If we take our standard series

$$\frac{1}{1^p} + \frac{1}{2^p} + \frac{1}{3^p} + \frac{1}{4^p} + \frac{1}{5^p} + \frac{1}{6^p} + \cdots$$

when $p = 2$, we get

$$\frac{1}{1^2} + \frac{1}{2^2} + \frac{1}{3^2} + \frac{1}{4^2} + \frac{1}{5^2} + \frac{1}{6^2} + \cdots$$

which we know to converge.

But $\dfrac{1}{1 \times 2} < \dfrac{1}{1^2}; \quad \dfrac{1}{2 \times 3} < \dfrac{1}{2^2}; \quad \dfrac{1}{3 \times 4} < \dfrac{1}{3^2}; \quad$ etc.

Each term of the given series is less than the corresponding term in the series known to converge.

Therefore ....

$$\boxed{\text{The given series converges}}$$

It is not always easy to devise a suitable comparison series, so we look for yet another test to apply, and here it is:

Test 3: D'Alembert's ratio test for positive terms

Let $u_1 + u_2 + u_3 + u_4 + \ldots u_n + \ldots$ be a series of *positive terms*. Find expressions for u_n and u_{n+1}, i.e. the nth term and the $(n + 1)$th term, and form the ratio $\dfrac{u_{n+1}}{u_n}$. Determine the limiting value of this ratio as $n \to \infty$.

If $\underset{n\to\infty}{Lim} \dfrac{u_{n+1}}{u_n} < 1$, the series converges

If $\underset{n\to\infty}{Lim} \dfrac{u_{n+1}}{u_n} > 1$, the series diverges

If $\underset{n\to\infty}{Lim} \dfrac{u_{n+1}}{u_n} = 1$, the series may converge or diverge and the test gives us no definite information.

Copy out D'Alembert's ratio test into your record book. Then on to Frame 44

44

Here it is again:

D'Alembert's ratio test for positive terms:

If $\underset{n\to\infty}{Lim}\dfrac{u_{n+1}}{u_n} < 1$, the series *converges*

If $\underset{n\to\infty}{Lim}\dfrac{u_{n+1}}{u_n} > 1$, the series *diverges*

If $\underset{n\to\infty}{Lim}\dfrac{u_{n+1}}{u_n} = 1$, the result is inconclusive.

For example: To test the series $\dfrac{1}{1} + \dfrac{3}{2} + \dfrac{5}{2^2} + \dfrac{7}{2^3} + \cdots$

We first of all decide on the pattern of the terms and hence write down the nth term. In this case $u_n = \dfrac{2n-1}{2^{n-1}}$. The $(n+1)$th term will then be the same with n replaced by $(n+1)$

i.e. $u_{n+1} = \dfrac{2n+1}{2^n}$

$\therefore \dfrac{u_{n+1}}{u_n} = \dfrac{2n+1}{2^n}\cdot\dfrac{2^{n-1}}{2n-1} = \dfrac{1}{2}\cdot\dfrac{2n+1}{2n-1}$

We now have to find the limiting value of this ratio as $n \to \infty$. From our prevous work on limiting values, we know that the next step, then, is to divide top and bottom by ....

45

Divide top and bottom by n

So $\underset{n\to\infty}{Lim}\dfrac{u_{n+1}}{u_n} = \underset{n\to\infty}{Lim}\dfrac{1}{2}\cdot\dfrac{2n+1}{2n-1} = \underset{n\to\infty}{Lim}\dfrac{1}{2}\cdot\dfrac{2+1/n}{2-1/n} = \dfrac{1}{2}\cdot\dfrac{2+0}{2-0} = \dfrac{1}{2}$

Because in this case, $\underset{n\to\infty}{Lim}\dfrac{u_{n+1}}{u_n} < 1$, we know that the given series is *convergent*.

Let us do another one in the same way:

Apply D'Alembert's ratio test to the series

$\dfrac{1}{2} + \dfrac{2}{3} + \dfrac{3}{4} + \dfrac{4}{5} + \dfrac{5}{6} + \cdots$

First of all, we must find an expression for u_n.

In this series, $u_n = $ ....

$$\frac{1}{2} + \frac{2}{3} + \frac{3}{4} + \frac{4}{5} + \dots \quad \boxed{u_n = \frac{n}{n+1}}$$

Then u_{n+1} is found by simply replacing n by $(n+1)$.

$$\therefore \ u_{n+1} = \frac{n+1}{n+2}$$

So that $\dfrac{u_{n+1}}{u_n} = \dfrac{n+1}{n+2} \cdot \dfrac{n+1}{n} = \dfrac{n^2 + 2n + 1}{n^2 + 2n}$

We now have to find $\underset{n \to \infty}{Lim}\,\dfrac{u_{n+1}}{u_n}$ and in order to do that we must divide top and bottom, in this case, by ....

$$\boxed{n^2}$$

$$\therefore \ \underset{n \to \infty}{Lim}\,\frac{u_{n+1}}{u_n} = \underset{n \to \infty}{Lim}\,\frac{n^2 + 2n + 1}{n^2 + 2n} = \underset{n \to \infty}{Lim}\,\frac{1 + 2/n + 1/n^2}{1 + 2/n}$$

$$= \frac{1 + 0 + 0}{1 + 0} = 1$$

$\therefore \ \underset{n \to \infty}{Lim}\,\dfrac{u_{n+1}}{u_n} = 1$, which is inconclusive and merely tells us that the series may be convergent or divergent. So where do we go from there?

We have of course, forgotten about Test 1, which states that:

(a) if $\underset{n \to \infty}{Lim}\,u_n = 0$, the series *may* be convergent

(b) if $\underset{n \to \infty}{Lim}\,u_n \neq 0$, the series is certainly *divergent*.

In our present series: $u_n = \dfrac{n}{n+1}$

$$\therefore \ \underset{n \to \infty}{Lim}\,u_n = \underset{n \to \infty}{Lim}\,\frac{n}{n+1} = \underset{n \to \infty}{Lim}\,\frac{1}{1 + 1/n} = 1$$

This is *not* zero. Therefore the series is *divergent*.

Now you do this one entirely on your own:

Test the series $\dfrac{1}{5} + \dfrac{2}{6} + \dfrac{2^2}{7} + \dfrac{2^3}{8} + \dfrac{2^4}{9} + \dots$

When you have finished, check your result with that in Frame 48

48

Here is the solution in detail: see if you agree with it.

$$\frac{1}{5} + \frac{2}{6} + \frac{2^2}{7} + \frac{2^3}{8} + \frac{2^4}{9} + \dots$$

$$u_n = \frac{2^{n-1}}{4+n}; \quad u_{n+1} = \frac{2^n}{5+n}$$

$$\therefore \frac{u_{n+1}}{u_n} = \frac{2^n}{5+n} \cdot \frac{4+n}{2^{n-1}}$$

The power 2^{n-1} cancels with the power 2^n to leave a single factor 2.

$$\therefore \frac{u_{n+1}}{u_n} = \frac{2(4+n)}{5+n}$$

$$\therefore \operatorname*{Lim}_{n\to\infty} \frac{u_{n+1}}{u_n} = \operatorname*{Lim}_{n\to\infty} \frac{2(4+n)}{5+n} = \operatorname*{Lim}_{n\to\infty} \frac{2(4/n+1)}{5/n+1} = \frac{2(0+1)}{0+1} = 2$$

$$\therefore \operatorname*{Lim}_{n\to\infty} \frac{u_{n+1}}{u_n} = 2$$

And since the limiting value is >1, we know the series is . . . .

49

$$\boxed{\text{divergent}}$$

Next frame

Series in general. Absolute convergence

50

So far we have considered series with positive terms only. Some series consist of alternate positive and negative terms.

For example, the series $1 - \frac{1}{2} + \frac{1}{3} - \frac{1}{4} + \dots$ is in fact convergent

while the series $1 + \frac{1}{2} + \frac{1}{3} + \frac{1}{4} + \dots$ is divergent.

If u_n denotes the nth term of a series in general, it may well be positive or negative. But $|u_n|$, or 'mod u_n' denotes the numerical value of u_n, so that if $u_1 + u_2 + u_3 + u_4 + \dots$ is a series of mixed terms, i.e. some positive, some negative, then the series $|u_1| + |u_2| + |u_3| + |u_4| + \dots$ will be a series of positive terms.

So if $\sum u_n = 1 - 3 + 5 - 7 + 9 - \dots$

Then $\sum |u_n| = $ . . . .

51

$$\sum |u_n| = 1 + 3 + 5 + 7 + 9 + \dots$$

Note: If a series $\sum u_n$ is convergent, then the series $\sum |u_n|$ may very well not be convergent, as in the example stated in the previous frame. But if $\sum |u_n|$ is found to be convergent, we can be sure that $\sum u_n$ is convergent.

If $\sum |u_n|$ converges, the series $\sum u_n$ is said to be *absolutely convergent*.

If $\sum |u_n|$ is not convergent, but $\sum u_n$ does converge, then $\sum u_n$ is said to be *conditionally convergent*.

So, if $\sum u_n = 1 - \dfrac{1}{2} + \dfrac{1}{3} - \dfrac{1}{4} + \dfrac{1}{5} - \dots$ converges

and $\sum |u_n| = 1 + \dfrac{1}{2} + \dfrac{1}{3} + \dfrac{1}{4} + \dfrac{1}{5} + \dots$ diverges

then $\sum u_n$ is .... convergent.

(absolutely or conditionally)

52

conditionally

As an example, find the range of values of x for which the following series is absolutely convergent:

$$\frac{x}{2 \times 5} - \frac{x^2}{3 \times 5^2} + \frac{x^3}{4 \times 5^3} - \frac{x^4}{5 \times 5^4} + \frac{x^5}{6 \times 5^5} - \dots$$

$$|u_n| = \frac{x^n}{(n+1)5^n}; \quad |u_{n+1}| = \frac{x^{n+1}}{(n+2)5^{n+1}}$$

$$\therefore \left| \frac{u_{n+1}}{u_n} \right| = \frac{x^{n+1}}{(n+2)5^{n+1}} \cdot \frac{(n+1)5^n}{x^n}$$

$$= \frac{x(n+1)}{5(n+2)} = \frac{x(1+1/n)}{5(1+2/n)}$$

$$\therefore \lim_{n \to \infty} \left| \frac{u_{n+1}}{u_n} \right| = \frac{x}{5}$$

For absolute convergence $\lim\limits_{n \to \infty} \left| \dfrac{u_{n+1}}{u_n} \right| < 1$. $\therefore$ Series convergent when $\left| \dfrac{x}{5} \right| < 1$, i.e. for $|x| < 5$.

You have now reached the end of this Program, except for the **Can You?** checklist and the **Test exercise** which follow. Before you work through them, here is a summary of the topics we have covered. Read through it carefully: it will refresh your memory on what we have been doing.

On to Frame 53

53 **Review summary**

1 *Arithmetic series:* $a + (a + d) + (a + 2d) + (a + 3d) + \ldots$

$u_n = a + (n-1)d \qquad S_n = \frac{n}{2}[2a + (n-1)d]]$

2 *Geometric series:* $a + ar + ar^2 + ar^3 + \ldots$

$u_n = ar^{n-1} \quad S_n = \frac{a(1 - r^n)}{1 - r}$

If $|r| < 1$, $S_\infty = \frac{a}{1 - r}$

3 *Powers of natural numbers*

$$\sum_{r=1}^{n} r = \frac{n(n+1)}{2} \qquad \sum_{r=1}^{n} r^2 = \frac{n(n+1)(2n+1)}{6}$$

$$\sum_{r=1}^{n} r^3 = \left\{ \frac{n(n+1)}{2} \right\}^2$$

4 *Infinite series:* $S_n = u_1 + u_2 + u_3 + u_4 + \ldots + u_n + \ldots$

If $\underset{n \to \infty}{Lim}\, S_n$ is a definite value, series is convergent

If $\underset{n \to \infty}{Lim}\, S_n$ is not a definite value, series is divergent.

5 *Tests for convergence*

(1) If $\underset{n \to \infty}{Lim}\, u_n = 0$, the series may be convergent

If $\underset{n \to \infty}{Lim}\, u_n \neq 0$ the series is certainly divergent.

(2) *Comparison test* – Useful standard series

$$\frac{1}{1^p} + \frac{1}{2^p} + \frac{1}{3^p} + \frac{1}{4^p} + \frac{1}{5^p} + \ldots + \frac{1}{n^p} \ldots$$

For $p > 1$, series converges: for $p \leq 1$, series diverges.

(3) *D'Alembert's ratio test for positive terms*

If $\underset{n \to \infty}{Lim} \frac{u_{n+1}}{u_n} < 1$, series converges

If $\underset{n \to \infty}{Lim} \frac{u_{n+1}}{u_n} > 1$, series diverges

If $\underset{n \to \infty}{Lim} \frac{u_{n+1}}{u_n} = 1$, inconclusive.

(4) *For general series*

(a) If $\sum |u_n|$ converges, $\sum u_n$ is absolutely convergent.

(b) If $\sum |u_n|$ diverges, but $\sum u_n$ converges,

then $\sum u_n$ is conditionally convergent.

Now you are ready for the **Can You?** checklist and the **Test exercise**.

So move on to Frame 54

☑ Can You?

Checklist 3

54

Check this list before and after you try the end of Program test.

On a scale of 1 to 5 how confident are you that you can: Frames

- Manipulate arithmetic and geometric series?
  Yes ☐ ☐ ☐ ☐ ☐ *No*   1 to 17

- Manipulate series of powers of the natural numbers?
  Yes ☐ ☐ ☐ ☐ ☐ *No*   18 to 23

- Determine the limiting values of arithmetic and geometric series?
  Yes ☐ ☐ ☐ ☐ ☐ *No*   24 to 27

- Determine the limiting values of simple indeterminate forms?   28 to 32
  Yes ☐ ☐ ☐ ☐ ☐ *No*

- Apply various convergence tests to infinite series?   33 to 49
  Yes ☐ ☐ ☐ ☐ ☐ *No*

- Distinguish between absolute and conditional convergence?   50 to 52
  Yes ☐ ☐ ☐ ☐ ☐ *No*

🚴 Test exercise 3

Take your time and work carefully.

55

1   The 3rd term of an AP is 34 and the 17th term is −8. Find the sum of the first 20 terms

2   For the series $1 + 1.2 + 1.44 + \ldots$ find the 6th term and the sum of the first 10 terms.

3   Evaluate $\displaystyle\sum_{n=1}^{8} n(3 + 2n + n^2)$.

▶

4 Determine whether each of the following series is convergent:

(a) $\dfrac{2}{2 \times 3} + \dfrac{2}{3 \times 4} + \dfrac{2}{4 \times 5} + \dfrac{2}{5 \times 6} + \ldots$

(b) $\dfrac{2}{1^2} + \dfrac{2^2}{2^2} + \dfrac{2^3}{3^2} + \dfrac{2^4}{4^2} + \ldots + \dfrac{2^n}{n^2} + \ldots$

(c) $u_n = \dfrac{1 + 2n^2}{1 + n^2}$

(d) $u_n = \dfrac{1}{n!}$

5 Find the range of values of x for which each of the following series is convergent or divergent:

(a) $1 + x + \dfrac{x^2}{2!} + \dfrac{x^3}{3!} + \dfrac{x^4}{4!} + \ldots$

(b) $\dfrac{x}{1 \times 2} + \dfrac{x^2}{2 \times 3} + \dfrac{x^3}{3 \times 4} + \dfrac{x^4}{4 \times 5} \ldots$

(c) $\displaystyle\sum_{n=1}^{\infty} \dfrac{(n+1)}{n^3} x^n$

Further problems 13

56

1 Find the sum of n terms of the series
$S_n = 1^2 + 3^2 + 5^2 + \ldots + (2n-1)^2$.

2 Find the sum to n terms of
$$\dfrac{1}{1 \cdot 2 \cdot 3} + \dfrac{3}{2 \cdot 3 \cdot 4} + \dfrac{5}{3 \cdot 4 \cdot 5} + \dfrac{7}{4 \cdot 5 \cdot 6} + \ldots$$

3 Sum to n terms, the series
$$1 \cdot 3 \cdot 5 + 2 \cdot 4 \cdot 6 + 3 \cdot 5 \cdot 7 + \ldots$$

4 Evaluate the following:

(a) $\displaystyle\sum_{r=1}^{n} r(r+3)$ (b) $\displaystyle\sum_{r=1}^{n} (r+1)^3$

5 Find the sum to infinity of the series
$$1 + \dfrac{4}{3!} + \dfrac{6}{4!} + \dfrac{8}{5!} + \ldots$$

6 For the series $5 - \dfrac{5}{2} + \dfrac{5}{4} - \dfrac{5}{8} + \ldots + \dfrac{(-1)^{n-1}5}{2^{n-1}}$
find an expression for S_n, the sum of the first n terms. Also if the series converges, find the sum to infinity.

7 Find the limiting values of:

(a) $\dfrac{3x^2 + 5x - 4}{5x^2 - x + 7}$ as $x \to \infty$

(b) $\dfrac{x^2 + 5x - 4}{2x^2 - 3x + 1}$ as $x \to \infty$

8 Determine whether each of the following series converges or diverges:

(a) $\displaystyle\sum_{n=1}^{\infty} \frac{n}{n+2}$ (b) $\displaystyle\sum_{n=1}^{\infty} \frac{n}{n^2+1}$

(c) $\displaystyle\sum_{n=1}^{\infty} \frac{1}{n^2+1}$ (d) $\displaystyle\sum_{n=0}^{\infty} \frac{1}{(2n+1)!}$

9 Find the range of values of x for which the series

$$\frac{x}{27} + \frac{x^2}{125} + \dots + \frac{x^n}{(2n+1)^3} + \dots$$

is absolutely convergent.

10 Show that the series

$$1 + \frac{x}{1 \times 2} + \frac{x^2}{2 \times 3} + \frac{x^3}{3 \times 4} + \dots$$

is absolutely convergent when $-1 < x < +1$.

11 Determine the range of values of x for which the following series is convergent:

$$\frac{x}{1 \cdot 2 \cdot 3} + \frac{x^2}{2 \cdot 3 \cdot 4} + \frac{x^3}{3 \cdot 4 \cdot 5} + \frac{x^4}{4 \cdot 5 \cdot 6} + \dots$$

12 Find the range of values of x for convergence for the series

$$x + \frac{2^4 x^2}{2!} + \frac{3^4 x^3}{3!} + \frac{4^4 x^4}{4!} + \dots$$

13 Investigate the convergence of the series

$$\frac{1}{1 \times 2} + \frac{x}{2 \times 3} + \frac{x^2}{3 \times 4} + \frac{x^3}{4 \times 5} + \dots \text{ for } x > 0.$$

14 Show that the following series is convergent: $2 + \dfrac{3}{2} \cdot \dfrac{1}{4} + \dfrac{4}{3} \cdot \dfrac{1}{4^2} + \dfrac{5}{4} \cdot \dfrac{1}{4^3} + \dots$

15 Prove that

$$\frac{1}{\sqrt{1}} + \frac{1}{\sqrt{2}} + \frac{1}{\sqrt{3}} + \frac{1}{\sqrt{4}} + \dots \text{ is divergent}$$

and that

$$\frac{1}{1^2} + \frac{1}{2^2} + \frac{1}{3^2} + \frac{1}{4^2} + \dots \text{ is convergent.}$$

▶

16 Determine whether each of the following series is convergent or divergent:

(a) $\displaystyle\sum_{n=1}^{\infty} \frac{1}{2n(2n+1)}$

(b) $\displaystyle\sum_{n=1}^{\infty} \frac{1+3n^2}{1+n^2}$

(c) $\displaystyle\sum_{n=1}^{\infty} \frac{n}{\sqrt{4n^2+1}}$

(d) $\displaystyle\sum_{n=1}^{\infty} \frac{3n+1}{3n^2-2}$

17 Show that the series

$$1 + \frac{2x}{5} + \frac{3x^2}{25} + \frac{4x^3}{125} + \ldots \text{ is convergent}$$

if $-5 < x < 5$ and for no other values of x.

18 Investigate the convergence of:

(a) $1 + \dfrac{3}{2 \times 4} + \dfrac{7}{4 \times 9} + \dfrac{15}{8 \times 16} + \dfrac{31}{16 \times 25} + \ldots$

(b) $\dfrac{1}{1 \times 2} + \dfrac{1}{2 \times 2^2} + \dfrac{1}{3 \times 2^3} + \dfrac{1}{4 \times 2^4} + \ldots$

19 Find the range of values of x for which the following series is convergent:

$$\frac{(x-2)}{1} + \frac{(x-2)^2}{2} + \frac{(x-2)^3}{3} + \ldots + \frac{(x-2)^n}{n} + \ldots$$

20 If $u_r = r(2r+1) + 2^{r+1}$, find the value of $\displaystyle\sum_{r=1}^{n} u_r$.

Series 2

Learning outcomes

When you have completed this Program you will be able to:

- Derive the power series for $\sin x$
- Use Maclaurin's series to derive series of common functions
- Use Maclaurin's series to derive the binomial series
- Derive power series expansions of miscellaneous functions using known expansions of common functions
- Use power series expansions in numerical approximations
- Use l'Hôpital's rule to evaluate limits of indeterminate forms
- Extend Maclaurin's series to Taylor's series

Power series

1

Introduction

In the first Program on series (no. 3), we saw how important it is to know something of the convergence properties of any infinite series we may wish to use and to appreciate the conditions in which the series is valid.

This is very important, since it is often convenient to represent a function as a series of ascending powers of the variable. This, in fact, is just how a computer finds the value of the sine of a given angle. Instead of storing the whole of the mathematical tables, it sums up the terms of a series representing the sine of an angle.

That is just one example. There are many occasions when we have need to express a function of x as an infinite series of powers of x. It is not at all difficult to express a function in this way, as you will soon see in this Program.

So make a start and move on to Frame 2

Suppose we wish to express $\sin x$ as a series of ascending powers of x. The series will be of the form

$$\sin x \equiv a + bx + cx^2 + dx^3 + ex^4 + \ldots$$

where a, b, c, etc. are constant coefficients, i.e. numerical factors of some kind. Notice that we have used the 'equivalent' sign and not the usual 'equals' sign. The statement is not an equation: it is an identity. The right-hand side does not *equal* the left-hand side: the RHS *is* the LHS expressed in a different form and the expression is therefore true for any value of x that we like to substitute.

Can you pick out an identity from these?

$$(x + 4)^2 = 3x^2 - 2x + 1$$
$$(2x + 1)^2 = 4x^2 + 4x - 3$$
$$(x + 2)^2 = x^2 + 4x + 4$$

When you have decided, move on to Frame 3

$$(x + 2)^2 \equiv x^2 + 4x + 4$$

Correct. This is the only identity of the three, since it is the only one in which the RHS is the LHS written in a different form. Right. Now back to our series:

$$\sin x \equiv a + bx + cx^2 + dx^3 + ex^4 + \dots$$

To establish the series, we have to find the values of the constant coefficients a, b, c, d, etc.

Suppose we substitute $x = 0$ on both sides.

Then $\sin 0 \equiv a + 0 + 0 + 0 + 0 + \dots$

and since $\sin 0 = 0$, we immediately get the value of a.

$a = \dots\dots\dots\dots$

$$a = 0$$

4

Now can we substitute some value for x, which will make all the terms disappear except the second? If we could, we should then find the value of b. Unfortunately, we cannot find any such substitution, so what is the next step?

Here is the series once again:

$$\sin x \equiv a + bx + cx^2 + dx^3 + ex^4 + \dots$$

and so far we know that $a = 0$.

The key to the whole business is simply this:

Differentiate both sides with respect to x.

On the left, we get $\cos x$.

On the right the terms are simply powers of x, so we get

$\cos x \equiv \dots\dots\dots\dots$

$$\cos x \equiv b + c \cdot 2x + d \cdot 3x^2 + c \cdot 4x^3 + \dots$$

5

This is still an identity, so we can substitute in it any value for x we like.

Notice that the a has now disappeared from the scene and that the constant term at the beginning of the expression is now b.

So what do you suggest that we substitute in the identity as it now stands, in order that all the terms except the first shall vanish?

We substitute $x = \dots\dots\dots\dots$ again.

6

$$\boxed{\text{Substitute } x = 0 \text{ again}}$$

Correct: because then all the terms will disappear except the first and we shall be able to find b.

$$\cos x \equiv b + c \cdot 2x + d \cdot 3x^2 + e \cdot 4x^3 + \dots$$

Put $x = 0$:

$$\therefore \quad \cos 0 = 1 = b + 0 + 0 + 0 + 0 + \dots \qquad \therefore \ b = 1$$

So far, so good. We have found the values of a and b. To find c and d and all the rest, we merely repeat the process over and over again at each successive stage.

i.e. *Differentiate both sides with respect to x and substitute* . . . .

7

$$\boxed{\text{Substitute } x = 0}$$

So we now get this, from the beginning:

$$\sin x \equiv a + bx + cx^2 + dx^3 + ex^4 + fx^5 + \dots$$

Put $x = 0$. $\quad \therefore \ \sin 0 = 0 = a + 0 + 0 + 0 + \dots \qquad\qquad\qquad \therefore \ a = 0$

$\begin{cases} \text{Differentiate.} \quad \cos x \equiv b + c \cdot 2x + d \cdot 3x^2 + e \cdot 4x^3 + f \cdot 5x^4 + \dots \\ \text{Put } x = 0. \quad \therefore \ \cos 0 = 1 + b + 0 + 0 + 0 + \dots \qquad\qquad \therefore \ b = 1 \end{cases}$

$\begin{cases} \text{Differentiate.} \quad -\sin x \equiv c \cdot 2 + d \cdot 3 \cdot 2x + e \cdot 4 \cdot 3x^2 + f \cdot 5 \cdot 4x^3 + \dots \\ \text{Put } x = 0. \quad \therefore \ -\sin 0 = 0 = c \cdot 2 + 0 + 0 + \dots \qquad\qquad \therefore \ c = 0 \end{cases}$

$\begin{cases} \text{Differentiate.} \quad -\cos x \equiv d \cdot 3 \cdot 2 \cdot 1 + e \cdot 4 \cdot 3 \cdot 2x + f \cdot 5 \cdot 4 \cdot 3x^2 + \dots \\ \text{Put } x = 0. \quad \therefore \ -\cos 0 = -1 = d \cdot 3! + 0 + 0 + \dots \qquad \therefore \ d = -\dfrac{1}{3!} \end{cases}$

$\begin{cases} \text{And again:} \quad \sin x \equiv e \cdot 4 \cdot 3 \cdot 2 \cdot 1 + f \cdot 5 \cdot 4 \cdot 3 \cdot 2x + \dots \\ \text{Put } x = 0. \quad \therefore \ \sin 0 = 0 = e \cdot 4! + 0 + 0 + \dots \qquad\qquad \therefore \ e = 0 \end{cases}$

$\begin{cases} \text{Once more:} \quad \cos x \equiv f \cdot 5 \cdot 4 \cdot 3 \cdot 2 \cdot 1 + \dots \\ \text{Put } x = 0. \quad \therefore \ \cos 0 = 1 = f \cdot 5! + 0 + \dots \qquad\qquad\qquad \therefore \ f = \dfrac{1}{5!} \end{cases}$

etc.

All that now remains is to put these values for the constant coefficients back into the original series.

$$\sin x \equiv 0 + 1 \cdot x + 0 \cdot x^2 + -\frac{1}{3!}x^3 + 0 \cdot x^4 + \frac{1}{5!}x^5 + \dots$$

i.e. $\sin x \equiv x - \dfrac{x^3}{3!} + \dfrac{x^5}{5!} - \dots$

Now we have obtained the first few terms of an infinite series representing the function $\sin x$, and you can see how the terms are likely to proceed.

Write down the first six terms of the series for $\sin x$.

When you have done so, move on to Frame 8

8

$$\sin x \equiv x - \frac{x^3}{3!} + \frac{x^5}{5!} - \frac{x^7}{7!} + \frac{x^9}{9!} - \frac{x^{11}}{11!} + \ldots$$

Provided we can differentiate a given function over and over again, and find the values of the derivatives when we put $x = 0$, then this method would enable us to express any function as a series of ascending powers of x.

However, it entails a considerable amount of writing, so we now establish a general form of such a series, which can be applied to most functions with very much less effort. This general series is known as *Maclaurin's series*.

So move on to Frame 9 and we will find out all about it

Maclaurin's series

9

To establish the series, we repeat the process of the previous example, but work with a general function, $f(x)$, instead of $\sin x$. The first derivative of $f(x)$ will be denoted by $f'(x)$; the second by $f''(x)$; the third by $f'''(x)$; and so on. Here it is then:

Let $f(x) = a + bx + cx^2 + dx^3 + ex^4 + fx^5 + \ldots$

Put $x = 0$. Then $f(0) = a + 0 + 0 + 0 + \ldots$ $\therefore\ a = f(0)$

   i.e. $a =$ the value of the function with x put equal to 0.

Differentiate. $f'(x) = b + c \cdot 2x + d \cdot 3x^2 + e \cdot 4x^3 + f \cdot 5x^4 + \ldots$

   Put $x = 0$. $\therefore\ f'(0) = b + 0 + 0 + \ldots$ $\therefore\ b = f'(0)$

Differentiate. $f''(x) = c \cdot 2 \cdot 1 + d \cdot 3 \cdot 2x + e \cdot 4 \cdot 3x^2 + f \cdot 5 \cdot 4x^3 \ldots$

   Put $x = 0$. $\therefore\ f''(0) = c \cdot 2! + 0 + 0 + \ldots$ $\therefore\ c = \dfrac{f''(0)}{2!}$

Now go on and find d and e, remembering that we denote

$\dfrac{\mathrm{d}}{\mathrm{d}x}\left\{f''(x)\right\}$ by $f'''(x)$ and $\dfrac{\mathrm{d}}{\mathrm{d}x}\left\{f'''(x)\right\}$ by $f^{iv}(x)$

   So, $d = \ldots\ldots\ldots\ldots$ and $e = \ldots\ldots\ldots\ldots$

10

$$d = \frac{f'''(0)}{3!}; \quad e = \frac{f^{iv}(0)}{4!}$$

Here it is. We had:

   $f''(x) = c \cdot 2 \cdot 1 + d \cdot 3 \cdot 2x + e \cdot 4 \cdot 3x^2 + f \cdot 5 \cdot 4x^3 + \ldots$

$\begin{cases} \text{Differentiate.} \quad \therefore\ f'''(x) = d \cdot 3 \cdot 2 \cdot 1 + e \cdot 4 \cdot 3 \cdot 2x + f \cdot 5 \cdot 4 \cdot 3x^2 + \ldots \\ \text{Put } x = 0. \quad\ \ \ \therefore\ f'''(0) = d \cdot 3! + 0 + 0 + \ldots \end{cases}$ $\therefore\ d = \dfrac{f'''(0)}{3!}$

$\begin{cases} \text{Differentiate.} \quad \therefore\ f^{iv}(x) = e \cdot 4 \cdot 3 \cdot 2 \cdot 1 + f \cdot 5 \cdot 4 \cdot 3 \cdot 2x + \ldots \\ \text{Put } x = 0. \quad\ \ \ \therefore\ f^{iv}(0) = e \cdot 4! + 0 + 0 + \ldots \end{cases}$ $\therefore\ e = \dfrac{f^{iv}(0)}{4!}$

   etc. So $a = f(0); b = f'(0); c = \dfrac{f''(0)}{2!}; d = \dfrac{f'''(0)}{3!}; e = \dfrac{f^{iv}(0)}{4!}; \ldots$

▶

Now, in just the same way as we did with our series for $\sin x$, we put the expressions for a, b, c, etc. back into the original series and get:

$$f(x) = \ldots\ldots\ldots$$

11

$$f(x) = f(0) + f'(0) \cdot x + \frac{f''(0)}{2!} \cdot x^2 + \frac{f'''(0)}{3!} \cdot x^3 + \ldots$$

and this is usually written as

$$f(x) = f(0) + x \cdot f'(0) + \frac{x^2}{2!} \cdot f''(0) + \frac{x^3}{3!} \cdot f'''(0) + \ldots \qquad \qquad \text{I}$$

This is *Maclaurin's series* and important!

Notice how tidy each term is.

The term in x^2 is divided by 2! and multiplied by $f''(0)$

The term in x^3 is divided by 3! and multiplied by $f'''(0)$

The term in x^4 is divided by 4! and multiplied by $f^{iv}(0)$

Copy the series into your record book for future reference.

Move on to Frame 12

12

Maclaurin's series

$$f(x) = f(0) + x \cdot f'(0) + \frac{x^2}{2!} \cdot f''(0) + \frac{x^3}{3!} \cdot f'''(0) + \ldots$$

Now we will use Maclaurin's series to find a series for $\sinh x$. We have to find the successive derivatives of $\sinh x$ and put $x = 0$ in each. Here goes, then:

$$f(x) = \sinh x \qquad\qquad f(0) = \sinh 0 = 0$$
$$f'(x) = \cosh x \qquad\qquad f'(0) = \cosh 0 = 1$$
$$f''(x) = \sinh x \qquad\qquad f''(0) = \sinh 0 = 0$$
$$f'''(x) = \cosh x \qquad\qquad f'''(0) = \cosh 0 = 1$$
$$f^{iv}(x) = \sinh x \qquad\qquad f^{iv}(0) = \sinh 0 = 0$$
$$f^{v}(x) = \cosh x \qquad\qquad f^{v}(0) = \cosh 0 = 1 \quad \text{etc.}$$

$$\therefore \ \sinh x = \cancel{0} + x \cdot 1 + \frac{\cancel{x^2}}{\cancel{2!}} \cdot (0) + \frac{x^3}{3!} \cdot (1) + \frac{\cancel{x^4}}{\cancel{4!}} \cdot (0) + \frac{x^5}{5!} \cdot (1) + \ldots$$

$$\therefore \ \sinh x = x + \frac{x^3}{3!} + \frac{x^5}{5!} + \frac{x^7}{7!} + \ldots$$

Move on to Frame 13

Now let us find a series for $\ln(1+x)$ in just the same way: **13**

$$f(x) = \ln(1+x) \qquad\qquad \therefore f(0) = \ldots\ldots\ldots$$

$$f'(x) = \frac{1}{1+x} = (1+x)^{-1} \qquad \therefore f'(0) = \ldots\ldots\ldots$$

$$f''(x) = -(1+x)^{-2} = \frac{-1}{(1+x)^2} \qquad \therefore f''(0) = \ldots\ldots\ldots$$

$$f'''(x) = 2(1+x)^{-3} = \frac{2}{(1+x)^3} \qquad \therefore f'''(0) = \ldots\ldots\ldots$$

$$f^{iv}(x) = -3\cdot 2(1+x)^{-4} = -\frac{3.2}{(1+x)^4} \quad \therefore f^{iv}(0) = \ldots\ldots\ldots$$

$$f^{v}(x) = 4\cdot 3\cdot 2(1+x)^{-5} = \frac{4!}{(1+x)^5} \quad \therefore f^{v}(0) = \ldots\ldots\ldots$$

You complete the work. Evaluate the derivatives when $x = 0$, remembering that $\ln 1 = 0$, and substitute back into Maclaurin's series to obtain the series for $\ln(1+x)$.

So, $\ln(1+x) = \ldots\ldots\ldots$

$f(0) = \ln 1 = 0; \quad f'(0) = 1; \quad f''(0) = -1; \quad f'''(0) = 2;$ **14**
$f^{iv}(0) = -3!; \quad f^{v}(0) = 4!; \quad \ldots$

Also
$$f(x) = f(0) + x\cdot f'(0) + \frac{x^2}{2!}f''(0) + \frac{x^3}{3!}f'''(0) + \ldots$$

$$\ln(1+x) = 0 + x\cdot 1 + \frac{x^2}{2!}(-1) + \frac{x^3}{3!}(2) + \frac{x^4}{4!}(-3!) + \ldots$$

$$\ln(1+x) = x - \frac{x^2}{2} + \frac{x^3}{3} - \frac{x^4}{4} + \frac{x^5}{5}$$

Note that in this series, the denominators are the natural numbers, not factorials!

Another example in Frame 15

Expand $\sin^2 x$ as a series of ascending powers of x. **15**

Maclaurin's series:

$$f(x) = f(0) + x\cdot f'(0) + \frac{x^2}{2!}\cdot f''(0) + \frac{x^3}{3!}\cdot f'''(0) + \ldots$$

$\therefore f(x) = \sin^2 x \qquad\qquad f(0) = \ldots\ldots\ldots$
$\quad f'(x) = 2\sin x \cos x = \sin 2x \qquad f'(0) = \ldots\ldots\ldots$
$\quad f''(x) = 2\cos 2x \qquad\qquad f''(0) = \ldots\ldots\ldots$
$\quad f'''(x) = -4\sin 2x \qquad\qquad f'''(0) = \ldots\ldots\ldots$
$\quad f^{iv}(x) = \ldots\ldots\ldots \qquad\qquad f^{iv}(0) = \ldots\ldots\ldots$

There we are! Finish it off: find the first three non-vanishing terms of the series.

Then move on to Frame 16

16

$$\sin^2 x = x^2 - \frac{x^4}{3} + \frac{2}{45}x^6 \ldots$$

Because

$f(x) = \sin^2 x$ $\therefore\; f(0) = 0$

$f'(x) = 2 \sin x \cos x = \sin 2x$ $\therefore\; f'(0) = 0$

$f''(x) = 2 \cos 2x$ $\therefore\; f''(0) = 2$

$f'''(x) = -4 \sin 2x$ $\therefore\; f'''(0) = 0$

$f^{iv}(x) = -8 \cos 2x$ $\therefore\; f^{iv}(0) = -8$

$f^v(x) = 16 \sin 2x$ $\therefore\; f^v(0) = 0$

$f^{vi}(x) = 32 \cos 2x$ $\therefore\; f^{vi}(0) = 32$ etc.

$$f(x) = f(0) + x \cdot f'(0) + \frac{x^2}{2!} \cdot f''(0) + \frac{x^3}{3!} \cdot f'''(0) + \ldots$$

$$\therefore\; \sin^2 x = 0 + x(0) + \frac{x^2}{2!}(2) + \frac{x^3}{3!}(0) + \frac{x^4}{4!}(-8) + \frac{x^5}{5!}(0) + \frac{x^6}{6!}(32) + \ldots$$

$$\therefore\; \sin^2 x = x^2 - \frac{x^4}{3} + \frac{2x^6}{45} + \ldots$$

Next we will find the series for $\tan x$. This is a little heavier but the method is always the same.

Move to Frame 17

17

Series for tan *x*

$f(x) = \tan x$ $\therefore\; f(0) = 0$

$\therefore\; f'(x) = \sec^2 x$ $\therefore\; f'(0) = 1$

$\therefore\; f''(x) = 2 \sec^2 x \tan x$ $\therefore\; f''(0) = 0$

$\therefore\; f'''(x) = 2 \sec^4 x + 4 \sec^2 x \tan^2 x$

$\qquad = 2 \sec^4 x + 4(1 + \tan^2 x) \tan^2 x$

$\qquad = 2 \sec^4 x + 4 \tan^2 x + 4 \tan^4 x$ $\therefore\; f'''(0) = 2$

$\therefore\; f^{iv}(x) = 8 \sec^4 x \tan x + 8 \tan x \sec^2 x + 16 \tan^3 x \sec^2 x$

$\qquad = 8(1 + t^2)^2 t + 8t(1 + t^2) + 16t^3(1 + t^2)$ [putting $t \equiv \tan x$]

$\qquad = 8(1 + 2t^2 + t^4)t + 8t + 8t^3 + 16t^3 + 16t^5$

$\qquad = 16t + 40t^3 + 24t^5$ $\therefore\; f^{iv}(0) = 0$

$\therefore\; f^v(x) = 16 \sec^2 x + 120t^2 \cdot \sec^2 x + 120t^4 \cdot \sec^2 x$ $\therefore\; f^v(0) = 16$

$\therefore\; \tan x = \ldots\ldots\ldots\ldots$

$$\therefore \quad \tan x = x + \frac{x^3}{3} + \frac{2x^5}{15} + \ldots$$

Move on to Frame 19

Standard series

19

Using Maclaurin's series we can build up a list of series representing many of the common functions – we have already found series for $\sin x$, $\sinh x$ and $\ln(1 + x)$.

To find a series for $\cos x$, we could apply the same technique all over again. However, let us be crafty about it. Suppose we take the series for $\sin x$ and differentiate both sides with respect to x just once, we get:

$$\sin x = x - \frac{x^3}{3!} + \frac{x^5}{5!} - \frac{x^7}{7!} + \ldots$$

Differentiate:

$$\cos x = 1 - \frac{3x^2}{3!} + \frac{5x^4}{5!} - \frac{7x^6}{7!} + \ldots$$

$$\therefore \quad \cos x = 1 - \frac{x^2}{2!} + \frac{x^4}{4!} - \frac{x^6}{6!} + \ldots$$

In the same way, we can obtain the series for $\cosh x$. We already know that

$$\sinh x = x + \frac{x^3}{3!} + \frac{x^5}{5!} + \frac{x^7}{7!} + \ldots$$

so if we differentiate both sides we shall establish a series for $\cosh x$.
What do we get?

20

We get:

$$\sinh x = x + \frac{x^3}{3!} + \frac{x^5}{5!} + \frac{x^7}{7!} + \frac{x^9}{9!} + \ldots$$

Differentiate:

$$\cosh x = 1 + \frac{3x^2}{3!} + \frac{5x^4}{5!} + \frac{7x^6}{7!} + \frac{9x^8}{9!} + \ldots$$

giving:

$$\cosh x = 1 + \frac{x^2}{2!} + \frac{x^4}{4!} + \frac{x^6}{6!} + \frac{x^8}{8!} + \ldots$$

Let us pause at this point and take stock of the series we have obtained.
We will make a list of them, so move on to Frame 21

21   **Summary**

Here are the standard series that we have established so far:

$$\sin x = x - \frac{x^3}{3!} + \frac{x^5}{5!} - \frac{x^7}{7!} + \frac{x^9}{9!} + \ldots \qquad \text{II}$$

$$\cos x = 1 - \frac{x^2}{2!} + \frac{x^4}{4!} - \frac{x^6}{6!} + \frac{x^8}{8!} + \ldots \qquad \text{III}$$

$$\tan x = x + \frac{x^3}{3} + \frac{2x^5}{15} + \ldots \qquad \text{IV}$$

$$\sinh x = x + \frac{x^3}{3!} + \frac{x^5}{5!} + \frac{x^7}{7!} + \ldots \qquad \text{V}$$

$$\cosh x = 1 + \frac{x^2}{2!} + \frac{x^4}{4!} + \frac{x^6}{6!} + \frac{x^8}{8!} + \ldots \qquad \text{VI}$$

$$\ln(1+x) = x - \frac{x^2}{2} + \frac{x^3}{3} - \frac{x^4}{4} + \frac{x^5}{5} + \ldots \qquad \text{VII}$$

Make a note of these six series in your record book.

Then move on to Frame 22

The binomial series

22

By the same method, we can apply Maclaurin's series to obtain the binomial series for $(1+x)^n$. Here it is:

$$f(x) = (1+x)^n \qquad\qquad\qquad f(0) = 1$$
$$f'(x) = n \cdot (1+x)^{n-1} \qquad\qquad f'(0) = n$$
$$f''(x) = n(n-1) \cdot (1+x)^{n-2} \qquad f''(0) = n(n-1)$$
$$f'''(x) = n(n-1)(n-2) \cdot (1+x)^{n-3} \qquad f'''(0) = n(n-1)(n-2)$$
$$f^{iv}(x) = n(n-1)(n-2)(n-3) \cdot (1+x)^{n-4} \qquad f^{iv}(0) = n(n-1)(n-2)(n-3)$$
$$\text{etc.} \qquad\qquad\qquad\qquad\qquad \text{etc.}$$

General Maclaurin's series:

$$f(x) = f(0) + x \cdot f'(0) + \frac{x^2}{2!} f''(0) + \frac{x^3}{3!} f'''(0) + \ldots$$

Therefore, in this case:

$$(1+x)^n = 1 + xn + \frac{x^2}{2!} n(n-1) + \frac{x^3}{3!} n(n-1)(n-2) + \ldots$$

$$(1+x)^n = 1 + nx + \frac{n(n-1)}{2!} x^2 + \frac{n(n-1)(n-2)}{3!} x^3 + \ldots \qquad \text{VIII}$$

By replacing x wherever it occurs by $(-x)$, determine the series for $(1-x)^n$.

When finished, move to Frame 23

23

$$(1-x)^n = 1 - nx + \frac{n(n-1)}{2!}x^2 - \frac{n(n-1)(n-2)}{3!}x^3 + \dots$$

Now we will work through another example. Here it is:

To find a series for $\tan^{-1} x$.

As before, we need to know the successive derivatives in order to insert them in Maclaurin's series.

$$f(x) = \tan^{-1} x \text{ and } f'(x) = \frac{1}{1+x^2}$$

If we differentiate again, we get $f''(x) = -\dfrac{2x}{(1+x^2)^2}$, after which the working becomes rather heavy, so let us be crafty and see if we can avoid unnecessary work.

We have $f(x) = \tan^{-1} x$ and $f'(x) = \dfrac{1}{1+x^2} = (1+x^2)^{-1}$. If we now expand $(1+x^2)^{-1}$ as a binomial series, we shall have a series of powers of x from which we can easily find the higher derivatives.

So see how it works out in the next frame

24

To find a series for $\tan^{-1} x$

$$f(x) = \tan^{-1} x \qquad\qquad\qquad \therefore f(0) = 0$$

$$\therefore f'(x) = \frac{1}{1+x^2} = (1+x^2)^{-1}$$

$$= 1 - x^2 + \frac{(-1)(-2)}{1\cdot 2}x^4 + \frac{(-1)(-2)(-3)}{1\cdot 2\cdot 3}x^6 + \dots$$

$$= 1 - x^2 + x^4 - x^6 + x^8 - \dots \qquad\qquad f'(0) = 1$$

$$\therefore f''(x) = -2x + 4x^3 - 6x^5 + 8x^7 - \dots \qquad\qquad f''(0) = 0$$

$$\therefore f'''(x) = -2 + 12x^2 - 30x^4 + 56x^6 - \dots \qquad\qquad f'''(0) = -2$$

$$\therefore f^{iv}(x) = 24x - 120x^3 + 336x^5 - \dots \qquad\qquad f^{iv}(0) = 0$$

$$\therefore f^{v}(x) = 24 - 360x^2 + 1680x^4 - \dots \qquad\qquad f^{v}(0) = 24 \text{ etc.}$$

$$\therefore \tan^{-1} x = f(0) + x\cdot f'(0) + \frac{x^2}{2!}f''(0) + \frac{x^3}{3!}f'''(0) + \dots \qquad\qquad \text{IX}$$

Substituting the values for the derivatives, gives us that $\tan^{-1} x = \dots\dots\dots$

Then on to Frame 25

25

$$\tan^{-1} x = 0 + x(1) + \frac{x^2}{2!}(0) + \frac{x^3}{3!}(-2) + \frac{x^4}{4!}(0) + \frac{x^5}{5!}(24) + \dots$$

$$\boxed{\tan^{-1} x = x - \frac{x^3}{3} + \frac{x^5}{5} - \frac{x^7}{7} + \dots} \qquad\qquad \text{X}$$

This is also a useful series, so make a note of it.

▶

Another series which you already know quite well is the series for e^x. Do you remember how it goes? Here it is anyway:

$$e^x = 1 + x + \frac{x^2}{2!} + \frac{x^3}{3!} + \frac{x^4}{4!} + \dots \qquad \text{XI}$$

and if we simply replace x by $(-x)$, we obtain the series for e^{-x}

$$e^{-x} = 1 - x + \frac{x^2}{2!} - \frac{x^3}{3!} + \frac{x^4}{4!} - \dots \qquad \text{XII}$$

So now we have quite a few. Add the last two to your list.

And then on to the next frame

26

Once we have established these standard series, we can of course combine them as necessary, as a couple of examples will show.

Example 1

Find the first three terms of the series for $e^x \cdot \ln(1 + x)$.

We know that $e^x = 1 + x + \dfrac{x^2}{2!} + \dfrac{x^3}{3!} + \dfrac{x^4}{4!} + \dots$

and that $\ln(1 + x) = x - \dfrac{x^2}{2!} + \dfrac{x^3}{3!} - \dfrac{x^4}{4!} + \dots$

So $e^x \cdot \ln(1+x) = \left\{ 1 + x + \dfrac{x^2}{2!} + \dfrac{x^3}{3!} + \dfrac{x^4}{4!} + \dots \right\} \left\{ x - \dfrac{x^2}{2!} + \dfrac{x^3}{3!} - \dots \right\}$

Now we have to multiply these series together. There is no constant term in the second series, so the lowest power of x in the product will be x itself. This can only be formed by multiplying the 1 in the first series by the x in the second.

The x^2 term is found by multiplying $1 \times \left(-\dfrac{x^2}{2} \right)$

and $x \times x$

$\left.\right\}$ $x^2 - \dfrac{x^2}{2} = \dfrac{x^2}{2}$

The x^3 term is found by multiplying $1 \times \dfrac{x^3}{3}$

and $x \times \left(-\dfrac{x^2}{2} \right)$

and $\dfrac{x^2}{2} \times x$

$\left.\right\}$ $\dfrac{x^3}{3} - \dfrac{x^3}{2} + \dfrac{x^3}{2} = \dfrac{x^3}{3}$

and so on.

27

$$\therefore\ e^x \cdot \ln(1+x) = x + \frac{x^2}{2} + \frac{x^3}{3} + \dots$$

It is not at all difficult, provided you are careful to avoid missing any of the products of the terms.

Here is one for you to do in the same way:

Example 2

Find the first four terms of the series for $e^x \sinh x$.

Take your time over it: then check your working with that in Frame 28

28

Here is the solution. Look through it carefully to see if you agree with the result.

$$e^x = 1 + x + \frac{x^2}{2!} + \frac{x^3}{3!} + \frac{x^4}{4!} + \dots$$

$$\sinh x = x + \frac{x^3}{3!} + \frac{x^5}{5!} + \frac{x^7}{7!} + \dots$$

$$e^x \cdot \sinh x = \left\{ 1 + x + \frac{x^2}{2!} + \frac{x^3}{3!} + \dots \right\} \left\{ x + \frac{x^3}{3!} + \frac{x^5}{5!} + \dots \right\}$$

Lowest power is x

Term in $x = 1 \cdot x = x$

Term in $x^2 = x \cdot x = x^2$

Term in $x^3 = 1 \cdot \dfrac{x^3}{3!} + \dfrac{x^2}{2!} \cdot x = x^3 \left(\dfrac{1}{6} + \dfrac{1}{2} \right) = \dfrac{2x^3}{3}$

Term in $x^4 = x \cdot \dfrac{x^3}{3!} + \dfrac{x^3}{3!} \cdot x = x^4 \left(\dfrac{1}{6} + \dfrac{1}{6} \right) = \dfrac{x^4}{3}$

$$\therefore\ e^x \cdot \sinh x = x + x^2 + \frac{2x^3}{3} + \frac{x^4}{3} + \dots$$

There we are. Now move on to Frame 29

Approximate values

29

This is a very obvious application of series and you will surely have done some examples on this topic at some time in the past. Here is just an example or two to refresh your memory.

▶

Example 1

Evaluate $\sqrt{1.02}$ correct to 5 decimal places.

$$1.02 = 1 + 0.02$$
$$\sqrt{1.02} = (1 + 0.02)^{\frac{1}{2}}$$

$$= 1 + \frac{1}{2}(0.02) + \frac{\frac{1}{2}\left(-\frac{1}{2}\right)}{1 \cdot 2}(0.02)^2 + \frac{\frac{1}{2}\left(-\frac{1}{2}\right)\left(-\frac{3}{2}\right)}{1 \cdot 2 \cdot 3}(0.02)^3 + \ldots$$

$$= 1 + 0.01 - \frac{1}{8}(0.0004) + \frac{1}{16}(0.000008) - \ldots$$

$$= 1 + 0.01 - 0.00005 + 0.0000005 - \ldots$$

$$= 1.010001 - 0.000050$$

$$= 1.009951 \qquad \therefore \ \sqrt{1.02} = 1.00995$$

Note: Whenever we substitute a value for x in any one of the standard series, we must be satisfied that the substitution value for x is within the range of values of x for which the series is valid.

The present series for $(1 + x)^n$ is valid for $|x| < 1$, so we are safe enough on this occasion.

Here is one for you to do.

Example 2

Evaluate $\tan^{-1} 0.1$ correct to 4 decimal places.

Complete the working and then check with the next frame

30

$$\boxed{\tan^{-1} 0.1 = 0.0997}$$

$$\tan^{-1} x = x - \frac{x^3}{3} + \frac{x^5}{5} - \frac{x^7}{7} + \ldots$$

$$\therefore \ \tan^{-1}(0.1) = 0.1 - \frac{0.001}{3} + \frac{0.00001}{5} - \frac{0.0000001}{7} + \ldots$$

$$= 0.1 - 0.00033 + 0.000002 - \ldots$$

$$= 0.0997 \qquad [\textit{Note}: x \text{ is measured in radians.}]$$

We will now consider a further use for series, so move now to Frame 31

Limiting values – indeterminate forms

31

In the first Program on series, we had occasion to find the limiting value of $\dfrac{u_{n+1}}{u_n}$ as $n \to \infty$. Sometimes, we have to find the limiting value of a function of x when $x \to 0$, or perhaps when $x \to a$.

e.g. $\underset{x \to 0}{Lim} \left\{ \dfrac{x^2 + 5x - 14}{x^2 - 5x + 8} \right\} = \dfrac{0 + 0 - 14}{0 - 0 + 8} = -\dfrac{14}{8} = -\dfrac{7}{4}$

That is easy enough, but suppose we have to find

$\underset{x \to 2}{Lim} \left\{ \dfrac{x^2 + 5x - 14}{x^2 - 5x + 6} \right\}$

Putting $x = 2$ in the function gives $\dfrac{4 + 10 - 14}{4 - 10 + 6} = \dfrac{0}{0}$ and what is the value of $\dfrac{0}{0}$?

Is it zero? Is it 1? Is it indeterminate?

When you have decided, move on to Frame 32

$\dfrac{0}{0}$, as it stands, is $\boxed{\text{indeterminate}}$

We can sometimes, however, use our knowledge of series to help us out of the difficulty. Let us consider an example or two.

Example 1

Find $\underset{x \to 0}{Lim} \left\{ \dfrac{\tan x - x}{x^3} \right\}$

If we substitute $x = 0$ in the function, we get the result $\dfrac{0}{0}$ which is indeterminate. So how do we proceed?

Well, we already know that $\tan x = x + \dfrac{x^3}{3} + \dfrac{2x^5}{15} + \dots$ So if we replace $\tan x$ by its series in the given function, we get:

$$\underset{x \to 0}{Lim} \left\{ \dfrac{\tan x - x}{x^3} \right\} = \underset{x \to 0}{Lim} \left\{ \dfrac{\left(\cancel{x} + \dfrac{x^3}{3} + \dfrac{2x^5}{15} + \dots \right) - \cancel{x}}{x^3} \right\}$$

$$= \underset{x \to 0}{Lim} \left\{ \dfrac{1}{3} + \dfrac{2x^2}{15} + \dots \right\} = \dfrac{1}{3}$$

$\therefore \underset{x \to 0}{Lim} \left\{ \dfrac{\tan x - x}{x^3} \right\} = \dfrac{1}{3}$ and the job is done!

Move on to Frame 33 for another example

33

Example 2

To find $\displaystyle \lim_{x \to 0} \left\{ \frac{\sinh x}{x} \right\}$

Direct substitution of $x = 0$ gives $\dfrac{\sin 0}{0}$ which is $\dfrac{0}{0}$ again. So we will express $\sinh x$ by its series, which is

$\sinh x = \ldots\ldots\ldots$

(If you do not remember, you will find it in your list of standard series which you have been compiling. Look it up.)

Then on to Frame 34

34

$$\sinh x = x + \frac{x^3}{3!} + \frac{x^5}{5!} + \frac{x^7}{7!} + \ldots$$

So $\displaystyle \lim_{x \to 0} \left\{ \frac{\sinh x}{x} \right\} = \lim_{x \to 0} \left\{ \frac{x + \dfrac{x^3}{3!} + \dfrac{x^5}{5!} + \dfrac{x^7}{7!} + \ldots}{x} \right\}$

$$= \lim_{x \to 0} \left\{ 1 + \frac{x^2}{3!} + \frac{x^4}{5!} + \ldots \right\}$$

$$= 1 + 0 + 0 + \ldots = 1$$

$\therefore \displaystyle \lim_{x \to 0} \left\{ \frac{\sinh x}{x} \right\} = 1$

Now, in very much the same way, you find $\displaystyle \lim_{x \to 0} \left\{ \frac{\sin^2 x}{x^2} \right\}$

Work it through: then check your result with that in the next frame

35

$$\lim_{x \to 0} \left\{ \frac{\sin^2 x}{x^2} \right\} = 1$$

Here is the working:

$$\lim_{x \to 0} \left\{ \frac{\sin^2 x}{x^2} \right\} = \lim_{x \to 0} \left\{ \frac{x^2 - \dfrac{x^4}{3} + \dfrac{2x^6}{45} - \ldots}{x^2} \right\} \quad \text{(see Frame 16)}$$

$$= \lim_{x \to 0} \left\{ 1 - \frac{x^2}{3} + \frac{2x^4}{45} - \ldots \right\} = 1$$

$\therefore \displaystyle \lim_{x \to 0} \left\{ \frac{\sin^2 x}{x^2} \right\} = 1$

▶

Here is one more for you to do in like manner.

Find $\displaystyle \lim_{x \to 0} \left\{ \frac{\sinh x - x}{x^3} \right\}$

Then on to Frame 36

$$\lim_{x \to 0} \left\{ \frac{\sinh x - x}{x^3} \right\} = \frac{1}{6}$$

Here is the working in detail:

$$\sinh x = x + \frac{x^3}{3!} + \frac{x^5}{5!} + \frac{x^7}{7!} + \cdots$$

$$\therefore \quad \frac{\sinh x - x}{x^3} = \frac{\cancel{x} + \dfrac{x^3}{3!} + \dfrac{x^5}{5!} + \dfrac{x^7}{7!} + \cdots - \cancel{x}}{x^3}$$

$$= \frac{1}{3!} + \frac{x^2}{5!} + \frac{x^4}{7!} + \cdots$$

$$\therefore \quad \lim_{x \to 0} \left\{ \frac{\sinh x - x}{x^3} \right\} = \lim_{x \to 0} \left\{ \frac{1}{3!} + \frac{x^2}{5!} + \frac{x^4}{7!} + \cdots \right\}$$

$$= \frac{1}{3!} = \frac{1}{6}$$

$$\therefore \quad \lim_{x \to 0} \left\{ \frac{\sinh x - x}{x^3} \right\} = \frac{1}{6}$$

So there you are: they are all done the same way:

(a) Express the given function in terms of power series.

(b) Simplify the function as far as possible.

(c) Then determine the limiting value – which should now be possible.

Of course, there may well be occasions when direct substitution gives the indeterminate form $\dfrac{0}{0}$ and we do not know the series expansion of the function concerned. What are we going to do then?

All is not lost! – for we do in fact have another method of finding limiting values which, in many cases, is quicker than the series method. It all depends upon the application of a rule which we must first establish, so move to the next frame for details.

Next frame

L'Hôpital's rule for finding limiting values

37

Suppose we have to find the limiting value of a function $F(x) = \dfrac{f(x)}{g(x)}$ at $x = a$,

when direct substitution of $x = a$ gives the indeterminate form $\dfrac{0}{0}$, i.e. at $x = a$,

$f(x) = 0$ and $g(x) = 0$.

If we represent the circumstances graphically, the diagram would look like this:

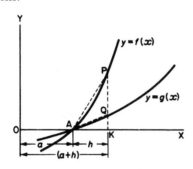

Note that at $x = a$, both of the graphs $y = f(x)$ and $y = g(x)$ cross the x-axis, so that at $x = a$, $f(x) = 0$ and $g(x) = 0$.

At a point K, i.e. $x = (a + h)$, $KP = f(a + h)$ and $KQ = g(a + h)$

$$\frac{f(a + h)}{g(a + h)} = \frac{KP}{KQ}$$

Now divide top and bottom by AK:

$$\frac{f(a + h)}{g(a + h)} = \frac{KP/AK}{KQ/AK} = \frac{\tan PAK}{\tan QAK}$$

Now $\displaystyle \lim_{x \to a} \frac{f(x)}{g(x)} = \lim_{h \to 0} \frac{f(a + h)}{g(a + h)} = \lim_{h \to 0} \frac{\tan PAK}{\tan QAK} = \frac{f'(a)}{g'(a)}$

i.e. the limiting value of $\dfrac{f(x)}{g(x)}$ as $x \to a$ (at which the function value by direct

substitution gives $\dfrac{0}{0}$) is given by the ratio of the derivatives of numerator and

denominator at $x = a$ (provided, of course, that $f'(a)$ and $g'(a)$ are not both zero themselves!). That is:

$$\lim_{x \to a} \left\{ \frac{f(x)}{g(x)} \right\} = \frac{f'(a)}{g'(a)} = \lim_{x \to a} \left\{ \frac{f'(x)}{g'(x)} \right\}$$

$$\therefore \ \lim_{x \to a} \left\{ \frac{f(x)}{g(x)} \right\} = \lim_{x \to a} \left\{ \frac{f'(x)}{g'(x)} \right\}$$

This is known as *l'Hôpital's rule* and is extremely useful for finding limiting values when the derivatives of the numerator and denominator can easily be found.

Copy the rule into your record book. Now we will use it

$$\underset{x \to a}{Lim}\Big\}\left\{\frac{f(x)}{g(x)}\right\} = \underset{x \to a}{Lim}\left\{\frac{f'(x)}{g'(x)}\right\}$$

Example 1

To find $\underset{x \to 1}{Lim}\left\{\dfrac{x^3 + x^2 - x - 1}{x^2 + 2x - 3}\right\}$

Note first that if we substitute $x = 1$, we get the indeterminate form $\dfrac{0}{0}$.

Therefore we apply l'Hôpital's rule.

We therefore differentiate numerator and denominator separately (*not* as a quotient):

$$\underset{x \to 1}{Lim}\left\{\frac{x^3 + x^2 - x - 1}{x^2 + 2x - 3}\right\} = \underset{x \to 1}{Lim}\left\{\frac{3x^2 + 2x - 1}{2x + 2}\right\}$$

$$= \frac{3 + 2 - 1}{2 + 2} = \frac{4}{4} = 1$$

$$\therefore \ \underset{x \to 1}{Lim}\left\{\frac{x^3 + x^2 - x - 1}{x^2 + 2x - 3}\right\} = 1$$

and that is all there is to it!

Let us do another example, so, on to the next frame

39

Example 2

Determine $\underset{x \to 0}{Lim}\left\{\dfrac{\cosh x - e^x}{x}\right\}$

We first of all try direct substitution, but we find that this leads us to the result $\dfrac{1 - 1}{0}$, i.e. $\dfrac{0}{0}$ which is indeterminate. Therefore, apply l'Hôpital's rule:

$$\underset{x \to a}{Lim}\left\{\frac{f(x)}{g(x)}\right\} = \underset{x \to a}{Lim}\left\{\frac{f'(x)}{g'(x)}\right\}$$

i.e. differentiate top and bottom separately and substitute the given value of x in the derivatives:

$$\therefore \ \underset{x \to 0}{Lim}\left\{\frac{\cosh x - e^x}{x}\right\} = \underset{x \to 0}{Lim}\left\{\frac{\sinh x - e^x}{1}\right\} = \frac{0 - 1}{1} = -1$$

$$\therefore \ \underset{x \to 0}{Lim}\left\{\frac{\cosh x - e^x}{x}\right\} = -1$$

Now you can do this one:

Determine $\underset{x \to 0}{Lim}\left\{\dfrac{x^2 - \sin 3x}{x^2 + 4x}\right\}$

40

$$\underset{x\to 0}{Lim}\left\{\frac{x^2-\sin 3x}{x^2+4x}\right\}=-\frac{3}{4}$$

The working is simply this:

Direct substitution gives $\dfrac{0}{0}$, so we apply l'Hôpital's rule which gives

$$\underset{x\to 0}{Lim}\left\{\frac{x^2-\sin 3x}{x^2+4x}\right\}=\underset{x\to 0}{Lim}\left\{\frac{2x-3\cos 3x}{2x+4}\right\}$$
$$=\frac{0-3}{0+4}=-\frac{3}{4}$$

Warning: l'Hôpital's rule applies only when the indeterminate form arises. If the limiting value can be found by direct substitution, the rule will not work. An example will soon show this.

Consider $\underset{x\to 2}{Lim}\left\{\dfrac{x^2+4x-3}{5-2x}\right\}$

By direct substitution, the limiting value $=\dfrac{4+8-3}{5-4}=9$. By l'Hôpital's rule

$\underset{x\to 2}{Lim}\left\{\dfrac{f'(x)}{g'(x)}\right\}=\underset{x\to 2}{Lim}\left\{\dfrac{2x+4}{-2}\right\}=-4$. As you will see, these results do not agree.

Before using l'Hôpital's rule, therefore, you must satisfy yourself that direct substitution gives the indeterminate form $\dfrac{0}{0}$. If it does, you may use the rule, but not otherwise.

41

Let us look at another example.

Example 3

Determine $\underset{x\to 0}{Lim}\left\{\dfrac{x-\sin x}{x^2}\right\}$

By direct substitution, limiting value $=\dfrac{0-0}{0}=\dfrac{0}{0}$.

Apply l'Hôpital's rule:

$$\underset{x\to 0}{Lim}\left\{\frac{x-\sin x}{x^2}\right\}=\underset{x\to 0}{Lim}\left\{\frac{1-\cos x}{2x}\right\}$$

We now find, with some horror, that substituting $x=0$ in the derivatives again produces the indeterminate form $\dfrac{0}{0}$. So what do you suggest we do now to find $\underset{x\to 0}{Lim}\left\{\dfrac{1-\cos x}{2x}\right\}$ (without bringing in the use of series)? Any ideas?

We . . . .

42

$$\boxed{\text{We apply the rule a second time}}$$

Correct, for our immediate problem now is to find $Lim_{x \to 0} \left\{ \dfrac{1 - \cos x}{2x} \right\}$.

If we do that, we get:

$$Lim_{x \to 0} \left\{ \frac{x - \sin x}{x^2} \right\} = \underbrace{Lim_{x \to 0} \left\{ \frac{1 - \cos x}{2x} \right\}}_{\text{First stage}} = \underbrace{Lim_{x \to 0} \left\{ \frac{\sin x}{2} \right\} = \frac{0}{2} = 0}_{\text{Second stage}}$$

$$\therefore \ Lim_{x \to 0} \left\{ \frac{x - \sin x}{x^2} \right\} = 0$$

So now we have the rule complete:

For limiting values when the indeterminate form $\left(\text{i.e. } \dfrac{0}{0}\right)$ exists, apply l'Hôpital's rule:

$$Lim_{x \to a} \left\{ \frac{f(x)}{g(x)} \right\} = Lim_{x \to a} \left\{ \frac{f'(x)}{g'(x)} \right\}$$

and continue to do so until a stage is reached where the numerator and/or the denominator are/is not zero.

Next frame

43

Just one more example to illustrate the point

Example 4

Determine $Lim_{x \to 0} \left\{ \dfrac{\sinh x - \sin x}{x^3} \right\}$

Direct substitution gives $\dfrac{0 - 0}{0}$, i.e. $\dfrac{0}{0}$. (indeterminate)

$$Lim_{x \to 0} \left\{ \frac{\sinh x - \sin x}{x^3} \right\} = Lim_{x \to 0} \left\{ \frac{\cosh x - \cos x}{3x^2} \right\}, \text{ giving } \frac{1 - 1}{0} = \frac{0}{0}$$

$$= Lim_{x \to 0} \left\{ \frac{\sinh x + \sin x}{6x} \right\}, \text{ giving } \frac{0 + 0}{0} = \frac{0}{0}$$

$$= Lim_{x \to 0} \left\{ \frac{\cosh x + \cos x}{6} \right\} = \frac{1 + 1}{6} = \frac{1}{3}$$

$$\therefore \ Lim_{x \to 0} \left\{ \frac{\sinh x - \sin x}{x^3} \right\} = \frac{1}{3}$$

Note: We apply l'Hôpital's rule again and again until we reach the stage where the numerator or the denominator (or both) is (are) *not* zero. We shall then arrive at a definite limiting value of the function.

Move on to frame 44

44

Here are three *Review Examples* for you to do. Work through all of them and then check your working with the results set out in the next frame. They are all straightforward and easy, so do not peep at the official solutions before you have done them all.

Determine

(a) $Lim\limits_{x \to 1} \left\{ \dfrac{x^3 - 2x^2 + 4x - 3}{4x^2 - 5x + 1} \right\}$

(b) $Lim\limits_{x \to 0} \left\{ \dfrac{\tan x - x}{\sin x - x} \right\}$

(c) $Lim\limits_{x \to 0} \left\{ \dfrac{x \cos x - \sin x}{x^3} \right\}$

Solutions in Frame 45

45

(a) $Lim\limits_{x \to 1} \left\{ \dfrac{x^3 - 2x^2 + 4x - 3}{4x^2 - 5x + 1} \right\}$ $\left(\text{Substitution gives } \dfrac{0}{0} \right)$

$= Lim\limits_{x \to 1} \left\{ \dfrac{3x^2 - 4x + 4}{8x - 5} \right\} = \dfrac{3}{3} = 1$

$\therefore Lim\limits_{x \to 1} \left\{ \dfrac{x^3 - 2x^2 + 4x - 3}{4x^2 - 5x + 1} \right\} = 1$

(b) $Lim\limits_{x \to 0} \left\{ \dfrac{\tan x - x}{\sin x - x} \right\}$ $\left(\text{Substitution gives } \dfrac{0}{0} \right)$

$= Lim\limits_{x \to 0} \left\{ \dfrac{\sec^2 x - 1}{\cos x - 1} \right\}$ $\left(\text{still gives } \dfrac{0}{0} \right)$

$= Lim\limits_{x \to 0} \left\{ \dfrac{2 \sec^2 x \tan x}{- \sin x} \right\}$ (and again!)

$= Lim\limits_{x \to 0} \left\{ \dfrac{2 \sec^2 x \sec^2 x + 4 \sec^2 x \tan^2 x}{- \cos x} \right\} = \dfrac{2 + 0}{-1} = -2$

$\therefore Lim\limits_{x \to 0} \left\{ \dfrac{\tan x - x}{\sin x - x} \right\} = -2$

(c) $Lim\limits_{x \to 0} \left\{ \dfrac{x \cos x - \sin x}{x^3} \right\}$ $\left(\text{Substitution gives } \dfrac{0}{0} \right)$

$= Lim\limits_{x \to 0} \left\{ \dfrac{-x \sin x + \cos x - \cos x}{3x^2} \right\}$

$= Lim\limits_{x \to 0} \left\{ \dfrac{- \sin x}{3x} \right\} = Lim\limits_{x \to 0} \left\{ \dfrac{- \cos x}{3} \right\} = -\dfrac{1}{3}$

$\therefore Lim\limits_{x \to 0} \left\{ \dfrac{x \cos x - \sin x}{x^3} \right\} = -\dfrac{1}{3}$

Next frame

Let us look at another useful series: Taylor's series. 46

Maclaurin's series $f(x) = f(0) + x.f'(0) + \dfrac{x^2}{2!}f''(0) + \ldots$ expresses a function in terms of its derivatives at $x = 0$, i.e. at the point K.

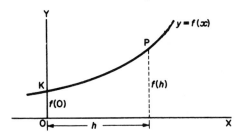

At P, $f(h) = f(0) + h \cdot f'(0) + \dfrac{h^2}{2!}f''(0) + \dfrac{h^3}{3!}f'''(0) \ldots$

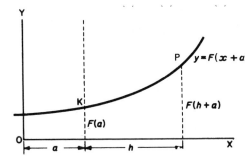

If we now move the y-axis a units to the left, the equation of the curve relative to the new axes now becomes $y = F(a + x)$ and the value at K is now $F(a)$.

At P, $F(a + h) = F(a) + h \cdot F'(a) + \dfrac{h^2}{2!}F''(a) + \dfrac{h^3}{3!}F'''(a) + \ldots$

This is, in fact, a general series and holds good when a and h are both variables. If we write $a = x$ in this result, we obtain:

$$f(x + h) = f(x) + h \cdot f'(x) + \dfrac{h^2}{2!}f''(x) + \dfrac{h^3}{3!}f'''(x) + \ldots$$

which is the usual form of Taylor's series.

Maclaurin's series and Taylor's series are very much alike in some respects. In 47 fact, Maclaurin's series is really a special case of Taylor's.

Maclaurin's series: $\quad f(x) = f(0) = x \cdot f'(0) + \dfrac{x^2}{2!}f''(0) + \dfrac{x^3}{3!}f'''(0) + \ldots$

Taylor's series: $\quad f(x + h) = f(x) + h \cdot f'(x) + \dfrac{h^2}{2!}f''(x) + \dfrac{h^3}{3!}f'''(x) + \ldots$

Copy the two series down together: it will help you remember them.

48

Example 1

Show that, if h is small, then

$$\tan^{-1}(x+h) = \tan^{-1}x + \frac{h}{1+x^2} - \frac{xh^2}{(1+x^2)^2} \quad \text{approximately.}$$

Taylor's series states:

$$f(x+h) = f(x) + h \cdot f'(x) + \frac{h^2}{2!}f''(x) + \frac{h^3}{3!}f'''(x) + \ldots$$

where $f(x)$ is the function obtained by putting $h=0$ in the function $f(x+h)$.
In this case then, $f(x) = \tan^{-1}x$

$$\therefore f'(x) = \frac{1}{1+x^2} \text{ and } f''(x) = -\frac{2x}{(1+x^2)^2}$$

Putting these expressions back into the series, we have:

$$\tan^{-1}(x+h) = \tan^{-1}x + h \cdot \frac{1}{1+x^2} - \frac{h^2}{2!} \cdot \frac{2x}{(1+x^2)^2} + \ldots$$

$$= \tan^{-1}x + \frac{h}{1+x^2} - \frac{xh^2}{(1+x^2)^2} \quad \text{approximately}$$

Why are we justified in omitting the terms that follow?

49

> The following terms contain higher powers of h which, by definition, are small. These terms will therefore be very small.

Example 2

Express $\sin(x+h)$ as a series of powers of h and evaluate $\sin 44°$ correct to 5 decimal places.

$$\sin(x+h) = f(x) + h \cdot f'(x) + \frac{h^2}{2!}f''(x) + \frac{h^3}{3!}f'''(x) + \ldots$$

$$f(x) = \sin x; \quad f'(x) = \cos x; \quad f''(x) = -\sin x$$

$$f'''(x) = -\cos x; \quad f^{iv}(x) = \sin x; \text{ etc.}$$

$$\therefore \sin(x+h) = \sin x + h\cos x - \frac{h^2}{2!}\sin x - \frac{h^3}{3!}\cos x + \ldots$$

$$\sin 44° = \sin(45° - 1°) = \sin\left(\frac{\pi}{4} - 0.01745\right) \text{ and } \sin\frac{\pi}{4} = \cos\frac{\pi}{4} = \frac{1}{\sqrt{2}}$$

$$\therefore \sin 44° = \frac{1}{\sqrt{2}}\left\{1 + h - \frac{h^2}{2} - \frac{h^3}{6} + \ldots\right\} \qquad h = -0.01745$$

$$= \frac{1}{\sqrt{2}}\left\{1 - 0.01745 - \frac{0.0003046}{2} + \frac{0.0000053}{6} + \ldots\right\}$$

$$= \frac{1}{\sqrt{2}}\{1 - 0.01745 - 0.0001523 + 0.0000009 - \ldots\}$$

$$= 0.7071(0.982395) = 0.69466$$

Remember that in all the trigonometric expansions, the angle *must* be in radians.

▶

You have now reached the end of the Program, except for the **Can You?** checklist and the **Test exercise** which follow.

Now move on to Frames 50 and 51

☑ Can You?

Checklist 4

50

Check this list before and after you try the end of Program test.

On a scale of 1 to 5 how confident are you that you can: Frames

- Derive the power series for $\sin x$?   [1] to [8]
  Yes ☐ ☐ ☐ ☐ ☐ *No*

- Use Maclaurin's series to derive series of common functions?   [9] to [21]
  Yes ☐ ☐ ☐ ☐ ☐ *No*

- Use Maclaurin's series to derive the binomial series?   [22] to [23]
  Yes ☐ ☐ ☐ ☐ ☐ *No*

- Derive power series expansions of miscellaneous functions
  using known expansions of common functions?   [24] to [28]
  Yes ☐ ☐ ☐ ☐ ☐ *No*

- Use power series expansions in numerical approximations?   [29] to [30]
  Yes ☐ ☐ ☐ ☐ ☐ *No*

- Use l'Hôpital's rule to evaluate limits of indeterminate forms?   [31] to [45]
  Yes ☐ ☐ ☐ ☐ ☐ *No*

- Extend Maclaurin's series to Taylor's series?   [46] to [49]
  Yes ☐ ☐ ☐ ☐ ☐ *No*

🚲 Test exercise 4

The questions are all straightforward and you will have no trouble with them. Work
through at your own speed. There is no need to hurry.

51

1 State Maclaurin's series.

2 Find the first 4 non-zero terms in the expansion of $\cos^2 x$.

3 Find the first 3 non-zero terms in the series for $\sec x$.

▶

4 Show that $\tan^{-1} x = x - \dfrac{x^3}{3} + \dfrac{x^5}{5} - \dfrac{x^7}{7} + \ldots$

5 Assuming the series for e^x and $\tan x$, determine the series for $e^x \cdot \tan x$ up to and including the term in x^4.

6 Evaluate $\sqrt{1.05}$ correct to 5 significant figures.

7 Find:

(a) $\displaystyle \lim_{x \to 0} \left\{ \frac{1 - 2\sin^2 x - \cos^3 x}{5x^2} \right\}$

(b) $\displaystyle \lim_{x \to 0} \left\{ \frac{\tan x \cdot \tan^{-1} x - x^2}{x^6} \right\}$

(c) $\displaystyle \lim_{x \to 0} \left\{ \frac{x - \sin x}{x - \tan x} \right\}$

8 Expand $\cos(x + h)$ as a series of powers of h and hence evaluate $\cos 31°$ correct to 5 decimal places.

You are now ready to start the next Program

🚴 **Further problems 4**

52

1 Prove that $\cos x = 1 - \dfrac{x^2}{2!} + \dfrac{x^4}{4!} - \dfrac{x^6}{6!} + \ldots$ and that the series is valid for all values of x. Deduce the power series for $\sin^2 x$ and show that, if x is small, $\dfrac{\sin^2 x - x^2 \cos x}{x^4} = \dfrac{1}{6} + \dfrac{x^2}{360}$ approximately.

2 Apply Maclaurin's series to establish a series for $\ln(1 + x)$. If $1 + x = \dfrac{b}{a}$, show that $(b^2 - a^2)/2ab = x - \dfrac{x^2}{2} + \dfrac{x^3}{2} - \ldots$

Hence show that, if b is nearly equal to a, then $(b^2 - a^2)/2ab$ exceeds $\ln\left(\dfrac{b}{a}\right)$ by approximately $(b - a)^3/6a^3$.

3 Evaluate:

(a) $\displaystyle \lim_{x \to 0} \left\{ \frac{\sin x - \cos x}{x^3} \right\}$

(b) $\displaystyle \lim_{x \to 0} \left\{ \frac{\tan x - \sin x}{x^3} \right\}$

(c) $\displaystyle \lim_{x \to 0} \left\{ \frac{\sin x - x}{x^3} \right\}$

(d) $\displaystyle \lim_{x \to 0} \left\{ \frac{\tan x - x}{x - \sin x} \right\}$

(e) $\displaystyle \lim_{x \to 0} \left\{ \frac{1 - 2\sin^2 x - \cos^3 x}{5x^2} \right\}$

▶

4 Write down the expansions of:

(a) $\cos x$ and   (b) $\dfrac{1}{1+x}$, and hence show that

$$\frac{\cos x}{1+x} = 1 - x + \frac{x^2}{2} - \frac{x^3}{2} + \frac{13x^4}{24} - \cdots$$

5 State the series for $\ln(1+x)$ and the range of values of x for which it is valid. Assuming the series for $\sin x$ and for $\cos x$, find the series for $\ln\left(\dfrac{\sin x}{x}\right)$ and $\ln(\cos x)$ as far as the term in x^4. Hence show that, if x is small, $\tan x$ is approximately equal to $x \cdot e^{\frac{x^2}{3}}$.

6 Use Maclaurin's series to obtain the expansion of e^x and of $\cos x$ in ascending powers of x and hence determine

$$\underset{x \to 0}{Lim} \left\{ \frac{e^x + e^{-x} - 2}{2\cos 2x - 2} \right\}.$$

7 Find the first four terms in the expansion of $\dfrac{x-3}{(1-x)^2(2+x^2)}$ in ascending powers of x.

8 Write down the series for $\ln(1+x)$ in ascending powers of x and state the conditions for convergence.
If a and b are small compared with x, show that

$$\ln(x+a) - \ln x = \frac{a}{b}\left(1 + \frac{b-a}{2x}\right)\{\ln(x+b) - \ln x\}.$$

9 Find the value of k for which the expansion of $(1+kx)\left(1+\dfrac{x}{6}\right)^{-1}\ln(1+x)$ contains no term in x^2.

10 Evaluate

(a) $\underset{x \to 0}{Lim} \left\{ \dfrac{\sinh x - \tanh x}{x^3} \right\}$ (b) $\underset{x \to 1}{Lim} \left\{ \dfrac{\ln x}{x^2 - 1} \right\}$

(c) $\underset{x \to 0}{Lim} \left\{ \dfrac{x + \sin x}{x^2 + x} \right\}$

11 If u_r and u_{r-1} indicate the rth term and the $(r-1)$th term respectively of the expansion of $(1+x)^n$, determine an expression, in its simplest form, for the ratio $\dfrac{u_r}{u_{r-1}}$. Hence show that in the binomial expansion of $(1+0.03)^{12}$ the rth term is less than one-tenth of the $(r-1)$th term if $r > 4$. Use the expansion to evaluate $(1.03)^{12}$ correct to three places of decimals.

▶

12 By the use of Maclaurin's series, show that

$$\sin^{-1} x = x + \frac{x^3}{6} + \frac{3x^5}{40} + \dots$$

Assuming the series for e^x, obtain the expansion of $e^x \sin^{-1} x$, up to and including the term in x^4. Hence show that, when x is small, the graph of $y = e^x \sin^{-1} x$ approximates to the parabola $y = x^2 + x$.

13 By application of Maclaurin's series, determine the first two non-vanishing terms of a series for $\ln \cos x$. Express $(1 + \cos \theta)$ in terms of $\cos \frac{\theta}{2}$ and show that, if θ is small, $\ln(1 + \cos \theta) = \ln 2 - \frac{\theta^2}{4} - \frac{\theta^4}{96}$ approximately.

14 If x is small, show that:

(a) $\sqrt{\dfrac{1+x}{1-x}} \approx 1 + x + \dfrac{x^2}{2}$
(b) $\dfrac{\sqrt{(1+3x^2)}e^x}{1-x} \approx 1 + \dfrac{3x}{2} + \dfrac{25x^2}{8}$

15 Prove that:

(a) $\dfrac{x}{e^x - 1} = 1 - \dfrac{x}{2} + \dfrac{x^2}{12} - \dfrac{x^4}{720} + \dots$
(b) $\dfrac{x}{e^x + 1} = \dfrac{x}{2} - \dfrac{x^2}{4} + \dfrac{x^4}{48} - \dots$

16 Find:

(a) $\underset{x \to 0}{Lim} \left\{ \dfrac{\sinh^{-1} x - x}{x^3} \right\}$
(b) $\underset{x \to 0}{Lim} \left\{ \dfrac{e^{\sin x} - 1 - x}{x^2} \right\}$

17 Find the first three terms in the expansion of $\dfrac{\sinh x \cdot \ln(1+x)}{x^2(1+x)^3}$.

18 The field strength of a magnet (H) at a point on the axis, distance x from its centre, is given by

$$H = \frac{M}{2l} \left\{ \frac{1}{(x-l)^2} - \frac{1}{(x+l)^2} \right\}$$

where $2l$ = length of magnet and M = moment. Show that if l is very small compared with x, then $H \approx \dfrac{2M}{x^3}$.

19 Expand $[\ln(1+x)]^2$ in powers of x up to and including the term in x^4. Hence determine whether $\cos 2x - [\ln(1+x)]^2$ has a maximum value, minimum value, or point of inflexion at $x = 0$.

20 If l is the length of a circular arc, a is the length of the chord of the whole arc, and b is the length of the chord of half the arc, show that:

(a) $a = 2r \sin \dfrac{l}{2r}$ and (b) $b = 2r \sin \dfrac{l}{4r}$, where r is the radius of the circle.

By expanding $\sin \dfrac{l}{2r}$ and $\sin \dfrac{l}{4r}$ as series, show that $l = \dfrac{8b - a}{3}$ approximately.

Power series solutions of ordinary differential equations

Frames
[1] to [128]

Learning outcomes

When you have completed this Program you will be able to:

- Obtain the nth derivative of the exponential and circular and hyperbolic functions
- Apply the Leibnitz theorem to derive the nth derivative of a product of expressions
- Use the Leibnitz–Maclaurin method of obtaining a series solution to a second-order homogeneous differential equation with constant coefficients
- Use Frobenius' method of obtaining a series solution to a second-order homogeneous differential equation for different cases of the indicial equation
- Apply Frobenius' method to Bessel's equation to derive Bessel functions of the first kind
- Apply Frobenius' method to Legendre's equation to derive Legendre polynomials
- Use Rodrigue's formula to derive Legendre polynomials and the generating function to obtain some of their properties
- Recognise a Sturm–Liouville system and the orthogonality properties of its eigenfunctions
- Write a polynomial in x as a finite series of Legendre polynomials

Prerequisite:
Programs 3 Series 1, 4 Series 2 and 2 Second-order differential equations

Higher derivatives

1

If $y = \sin x$ $\quad \dfrac{dy}{dx} = \cos x = \sin\left(x + \dfrac{\pi}{2}\right)$

$$\dfrac{d^2y}{dx^2} = -\sin x = \sin(x + \pi) = \sin\left(x + \dfrac{2\pi}{2}\right)$$

$$\dfrac{d^3y}{dx^3} = -\cos x = \sin\left(x + \dfrac{3\pi}{2}\right) \quad \text{etc.}$$

We see a pattern developing. In general $\quad \dfrac{d^ny}{dx^n} = \sin\left(x + \dfrac{n\pi}{2}\right)$. Before we go further, we introduce a shorthand notation for the nth derivative of y as $y^{(n)} = \dfrac{d^ny}{dx^n}$. Note, however, we still use the 'prime' notation y', y'' and y''' to represent the first, second and third derivatives respectively.
The results above can therefore be written

If $y = \sin x$ $\quad \therefore \ y' = \cos x = \sin\left(x + \dfrac{\pi}{2}\right)$

$$y'' = -\sin x = \sin\left(x + \dfrac{2\pi}{2}\right)$$

$$y''' = -\cos x = \sin\left(x + \dfrac{3\pi}{2}\right)$$

and, in general, $y^{(n)} = \sin\left(x + \dfrac{n\pi}{2}\right)$

It is therefore possible to write down any particular derivative of $\sin x$ without calculating all the previous derivatives. For example

$$\dfrac{d^7y}{dx^7} = y^{(7)} = \sin\left(x + \dfrac{7\pi}{2}\right) = -\cos x$$

Similarly, starting with $y = \cos x$, we can determine an expression for the nth derivative of y which is ....

2

$$\boxed{y^{(n)} = \cos\left(x + \dfrac{n\pi}{2}\right)}$$

Because

$y = \cos x$ $\quad \therefore \ y' = -\sin x = \cos\left(x + \dfrac{\pi}{2}\right)$

$$y'' = -\cos x = \cos\left(x + \dfrac{2\pi}{2}\right)$$

$$y''' = \sin x \quad = \cos\left(x + \dfrac{3\pi}{2}\right) \quad \text{etc.}$$

$$\therefore \ y^{(n)} = \cos\left(x + \dfrac{n\pi}{2}\right)$$

Many of the standard functions can be treated in a similar manner.
For example, if $y = e^{ax}$, then $y^{(n)} = $ ....

$$y^{(n)} = a^n e^{ax}$$

Because

$$y = e^{ax}, \quad y' = ae^{ax}, \quad y'' = a^2 e^{ax}, \quad y''' = a^3 e^{ax}, \quad \text{etc.}$$

In general, $y^{(n)} = a^n e^{ax}$.

With no great effort, we can now write down expressions for the following

If $\quad y = \sin ax, \quad y^{(n)} = \ldots\ldots\ldots\ldots$

If $\quad y = \cos ax, \quad y^{(n)} = \ldots\ldots\ldots\ldots$

$$y = \sin ax, \quad y^{(n)} = a^n \sin\left(ax + \frac{n\pi}{2}\right)$$

$$y = \cos ax, \quad y^{(n)} = a^n \cos\left(ax + \frac{n\pi}{2}\right)$$

Now one more.

$$\text{If } y = \ln x, \quad y^{(n)} = \ldots\ldots\ldots\ldots$$

$$y^{(n)} = (-1)^{n-1} \cdot \frac{(n-1)!}{x^n}$$

Because

$$y = \ln x \quad \therefore \ y' = \frac{1}{x}$$

$$y'' = -\frac{1}{x^2}$$

$$y''' = \frac{2}{x^3}$$

$$y^{(4)} = -\frac{3!}{x^4} \qquad \therefore \ y^{(n)} = (-1)^{n-1} \cdot \frac{(n-1)!}{x^n}$$

We already know that, if $y = \ln x$, $\dfrac{dy}{dx} = y' = \dfrac{1}{x} = x^{-1}$.

Therefore, if the result obtained for $y^{(n)}$ is to be valid for $n = 1$, then

$$y' = (-1)^0 \cdot \frac{0!}{x} = \frac{0!}{x}$$

But $\quad y' = x^{-1} \qquad \qquad \therefore \ 0! = \ldots\ldots\ldots\ldots$

6

$$\boxed{0! = 1}$$

Now let us consider the derivatives of $\sinh ax$ and $\cosh ax$.

Next frame

7

If $y = \sinh ax$, $y' = a \cosh ax$

$$y'' = a^2 \sinh ax$$

$$y''' = a^3 \cosh ax \qquad \text{etc.}$$

Because $\sinh ax$ is not periodic, we cannot proceed as we did with $\sin ax$. We need to find a general statement for $y^{(n)}$ containing terms in $\sinh ax$ and in $\cosh ax$, such that, when n is even, the term in $\cosh ax$ disappears and, when n is odd, the term in $\sinh ax$ disappears.

This we can do by writing $y^{(n)}$ in the form

$$y^{(n)} = \frac{a^n}{2} \{ [1 + (-1)^n] \sinh ax + [1 - (-1)^n] \cosh ax \}$$

In very much the same way, we can determine the nth derivative of $y = \cosh ax$ as ....

8

$$\boxed{y^{(n)} = \frac{a^n}{2} \{ [1 - (-1)^n] \sinh ax + [1 + (-1)^n] \cosh ax \}}$$

Finally, let us deal with $y = x^a$.

$$y = x^a \quad \therefore \quad y' = ax^{a-1}$$

$$y'' = a(a-1)x^{a-2}$$

$$y''' = a(a-1)(a-2)x^{a-3}$$

$$............$$

$$\therefore \quad y^{(n)} = a(a-1)(a-2)\ldots(a-n+1)x^{a-n}$$

$$\therefore \quad y^{(n)} = \frac{a!}{(a-n)!} x^{a-n} \qquad (a \text{ is a positive integer})$$

So, collecting our results together, we have

$y = x^a$	$y^{(n)} = \dfrac{a!}{(a-n)!} x^{a-n}$
$y = e^{ax}$	$y^{(n)} = a^n e^{ax}$
$y = \sin ax$	$y^{(n)} = a^n \sin\left(ax + \dfrac{n\pi}{2}\right)$
$y = \cos ax$	$y^{(n)} = a^n \cos\left(ax + \dfrac{n\pi}{2}\right)$
$y = \sinh ax$	$y^{(n)} = \dfrac{a^n}{2} \{ [1 + (-1)^n] \sinh ax + [1 - (-1)^n] \cosh ax \}$
$y = \cosh ax$	$y^{(n)} = \dfrac{a^n}{2} \{ [1 - (-1)^n] \sinh ax + [1 + (-1)^n] \cosh ax \}$

Make a note of these, as a set, and then move on to the next frame

Exercise

9

Determine the following derivatives

1　$y = \sin 4x$　　　　　$y^{(5)} = \ldots\ldots\ldots$

2　$y = e^{x/2}$　　　　　$y^{(8)} = \ldots\ldots\ldots$

3　$y = \cosh 3x$　　　　$y^{(12)} = \ldots\ldots\ldots$

4　$y = \cos(x\sqrt{2})$　　$y^{(10)} = \ldots\ldots\ldots$

5　$y = x^8$　　　　　　$y^{(6)} = \ldots\ldots\ldots$

6　$y = \sinh 2x$　　　　$y^{(7)} = \ldots\ldots\ldots$

Finish them all; then check with the next frame

Here are the solutions

10

1　$y^{(5)} = 4^5 \sin\left(4x + \dfrac{5\pi}{2}\right) = 1024 \sin\left(4x + \dfrac{\pi}{2}\right) = 1024 \cos 4x$

2　$y^{(8)} = \left(\dfrac{1}{2}\right)^8 e^{x/2} = \dfrac{1}{256} e^{x/2} = e^{x/2}/256$

3　$y^{(12)} = \dfrac{3^{12}}{2}\{0 \sinh 3x + 2 \cosh 3x\} = 3^{12} \cosh 3x$

4　$y^{(10)} = (\sqrt{2})^{10} \cos\left(x\sqrt{2} + \dfrac{10\pi}{2}\right)$

　　　$= 32 \cos(x\sqrt{2} + 5\pi) = -32 \cos(x\sqrt{2})$

5　$y^{(6)} = \dfrac{8!}{2!} x^2 = 20\,160\, x^2$

6　$y^{(7)} = \dfrac{2^7}{2}\left\{[1 + (-1)^7] \sinh 2x + [1 - (-1)^7] \cosh 2x\right\}$

　　　$= 2^7 \cosh 2x$

Leibnitz theorem – *n*th derivative of a product of two functions

If $y = uv$, where u and v are functions of x, then

$$y' = uv' + vu' \quad \text{where} \quad v' = \frac{dv}{dx} \quad \text{and} \quad u' = \frac{du}{dx}$$

and　$y'' = uv'' + v'u' + vu'' + u'v' = u''v + 2u'v' + uv''$

If we differentiate the last result and collect like terms, we obtain

$$y''' = \ldots\ldots\ldots$$

11

$$y''' = u'''v + 3u''v' + 3u'v'' + uv'''$$

A further stage of differentiation would give

$$y^{(4)} = u^{(4)}v + 4u^{(3)}v^{(1)} + 6u^{(2)}v^{(2)} + 4u^{(1)}v^{(3)} + uv^{(4)}$$

These results can therefore be written

$$y = uv$$
$$y' = u'v + uv'$$
$$y'' = u''v + 2u'v' + uv''$$
$$y''' = u'''v + 3u''v' + 3u'v'' + uv'''$$
$$y^{(4)} = u^{(4)}v + 4u^{(3)}v^{(1)} + 6u^{(2)}v^{(2)} + 4u^{(1)}v^{(3)} + uv^{(4)}$$

Notice that in each case

(a) the superscript of u decreases regularly by 1

(b) the superscript of v increases regularly by 1

(c) the numerical coefficients are the normal binomial coefficients.

Indeed, $(uv)^{(n)}$ can be obtained by expanding $(u+v)^{(n)}$ using the binomial theorem where the 'powers' are interpreted as derivatives. So the expression for the nth derivative can therefore be written as

$$y^{(n)} = u^{(n)}v + nu^{(n-1)}v^{(1)} + \frac{n(n-1)}{1 \times 2}u^{(n-2)}v^{(2)}$$
$$+ \frac{n(n-1)(n-2)}{1 \times 2 \times 3}u^{(n-3)}v^{(3)} + \dots$$
$$= u^{(n)}v + nu^{(n-1)}v^{(1)} + \frac{n(n-1)}{2!}u^{(n-2)}v^{(2)}$$
$$+ \frac{n(n-1)(n-2)}{3!}u^{(n-3)}v^{(3)} + \dots$$

i.e. $y^{(n)} = u^{(n)}v + {}^nC_1 u^{(n-1)}v^{(1)} + {}^nC_2 u^{(n-2)}v^{(2)} + \dots$
$$+ {}^nC_{n-1}u^{(1)}v^{(n-1)} + uv^{(n)}$$

where ${}^nC_r = \dfrac{n!}{r!(n-r)!}$

If $y = uv$ $y^{(n)} = \displaystyle\sum_{r=0}^{n} {}^nC_r u^{(n-r)}v^{(r)}$ where $u^{(0)} \equiv u$

This is the *Leibnitz theorem*. We shall certainly be using it often in the work ahead, so make a note of it for future reference. Then we can see it in use.

Choice of function for *u* and *v*   [12]

For the product $y = uv$ the function taken as

(a) u is the one whose nth derivative can readily be obtained

(b) v is the one whose derivatives reduce to zero after a small number of stages of differentiation.

Example 1

To find $y^{(n)}$ when $y = x^3 e^{2x}$.

Here we choose $v = x^3$ – whose fourth derivative is zero

$\qquad\qquad\qquad u = e^{2x}$ – because we know that the nth derivative

$\qquad\qquad\qquad u^{(n)} = \ldots\ldots\ldots\ldots$

[13]

$$\boxed{u^{(n)} = 2^n e^{2x}}$$

Using the Leibnitz theorem:

$$y^{(n)} = u^{(n)}v + nu^{(n-1)}v^{(1)} + \frac{n(n-1)}{2!}u^{(n-2)}v^{(2)}$$
$$+ \frac{n(n-1)(n-2)}{3!}u^{(n-3)}v^{(3)} + \ldots$$

$v = x^3; \quad v^{(1)} = 3x^2; \quad v^{(2)} = 6x; \quad v^{(3)} = 6; \quad v^{(4)} = 0$

$u = e^{2x}; \quad u^{(n)} = 2^n e^{2x}$

$$\therefore \ y^{(n)} = \ldots\ldots\ldots\ldots$$

[14]

$$\boxed{y^{(n)} = e^{2x}\, 2^{n-3}\left\{8x^3 + 12nx^2 + n(n-1)\,6x + n(n-1)(n-2)\right\}}$$

Example 2

If $x^2 y'' + xy' + y = 0$, show that

$$x^2 y^{(n+2)} + (2n+1)\, xy^{(n+1)} + (n^2 + 1)\, y^{(n)} = 0.$$

We take the given equation $x^2 y'' + xy' + y = 0$ and differentiate n times, treating each term in turn.

$$\begin{aligned}
\text{If } \ w &= x^2 y'' & w^{(n)} &= \ldots\ldots\ldots\ldots \\
\text{If } \ w &= xy' & w^{(n)} &= \ldots\ldots\ldots\ldots \\
\text{If } \ w &= y & w^{(n)} &= \ldots\ldots\ldots\ldots
\end{aligned}$$

15

$$w = x^2 y'' \qquad \therefore \ w^{(n)} = y^{(n+2)} x^2 + n y^{(n+1)} 2x + \frac{n(n-1)}{2!} y^{(n)} 2 + 0 \ldots$$

$$w = xy' \qquad \therefore \ w^{(n)} = y^{(n+1)} x + n y^{(n)} 1 + 0 + \ldots$$

$$w = y \qquad \therefore \ w^{(n)} = y^{(n)}$$

Then $\left[x^2 y'' + xy' + y \right]^{(n)} = 0$ becomes

....

16

$$x^2 y^{(n+2)} + (2n+1)xy^{(n+1)} + (n^2 + 1)y^{(n)} = 0$$

which is what we had to show.

Example 3

Differentiate n times

$$(1 + x^2)y'' + 2xy' - 5y = 0.$$

The result ....

17

$$(1 + x^2)\, y^{(n+2)} + 2(n+1)\, xy^{(n+1)} + (n^2 + n - 5)\, y^{(n)} = 0$$

Because, by the Leibnitz theorem

$$\left\{ y^{(n+2)}(1 + x^2) + n y^{(n+1)} 2x + \frac{n(n-1)}{2!} y^{(n)} 2 \right\}$$

$$+ 2\left\{ xy^{(n+1)} + n y^{(n)} \cdot 1 \right\} - 5y^{(n)} = 0$$

$$(1 + x^2)\, y^{(n+2)} + 2(n+1)\, xy^{(n+1)} + \{n(n-1) + 2n - 5\}\, y^{(n)} = 0$$

$$(1 + x^2)\, y^{(n+2)} + 2(n+1)\, xy^{(n+1)} + (n^2 + n - 5)\, y^{(n)} = 0$$

We shall be using the Leibnitz theorem in the rest of this Program, so let us move on to see some of its applications.

Power series solutions

18

Second-order linear differential equations with constant coefficients of the form $a \dfrac{\mathrm{d}^2 y}{\mathrm{d}x^2} + b \dfrac{\mathrm{d}y}{\mathrm{d}x} + cy = 0$ can be solved by algebraic methods giving solutions in terms of the normal elementary functions such as exponentials, trigonometric and polynomial functions.

▶

In general, equations of the form $\dfrac{d^2y}{dx^2} + P(x)\dfrac{dy}{dx} + Q(x)y = 0,$ where $P(x)$ and $Q(x)$ are functions of x, cannot be solved in this way. However, it is often possible to obtain solutions in the form of infinite series of powers of x – and the next section of work investigates some of the methods which make this possible.

1 Leibnitz–Maclaurin method

As the title suggests, for this we need to be familiar with the Leibnitz theorem and with Maclaurin's series.

The *Leibnitz theorem* states that, if $y = uv$, where u and v are functions of x, then

$$y^{(n)} = \ldots\ldots\ldots\ldots$$

19

$$y^{(n)} = u^{(n)}v + nu^{(n-1)}v^{(1)} + \frac{n(n-1)}{2!}u^{(n-2)}v^{(2)} + \ldots$$
$$+ \frac{n(n-1)\ldots(n-r+1)}{r!}u^{(n-r)}v^{(r)} + \ldots + uv^{(n)}$$

where $u^{(r)}$ and $v^{(r)}$ denote $\dfrac{d^r u}{dx^r}$ and $\dfrac{d^r v}{dx^r}$ respectively.

Maclaurin's series for $y = f(x)$ can be stated as

$$y = \ldots\ldots\ldots\ldots$$

20

$$y = (y)_0 + x(y')_0 + \frac{x^2}{2!}(y'')_0 + \ldots + \frac{x^n}{n!}(y^{(n)})_0 + \ldots$$

where $(y^{(n)})_0$ denotes the value of the nth derivative of y at $x = 0$.

On to the next frame

21

Example 1

Find the power series solution of the equation

$$x\frac{d^2y}{dx^2} + \frac{dy}{dx} + xy = 1.$$

The equation can be written

$$xy'' + y' + xy = 1$$

In the first product term xy'', treat y'' as u and x as v. Then, differentiating the equation n times by the Leibnitz theorem, gives

$$\ldots\ldots\ldots\ldots$$

22

$$\left(xy^{(n+2)} + n \cdot 1 \cdot y^{(n+1)}\right) + y^{(n+1)} + \left(xy^{(n)} + n \cdot 1 \cdot y^{(n-1)}\right) = 0$$
$$\text{i.e.} \quad xy^{(n+2)} + (n+1)y^{(n+1)} + xy^{(n)} + ny^{(n-1)} = 0$$

At $x = 0$, this becomes

$$(n+1)\left(y^{(n+1)}\right)_0 + n\left(y^{(n-1)}\right)_0 = 0$$

$$\therefore \quad \left(y^{(n+1)}\right)_0 = -\frac{n}{n+1}\left(y^{(n-1)}\right)_0 \qquad n \geq 1$$

This relationship is called a *recurrence relation*.

We can now substitute $n = 1, 2, 3, \ldots$ and get a set of relationships between the various coefficients.

$$n = 1 \qquad (y'')_0 = -\tfrac{1}{2}(y)_0$$
$$n = 2 \qquad (y''')_0 = -\tfrac{2}{3}(y')_0$$
$$n = 3 \qquad (y^{(4)})_0 = -\tfrac{3}{4}(y'')_0 = \left(-\tfrac{3}{4}\right)\left(-\tfrac{1}{2}\right)(y)_0$$

Continuing in the same way,

$$(y^{(5)})_0 = \ldots\ldots\ldots$$
$$(y^{(6)})_0 = \ldots\ldots\ldots$$
$$(y^{(7)})_0 = \ldots\ldots\ldots$$
$$(y^{(8)})_0 = \ldots\ldots\ldots$$

23

$$n = 4 \qquad (y^{(5)})_0 = -\tfrac{4}{5}(y^{(3)})_0 = \left(-\tfrac{4}{5}\right)\left(-\tfrac{2}{3}\right)(y^{(1)})_0$$
$$n = 5 \qquad (y^{(6)})_0 = -\tfrac{5}{6}(y^{(4)})_0 = \left(-\tfrac{5}{6}\right)\left(-\tfrac{3}{4}\right)\left(-\tfrac{1}{2}\right)(y)_0$$
$$n = 6 \qquad (y^{(7)})_0 = -\tfrac{6}{7}(y^{(5)})_0 = \left(-\tfrac{6}{7}\right)\left(-\tfrac{4}{5}\right)\left(-\tfrac{2}{3}\right)(y^{(1)})_0$$
$$n = 7 \qquad (y^{(8)})_0 = -\tfrac{7}{8}(y^{(6)})_0 = \left(-\tfrac{7}{8}\right)\left(-\tfrac{5}{6}\right)\left(-\tfrac{3}{4}\right)\left(-\tfrac{1}{2}\right)(y)_0$$

Notice that, by this means, the values of all the derivatives at $x = 0$ can be expressed in terms of $(y)_0$ and $(y')_0$.

If we now substitute these values for $(y^{(r)})_0$ in the Maclaurin series

$$y = (y)_0 + x(y')_0 + \frac{x^2}{2!}(y'')_0 + \frac{x^3}{3!}(y''')_0 + \ldots + \frac{x^r}{r!}(y^{(r)})_0 + \ldots$$

we obtain $\ldots\ldots\ldots$

24

$$y = (y)_0 + x(y')_0 + \frac{x^2}{2!}\left(-\frac{1}{2}\right)(y)_0 + \frac{x^3}{3!}\left(-\frac{2}{3}\right)(y')_0$$

$$+ \frac{x^4}{4!}\left(-\frac{3}{4}\right)\left(-\frac{1}{2}\right)(y)_0 + \frac{x^5}{5!}\left(-\frac{4}{5}\right)\left(-\frac{2}{3}\right)(y')_0$$

$$+ \frac{x^6}{6!}\left(-\frac{5}{6}\right)\left(-\frac{3}{4}\right)\left(-\frac{1}{2}\right)(y)_0 + \ldots\ldots\ldots$$

Simplifying, this gives

$$y = (y)_0\left\{1 - \frac{x^2}{2^2} + \frac{x^4}{2^2 \times 4^2} - \frac{x^6}{2^2 \times 4^2 \times 6^2} + \ldots\right\}$$

$$+ (y')_0\left\{x - \frac{x^3}{3^2} + \frac{x^5}{3^2 \times 5^2} + \ldots\right\}$$

The values of $(y)_0$ and $(y')_0$ provide the two arbitrary constants for the second-order equation and are obtained from the given initial conditions.

For example, if at $x = 0$, $y = 2$ and $\dfrac{dy}{dx} = 1$, then the relevant particular

solution is ....

25

$$y = 2\left\{1 - \frac{x^2}{2^2} + \frac{x^4}{2^2 \times 4^2} - \frac{x^6}{2^2 \times 4^2 \times 6^2} + \ldots\right\}$$

$$+ \left\{x - \frac{x^3}{3^2} + \frac{x^5}{3^2 \times 5^2} + \ldots\right\}$$

Because at $x = 0$, $y = 2$ i.e. $(y)_0 = 2$

$$\frac{dy}{dx} = 1 \quad \text{i.e. } (y')_0 = 1.$$

To be a valid solution, the series obtained must converge. Application of the ratio test will normally indicate any restrictions on the values that x may have.

The Leibnitz–Maclaurin (power series) method therefore involves the following main steps:

(a) Differentiate the given equation n times, using the Leibnitz theorem.

(b) Rearrange the result to obtain the recurrence relation at $x = 0$.

(c) Determine the values of the derivatives at $x = 0$, usually in terms of $(y)_0$ and $(y')_0$.

(d) Substitute in the Maclaurin expansion for $y = f(x)$.

(e) Simplify the result where possible and apply boundary conditions if provided.

That is all there is to it. Let us go through the various steps with another example.

▶

Example 2

Determine a series solution of the equation

$$\frac{d^2y}{dx^2} + x\frac{dy}{dx} + y = 0.$$

The equation can be written $y'' + xy' + y = 0$

(a) Differentiate n times using the Leibnitz theorem, which gives

. . . .

26

$$y^{(n+2)} + xy^{(n+1)} + (n+1)y^{(n)} = 0$$

Because $y'' + xy' + y = 0$

$$\therefore \ y^{(n+2)} + \left\{xy^{(n+1)} + n \cdot 1 \cdot y^{(n)}\right\} + y^{(n)} = 0$$

$$\therefore \ y^{(n+2)} + xy^{(n+1)} + (n+1)y^{(n)} = 0.$$

(b) Determine the recurrence relation at $x = 0$, which is

. . . .

27

$$y^{(n+2)} = -(n+1)\,y^{(n)}$$

(c) Now taking $n = 0, 1, 2, 3, 4, 5$, determine the derivatives at $x = 0$ in terms of $(y)_0$ and $(y')_0$. List them, as we did before, in table form.

28

$$
\begin{aligned}
n = 0 \quad & (y'')_0 = -(y)_0 & &= -(y)_0 \\
1 \quad & (y''')_0 = -2\,(y')_0 & &= -2\,(y')_0 \\
2 \quad & (y^{(4)})_0 = -3\,(y'')_0 = (-3)[-(y)_0] & &= 3\,(y)_0 \\
3 \quad & (y^{(5)})_0 = -4\,(y''')_0 = (-4)[-2(y')_0] & &= 2 \times 4\,(y')_0 \\
4 \quad & (y^{(6)})_0 = -5\,(y^{(4)})_0 = (-5)[-3(y'')_0] & &= -3 \times 5\,(y)_0 \\
5 \quad & (y^{(7)})_0 = -6\,(y^{(5)})_0 = (-6)[-4(y''')_0] & &= -2 \times 4 \times 6\,(y')_0
\end{aligned}
$$

(d) Substitute these expressions for the derivatives in terms of $(y)_0$ and $(y')_0$ in Maclaurin's expansion

$$y = (y)_0 + x\,(y')_0 + \frac{x^2}{2!}\,(y'')_0 + \frac{x^3}{3!}\,(y''')_0 + \frac{x^4}{4!}\,(y^{(4)})_0 + \cdots$$

Then $\quad y = \ldots \ldots \ldots \ldots$

29

$$y = (y)_0 + x(y')_0 + \frac{x^2}{2!}(-y)_0 + \frac{x^3}{3!}(-2y')_0 + \frac{x^4}{4!}(3y)_0 + \frac{x^5}{5!}(8y')_0$$
$$+ \frac{x^6}{6!}(-15y)_0 + \frac{x^7}{7!}(-48y')_0 + \cdots$$

Collecting now the terms in $(y)_0$ and $(y')_0$, we finally obtain

$$y = (y)_0 \left\{ 1 - \frac{x^2}{2} + \frac{x^4}{2 \times 4} - \frac{x^6}{2 \times 4 \times 6} + \cdots \right\}$$
$$+ (y')_0 \left\{ x - \frac{x^3}{3} + \frac{x^5}{3 \times 5} - \frac{x^7}{3 \times 5 \times 7} + \cdots \right\}$$

They are all done in very much the same way. Here is another.

Example 3

Solve the equation $\dfrac{d^2 y}{dx^2} + \dfrac{dy}{dx} + 2xy = 0$ given that at $x = 0$, $y = 0$ and $\dfrac{dy}{dx} = 1$.

First write the equation as $y'' + y' + 2xy = 0$, differentiate n times by the Leibnitz theorem and obtain the recurrence relation at $x = 0$, which is
. . . .

30

$$y^{(n+2)} = -\left\{ y^{(n+1)} + 2ny^{(n-1)} \right\} \quad n \geq 1$$

Because $y'' + y' + 2xy = 0$
$$\therefore \ y^{(n+2)} + y^{(n+1)} + 2xy^{(n)} + n2y^{(n-1)} = 0$$
At $x = 0$, $\qquad\qquad y^{(n+2)} + y^{(n+1)} + 2ny^{(n-1)} = 0$
$$\therefore \ y^{(n+2)} = -\left\{ y^{(n+1)} + 2ny^{(n-1)} \right\}$$

Since we have a term in $y^{(n-1)}$, then n must start at 1 to give $(y)_0$. Therefore the recurrence relation applies for $n \geq 1$.

We now take $n = 1, 2, 3, \ldots$ to obtain the relationships between the coefficients up to $(y^{(6)})_0$. Complete the table and check with the next frame.

31

$$
\begin{array}{ll}
n = 1 & (y^{(3)})_0 = -\{(y^{(2)})_0 + 2(y)_0\} \\
n = 2 & (y^{(4)})_0 = -\{(y^{(3)})_0 + 4(y')_0\} \\
n = 3 & (y^{(5)})_0 = -\{(y^{(4)})_0 + 6(y^{(2)})_0\} \\
n = 4 & (y^{(6)})_0 = -\{(y^{(5)})_0 + 8(y^{(3)})_0\}
\end{array}
$$

We therefore have expressions for $(y''')_0, (y^{(4)})_0, (y^{(5)})_0, (y^{(6)})_0$, but what about $(y'')_0$?

If we refer to the initial conditions, we know that at $x = 0$, $y = 0$ and $y' = 1$. $\therefore$ $(y)_0 = 0$ and $(y')_0 = 1$.

We can find $(y'')_0$ by reference to the given equation itself, because

$$y'' + y' + 2xy = 0$$

Therefore, at $x = 0$, $(y'')_0 + (y')_0 = 0$ $\therefore$ $(y'')_0 = -(y')_0 = -1$.

So now we have $(y)_0 = 0$

$$
\begin{aligned}
(y')_0 &= 1 \\
(y'')_0 &= -1 \\
(y''')_0 &= -\{(y'')_0 + 2(y)_0\} & &= -\{(-1) + 0\} = 1 \\
(y^{(4)})_0 &= -\{(y''')_0 + 4(y')_0\} & &= -\{1 + 4\} & &= -5 \\
(y^{(5)})_0 &= -\{(y^{(4)})_0 + 6(y'')_0\} & &= -\{(-5) - 6\} = 11 \\
(y^{(6)})_0 &= -\{(y^{(5)})_0 + 8(y''')_0\} & &= -\{11 + 8\} & &= -19
\end{aligned}
$$

The required series solution is therefore

$$y = \ldots\ldots\ldots\ldots$$

32

$$
y = x - \frac{x^2}{2!} + \frac{x^3}{3!} - \frac{5x^4}{4!} + \frac{11x^5}{5!} - \frac{19x^6}{6!} + \ldots
$$

Because

$$
\begin{aligned}
y &= (y)_0 + x(y')_0 + \frac{x^2}{2!}(y'')_0 + \frac{x^3}{3!}(y''')_0 + \frac{x^4}{4!}(y^{(4)})_0 + \ldots \\
&= 0 + x(1) + \frac{x^2}{2!}(-1) + \frac{x^3}{3!}(1) + \frac{x^4}{4!}(-5) + \frac{x^5}{5!}(11) + \frac{x^6}{6!}(-19)
\end{aligned}
$$

$$
\therefore \ y = x - \frac{x^2}{2!} + \frac{x^3}{3!} - \frac{5x^4}{4!} + \frac{11x^5}{5!} - \frac{19x^6}{6!} + \ldots
$$

33

One more of the same kind.

Example 4

Determine the general series solution of the equation

$$(x^2 + 1)y'' + xy' - 4y = 0$$

As usual, establish the recurrence relation at $x = 0$, which is

$$\ldots\ldots\ldots\ldots$$

$$\boxed{y^{(n+2)} = (4 - n^2)y^{(n)}}$$

Because

$(x^2 + 1)y'' + xy' - 4y = 0$ therefore

$$\left\{(x^2 + 1)y^{(n+2)} + 2xny^{(n+1)} + 2\frac{n(n-1)}{2!}y^{(n)}\right\} + \left\{xy^{(n+1)} + ny^{(n)}\right\} - 4y^{(n)}$$

$= 0$

At $x = 0$, this becomes

$y^{(n+2)} + n(n-1)y^{(n)} + ny^{(n)} - 4y^{(n)} = 0$ that is $y^{(n+2)} = (4 - n^2)y^{(n)}$

Then, starting with $n = 0$, determine expressions for $(y^{(n)})_0$ as far as $n = 7$.

They are ....

$$
\begin{array}{llll}
n = 0 & (y'')_0 = 4(y)_0 & = 4(y)_0 \\
n = 1 & (y''')_0 = 3(y')_0 & = 3(y')_0 \\
n = 2 & (y^{(4)})_0 = 0 & = 0 \\
n = 3 & (y^{(5)})_0 = -5(y''')_0 & = -15(y')_0 \\
n = 4 & (y^{(6)})_0 = -12(y^{(4)})_0 = 0 \\
n = 5 & (y^{(7)})_0 = -21(y^{(5)})_0 = (-21)(-15)(y')_0 \\
\end{array}
$$

Now substitute in Maclaurin's expansion and simplify the result.

$y = $ ....

$$\boxed{y = A(1 + 2x^2) + B\left\{x + \frac{x^3}{2} - \frac{x^5}{8} + \frac{x^7}{16} + \cdots\right\}}$$

Because

$$y = (y)_0 + x(y')_0 + \frac{x^2}{2!}(y'')_0 + \frac{x^3}{3!}(y''')_0 + \frac{x^4}{4!}(y^{(4)})_0 + \cdots$$

$$= (y)_0 + x(y')_0 + \frac{x^2}{2!}4(y)_0 + \frac{x^3}{3!}3(y')_0 + \frac{x^4}{4!}(0) + \frac{x^5}{5!}(-15)(y')_0 + \text{etc.}$$

$$= (y)_0\{1 + 2x^2\} + (y')_0\left\{x + \frac{x^3}{2} - \frac{x^5}{8} + \frac{x^7}{16} + \cdots\right\}$$

Putting $(y)_0 = A$ and $(y')_0 = B$, we have the result stated.

Now to something slightly different

37 2 Frobenius' method

In each of the previous examples, we established the solution as a power series in integral powers of x. Such a solution is not always possible and a more general method is to assume a trial solution of the form

$$y = x^c \{a_0 + a_1 x + a_2 x^2 + a_3 x^3 + \ldots + a_r x^r + \ldots\}$$

where a_0 is the first coefficient that is not zero.

The type of equation that can be solved by this method is of the form

$$y'' + Py' + Qy = 0$$

where P and Q are functions of x.

However, certain conditions have to be satisfied.

(a) If the functions P and Q are such that both are finite when x is put equal to zero, $x = 0$ is called an *ordinary point* of the equation.

(b) If xP and $x^2 Q$ remain finite at $x = 0$, then $x = 0$ is called a *regular singular point* of the equation.

In both of these cases, the method of Frobenius can be applied.

(c) If, however, P and Q do not satisfy either of these conditions stated in (a) or (b), then $x = 0$ is called an *irregular singular point* of the equation and the method of Frobenius cannot be applied.

Solution of differential equations by the method of Frobenius

To solve a given equation, we have to find the coefficients $a_0, a_1, a_2, \ldots$ and also the index c in the trial solution. Basically, the steps in the method are as follows

(a) Differentiate the trial series as required.

(b) Substitute the results in the given differential equation.

(c) Equate coefficients of corresponding powers of x on each side of the equation.

The following examples will demonstrate the method – so move on

Example 1

38

Find a series solution for the equation

$$2x\frac{d^2y}{dx^2} + \frac{dy}{dx} + y = 0.$$

The equation can be written as $2xy'' + y' + y = 0$.

Assume a solution of the form

$$y = x^c\{a_0 + a_1x + a_2x^2 + a_3x^3 + \ldots + a_rx^r + \ldots\} \qquad a_0 \neq 0.$$
$$\therefore \ y = a_0x^c + a_1x^{c+1} + a_2x^{c+2} + \ldots + a_rx^{c+r} + \ldots$$

Differentiating term by term, we get

$$y' = \ldots\ldots\ldots\ldots$$

39

$$y' = a_0cx^{c-1} + a_1(c+1)x^c + a_2(c+2)x^{c+1} + \ldots + a_r(c+r)x^{c+r-1} + \ldots$$

Repeating the process one stage further, we have

$$y'' = \ldots\ldots\ldots\ldots \qquad \text{(give yourself plenty of room)}$$

40

$$y'' = a_0c(c-1)x^{c-2} + a_1c(c+1)x^{c-1} + a_2(c+1)(c+2)x^c + \ldots$$
$$+ a_r(c+r-1)(c+r)x^{c+r-2} + \ldots$$

So far, we have $\qquad 2xy'' + y' + y = 0$

$$y = a_0x^c + a_1x^{c+1} + a_2x^{c+2} + \ldots + a_rx^{c+r} + \ldots$$
$$y' = a_0cx^{c-1} + a_1(c+1)x^c + a_2(c+2)x^{c+1} + \ldots$$
$$\qquad + a_r(c+r)x^{c+r-1} + \ldots$$
$$y'' = a_0c(c-1)x^{c-2} + a_1c(c+1)x^{c-1} + a_2(c+1)(c+2)x^c + \ldots$$
$$\qquad + a_r(c+r-1)(c+r)x^{c+r-2} + \ldots$$

Considering each term of the equation in turn

$$2xy'' = 2a_0c(c-1)x^{c-1} + 2a_1c(c+1)x^c + 2a_2(c+1)(c+2)x^{c+1}$$
$$\qquad + \ldots + a_r(c+r-1)(c+r)x^{c+r-1} + \ldots$$
$$y' = a_0cx^{c-1} + a_1(c+1)x^c + a_2(c+2)x^{c+1} + \ldots$$
$$\qquad + a_r(c+r)x^{c+r-1} + \ldots$$
$$y = a_0x^c + a_1x^{c+1} + \ldots + a_rx^{c+r} + \ldots$$

Adding these three lines to form the left-hand side of the equation, we can equate the total coefficient of each power of x to zero, since the right-hand side is zero.

$$[x^{c-1}] \text{ gives } \ldots\ldots\ldots\ldots$$

41

$$[x^{c-1}]: \qquad 2a_0c(c-1) + a_0c = 0$$
$$\therefore \ a_0c(2c-1) = 0$$

So, $[x^{c-1}]$ gives $\qquad a_0c(2c-1) = 0$ \hfill (1)
Similarly, $[x^c]$ gives ....

42

$$2a_1c\,(c+1) + a_1(c+1) + a_0 = 0$$

Simplifying, this becomes

$$a_1(2c^2 + 3c + 1) + a_0 = 0$$
i.e. $\qquad a_1(c+1)(2c+1) + a_0 = 0$ \hfill (2)

Also $[x^{c+1}]$ gives ....

43

$$2a_2(c+1)(c+2) + a_2(c+2) + a_1 = 0$$

and this simplifies straight away to

$$a_2(c+2)(2c+3) + a_1 = 0$$ \hfill (3)

Note that the coefficient of x^c involves all three lines of the expressions and, from then on, a general relationship can be obtained for x^{c+r}, $r \geq 0$.

In the expression for $2xy''$ and y' we have terms in x^{c+r-1}. If we replace r by $(r+1)$, we shall obtain the corresponding terms in x^{c+r}.

In the series for $2xy''$, this is $\quad 2a_{r+1}(c+r)(c+r+1)x^{c+r}$
In the series for y', this is $\quad a_{r+1}(c+r+1)x^{c+r}$
In the series for y, this is $\quad a_rx^{c+r}$

Therefore, equating the total coefficient of x^{c+r} to zero, we have

....

44

$$2a_{r+1}(c+r)(c+r+1) + a_{r+1}(c+r+1) + a_r = 0$$

and this tidies up to

$$a_{r+1}\{(c+r+1)(2c+2r+1)\} + a_r = 0$$ \hfill (4)

Make a note of results (1), (2), (3) and (4): we shall return to them in due course.

Then move on

Indicial equation

45

Equation (1), formed from the coefficient of the lowest power of x, that is x^{c-1}, is called the *indicial equation* from which the values of c can be obtained. In the present example $a_0 c(2c - 1) = 0$

$$\therefore \ c = \ldots\ldots\ldots\ldots$$

46

$$\boxed{c = 0 \ \text{ or } \ \frac{1}{2}, \text{ since } a_0 \neq 0, \text{ by definition}}$$

Both values of c are valid, so that we have two possible solutions of the given equation. We will consider each in turn.

(a) *Using* $c = 0$

 (2) gives $a_1(1)(1) + a_0 = 0 \quad \therefore \ a_1 = -a_0$

 Similarly

 (3) gives $\ldots\ldots\ldots\ldots$

47

$$\boxed{a_2(2)(3) + a_1 = 0}$$

$a_1 = -a_0$ and $a_2 = -\dfrac{a_1}{2 \times 3} = \dfrac{a_0}{2 \times 3}$

and from (4) $\quad a_{r+1} = \dfrac{-a_r}{(r + 1)(2r + 1)} \qquad r \geq 0$

From the combined series, the term in x^c and all subsequent terms involve all three lines and the coefficient of the general term can be used.

So we have $a_1 = -a_0$ and $a_{r+1} = \dfrac{-a_r}{(r + 1)(2r + 1)}$ for $r = 0, 1, 2, \ldots$

$$\therefore \ a_2 = \frac{-a_1}{2 \times 3} = \frac{a_0}{2 \times 3}$$

$$a_2 = \frac{-a_2}{3 \times 5} = \frac{-a_0}{(2 \times 3)(3 \times 5)}$$

$$a_4 = \frac{-a_3}{4 \times 7} = \frac{a_0}{(2 \times 3 \times 4)(3 \times 5 \times 7)} \qquad \text{etc.}$$

$$\therefore \ y = x^0 \left\{ a_0 - a_0 x + \frac{a_0}{(2 \times 3)} x^2 - \frac{a_0}{(2 \times 3)(3 \times 5)} x^3 + \ldots \right\}$$

$$\therefore \ y = a_0 \left\{ 1 - x + \frac{x^2}{(2)(3)} - \frac{x^3}{(2 \times 3)(3 \times 5)} + \frac{x^4}{(2 \times 3 \times 4)(3 \times 5 \times 7)} + \ldots \right\}$$

Now we go through the same steps using our second value for c, i.e. $c = \frac{1}{2}$.

Next frame

48

(b) *Using $c = \frac{1}{2}$*

Our equations relating the coefficients were

$$a_0 c(2c - 1) = 0 \quad \text{which gave} \quad c = 0 \text{ or } c = \frac{1}{2} \tag{1}$$

$$a_1(c + 1)(2c + 1) + a_0 = 0 \tag{2}$$

$$a_2(c + 2)(2c + 3) + a_1 = 0 \tag{3}$$

$$a_{r+1}(c + r + 1)(2c + 2r + 1) + a_r = 0 \tag{4}$$

Putting $c = \frac{1}{2}$ in (2) gives ....

49

$$\boxed{a_1 = -\frac{a_0}{3}}$$

Similarly (3) gives $\quad a_2 = -\dfrac{a_1}{10} = \dfrac{a_0}{3 \times 10}$

and from the general relationship, (4), we have ....

50

$$\boxed{a_{r+1} = \frac{-a_r}{(r + 1)(2r + 3)}}$$

So $\quad a_1 = -\dfrac{a_0}{3}$

$a_2 = -\dfrac{a_1}{2 \times 5} = \dfrac{a_0}{(1 \times 2)(3 \times 5)}$

$a_3 = -\dfrac{a_2}{3 \times 7} = \dfrac{-a_0}{(1 \times 2 \times 3)(3 \times 5 \times 7)}$

$a_4 = -\dfrac{a_3}{4 \times 9} = \dfrac{a_0}{(1 \times 2 \times 3 \times 4)(3 \times 5 \times 7 \times 9)}$ etc.

$y = x^c\{a_0 + a_1 x + a_2 x^2 + a_3 x^3 + \ldots + a_r x^r + \ldots\}$

i.e. $\quad y = \ldots\ldots\ldots\ldots$

51

$$y = x^{\frac{1}{2}}\left\{a_0 - \frac{a_0}{3}x + \frac{a_0}{(1 \times 2)(3 \times 5)}x^2 - \frac{a_0}{(1 \times 2 \times 3)(3 \times 5 \times 7)}x^3 + \dots\right\}$$

i.e. $y = a_0 x^{\frac{1}{2}}\left\{1 - \frac{x}{(1 \times 3)} + \frac{x^2}{(1 \times 2)(3 \times 5)} - \frac{x^3}{(1 \times 2 \times 3)(3 \times 5 \times 7)} + \dots\right\}$

Since a_0 is an arbitrary (non-zero) constant in each solution, its values may well be different, A and B say. If we denote the first solution by $u(x)$ and the second by $v(x)$, then

$$u = A\left\{1 - x + \frac{x^2}{(2 \times 3)} - \frac{x^3}{(2 \times 3)(3 \times 5)} + \frac{x^4}{(2 \times 3 \times 4)(3 \times 5 \times 7)} + \dots\right\}$$

and

$$v = B x^{\frac{1}{2}}\left\{1 - \frac{x}{(1 \times 3)} + \frac{x^2}{(1 \times 2)(3 \times 5)} - \frac{x^3}{(1 \times 2 \times 3)(3 \times 5 \times 7)} + \dots\right\}$$

The general solution $y = u + v$ is therefore ....

52

$$y = A\left\{1 - x + \frac{x^2}{(2 \times 3)} - \frac{x^3}{(2 \times 3)(3 \times 5)} + \dots\right\} + B x^{\frac{1}{2}}\left\{1 - \frac{x}{(1 \times 3)}\right.$$
$$\left. + \frac{x^2}{(1 \times 2)(3 \times 5)} - \frac{x^3}{(1 \times 2 \times 3)(3 \times 5 \times 7)} + \dots\right\}$$

The method may seem somewhat lengthy, but we have set it out in detail. It is a straightforward routine. Here is another example with the same steps.

Example 2

Find the series solution for the equation

$$3x^2y'' - xy' + y - xy = 0.$$

We proceed in just the same way as in the previous example.

Assume $\quad y = x^c\{a_0 + a_1x + a_2x^2 + a_3x^3 + \dots + a_rx^r + \dots\}$

i.e. $\quad y = a_0x^c + a_1x^{c+1} + a_2x^{c+2} + \dots + a_rx^{c+r} + \dots$

$\therefore y' = a_0cx^{c-1} + a_1(c+1)x^c + a_2(c+2)x^{c+1} + \dots$
$\quad + a_r(c+r)x^{c+r-1} + \dots$

and $\quad y'' = \dots\dots\dots$

53

$$y'' = a_0 c(c-1)x^{c-2} + a_1(c+1)cx^{c-1} + a_2(c+2)(c+1)x^c + \ldots$$
$$+ a_r(c+r)(c+r-1)x^{c+r-2} + \ldots$$

Now we build up the terms in the given equation.

$$3x^2 y'' = 3a_0 c(c-1)x^c + 3a_1(c+1)cx^{c+1} + 3a_2(c+2)(c+1)x^{c+2} + \ldots$$
$$+ 3a_r(c+r)(c+r-1)x^{c+r} + \ldots$$
$$-xy' = -a_0 cx^c - a_1(c+1)x^{c+1} - a_2(c+2)x^{c+2} - \ldots - a_r(c+r)x^{c+r} - \ldots$$
$$y = a_0 x^c + a_1 x^{c+1} + a_2 x^{c+2} + \ldots + a_r x^{c+r} + \ldots$$
$$-xy = -a_0 x^{c+1} - a_1 x^{c+2} - \ldots - a_r x^{c+r+1} \ldots$$

The *indicial equation*, i.e. equating the coefficient of the lowest power of x to zero, gives the values of c. Thus, in this case

$$c = \ldots\ldots\ldots\ldots$$

54

$$\boxed{c = 1 \text{ or } \frac{1}{3}}$$

Because the lowest power is x^c and the coefficient of x^c equated to zero gives

$$3a_0 c(c-1) - a_0 c + a_0 = 0$$

$$\therefore \ a_0(3c^2 - 4c + 1) = 0 \quad \therefore \ (3c-1)(c-1) = 0 \text{ since } a_0 \neq 0$$

$$\therefore \ c = 1 \quad \text{or} \quad \frac{1}{3}$$

The coefficient of the general term, i.e. x^{c+r} gives

$$3a_r(c+r)(c+r-1) - a_r(c+r) + a_r - a_{r-1} = 0$$

$$\therefore \ a_r = \ldots\ldots\ldots\ldots$$

55

$$\boxed{a_r = \frac{a_{r-1}}{3(c+r)^2 - 4(c+r) + 1} = \frac{a_{r-1}}{(c+r-1)(3c+3r-1)}}$$

(a) *Using* $c = 1$ the recurrence relation becomes

$$a_r = \frac{a_{r-1}}{r(3r+2)}$$

$$\therefore \ r = 1 \quad a_1 = \frac{a_0}{1 \times 5}$$

$$r = 2 \quad a_2 = \frac{a_1}{2 \times 8} = \frac{a_0}{(1 \times 2)(5 \times 8)}$$

$$r = 3 \quad a_3 = \frac{a_2}{3 \times 11} = \frac{a_0}{(1 \times 2 \times 3)(5 \times 8 \times 11)}$$

Our first solution is therefore

$$y = \ldots\ldots\ldots\ldots$$

56

$$y = x^1 \left\{ a_0 + \frac{a_0 x}{(1 \times 5)} + \frac{a_0 x^2}{(1 \times 2)(5 \times 8)} + \frac{a_0 x^3}{(1 \times 2 \times 3)(5 \times 8 \times 11)} + \cdots \right\}$$

$$\therefore \ y = Ax \left\{ 1 + \frac{x}{(1 \times 5)} + \frac{x^2}{(1 \times 2)(5 \times 8)} + \frac{x^3}{(1 \times 2 \times 3)(5 \times 8 \times 11)} + \cdots \right\}$$

(b) *For the second solution*, we put $c = \frac{1}{3}$. The recurrence relation then becomes

$$a_r = \ldots\ldots\ldots\ldots$$

57

$$a_r = \frac{a_{r-1}}{r(3r - 2)}$$

Therefore we can now determine the coefficients for $r = 1, 2, 3, \ldots$ and complete the second solution.

$$y = \ldots\ldots\ldots\ldots$$

58

$$y = Bx^{\frac{1}{3}} \left\{ 1 + x + \frac{x^2}{(2 \times 4)} + \frac{x^3}{(2 \times 3)(4 \times 7)} \right.$$
$$\left. + \frac{x^4}{(2 \times 3 \times 4)(4 \times 7 \times 10)} + \cdots \right\}$$

Because

$$a_1 = \frac{a_0}{1 \times 1}; \qquad a_2 = \frac{a_1}{2 \times 4} = \frac{a_0}{(1 \times 2)(2 \times 4)}$$

$$a_3 = \frac{a_2}{3 \times 7} = \frac{a_0}{(2 \times 3)(4 \times 7)}$$

$$a_4 = \frac{a_3}{4 \times 10} = \frac{a_0}{(2 \times 3 \times 4)(4 \times 7 \times 10)}$$

$$\therefore \ y = a_0 x^{\frac{1}{3}} \left\{ 1 + x + \frac{x^2}{(2 \times 4)} + \frac{x^3}{(2 \times 3)(4 \times 7)} \right.$$

$$\left. + \frac{x^4}{(2 \times 3 \times 4)(4 \times 7 \times 10)} + \cdots \right\}$$

Therefore, the general solution is

$$y = \ldots\ldots\ldots\ldots$$

59

$$y = Ax \left\{ 1 + \frac{x}{(1 \times 5)} + \frac{x^2}{(1 \times 2)(5 \times 8)} + \frac{x^3}{(1 \times 2 \times 3)(5 \times 8 \times 11)} + \ldots \right\}$$
$$+ Bx^{\frac{1}{3}} \left\{ 1 + x + \frac{x^2}{(2 \times 4)} + \frac{x^3}{(2 \times 3)(4 \times 7)} + \frac{x^4}{(2 \times 3 \times 4)(4 \times 7 \times 10)} + \ldots \right\}$$

Example 3

Find the series solution for the equation

$$\frac{d^2 y}{dx^2} - y = 0 \quad \text{i.e.} \quad y'' - y = 0.$$

As usual, we start off with the assumed solution

$$y = x^c \{ a_0 + a_1 x + a_2 x^2 + \ldots + a_r x^r + \ldots \}$$

i.e. $y = a_0 x^c + a_1 x^{c+1} + a_2 x^{c+2} + \ldots + a_r x^{c+r} + \ldots$

$\therefore \quad y' = a_0 c x^{c-1} + a_1 (c+1) x^c + a_2 (c+2) x^{c+1} + \ldots$
$\qquad + a_r (c+r) x^{c+r-1} + \ldots$

$y'' = a_0 c(c-1) x^{c-2} + a_1 (c+1) c x^{c-1} + a_2 (c+2)(c+1) x^c + \ldots$
$\qquad + a_r (c+r)(c+r-1) x^{c+r-2} + \ldots$

These three expansions are required regularly, so make a note of them

60

Now we build up the terms in the left-hand side of the equation.

$y'' = a_0 c(c-1) x^{c-2} + a_1 (c+1) c x^{c-1} + a_2 (c+2)(c+1) x^c + \ldots$
$\qquad + a_r (c+r)(c+r-1) x^{c+r-2} + \ldots$

$y = a_0 x^c + a_1 x^{c+1} + \ldots + a_r x^{c+r} + \ldots$

The term in x^{c+r} in the first of these expansions is

. . . .

61

$$\boxed{a_{r+2}(c+r+2)(c+r+1)x^{c+r}}$$

Because replacing r by $(r+2)$ in $a_r(c+r)(c+r+1)x^{c+r-2}$ gives this result.

Then $y'' - y = \ldots \ldots \ldots$

62

$$y'' - y = a_0 c(c-1)x^{c-2} + a_1(c+1)cx^{c-1} + [a_2(c+2)(c+1) - a_0]x^c$$
$$+ \ldots + [a_{r+2}(c+r+2)(c+r+1) - a_r]x^{c+r} + \ldots$$

We now equate each coefficient in turn to zero, since the right-hand side of the equation is zero. The coefficient of the lowest power of x gives the *indicial equation* from which we obtain the values of c.

So, in this case, $c = \ldots\ldots\ldots\ldots$

63

$$c = 0 \quad \text{or} \quad 1$$

For the term in x^{c-1}, we have

$[x^{c-1}]$: $a_1(c+1)c = 0$.

With $c = 1$, $a_1 = 0$.

But with $c = 0$, a_1 is indeterminate, because any value of a_1 combined with the zero value of c would make the product zero.

$[x^c]$: $a_2(c+2)(c+1) - a_0 = 0 \quad \therefore a_2 = \dfrac{a_0}{(c+1)(c+2)}$

For the general term

$[x^{c+r}]$: $\ldots\ldots\ldots\ldots$

64

$$a_{r+2} = \frac{a_r}{(c+r+1)(c+r+2)}$$

Because $a_{r+2}(c+r+2)(c+r+1) - a_r = 0$. Hence the result above.

From the indicial equation, $c = 0$ or $c = 1$.

(a) When $c = 0$ a_1 is indeterminate
$$a_2 = \frac{a_0}{2}$$

In general $a_{r+2} = \dfrac{a_r}{(r+1)(r+2)}$

$r = 1$ $\therefore a_3 = \dfrac{a_1}{2 \times 3}$

$r = 2$ $a_4 = \dfrac{a_2}{3 \times 4} = \dfrac{a_0}{4!}$

Therefore, one solution is $\ldots\ldots\ldots\ldots$

65

$$y = x^0 \left\{ a_0 + a_1 x + \frac{a_0}{2!} x^2 + \frac{a_1}{3!} x^3 + \frac{a_0}{4!} x^4 \ldots \right\}$$

i.e. $y = a_0 \left\{ 1 + \frac{x^2}{2!} + \frac{x^4}{4!} + \ldots \right\} + a_1 \left\{ x + \frac{x^3}{3!} + \frac{x^5}{5!} + \ldots \right\}$

a_0 and a_1 are arbitrary constants depending on the boundary conditions.

$$\therefore \ y = A \left\{ 1 + \frac{x^2}{2!} + \frac{x^4}{4!} + \ldots \right\} + B \left\{ x + \frac{x^3}{3!} + \frac{x^5}{5!} + \ldots \right\}$$

Notice that these two series are the Maclaurin series expansions of the hyperbolic functions, so that

$$y = A \cosh x + B \sinh x$$

It is not very often the case that the series solution is so easily expressible in terms of known functions.

(b) Similarly,

$$\text{when } c = 1 \qquad a_1 = 0$$

$$a_2 = \frac{a_0}{2 \times 3}$$

$$a_{r+2} = \ldots \ldots \ldots$$

66

$$a_{r+2} = \frac{a_r}{(r+2)(r+3)}$$

$$\therefore \ a_1 = 0$$

$$a_2 = \frac{a_0}{3!}$$

$$r = 1 \qquad a_3 = \frac{a_1}{3 \times 4} = 0$$

$$r = 2 \qquad a_4 = \frac{a_2}{4 \times 5} = \frac{a_0}{5!}$$

$$r = 3 \qquad a_5 = \frac{a_3}{5 \times 6} = 0 \quad \text{etc.}$$

A second solution with $c = 1$ is therefore

$$y = \ldots \ldots \ldots$$

$$y = a_0 \left\{ x + \frac{x^3}{3!} + \frac{x^5}{5!} + \cdots \right\}$$

and, because a_0 is an arbitrary constant

$$y = C \left\{ x + \frac{x^3}{3!} + \frac{x^5}{5!} + \frac{x^7}{7!} + \cdots \right\}$$

Note: This is not, in fact, a separate solution, since it already forms the second series in the solution for $c = 0$ obtained previously. Therefore, the first solution, with its two arbitrary constants, A and B, gives the general solution. This happens when the two values of c differ by an integer.

Make a note of the following:

> If the two values of c, i.e. c_1 and c_2, differ by an integer, and if $c = c_1$ results in a_1 being indeterminate, then this value of c gives the general solution.

> The solution resulting from $c = c_2$ is then merely a multiple of one of the series forming the first solution.

Our last problem was an example of this.
So far, we have met two distinct cases concerning the two roots $c = c_1$ and $c = c_2$ of the indicial equation.

(a) *If c_1 and c_2 differ by a quantity NOT an integer* then two independent solutions, $y = u(x)$ and $y = v(x)$, are obtained. The general solution is then $y = Au + Bv$.

(b) *If c_1 and c_2 differ by an integer*, i.e. $c_2 = c_1 + n$, and if one coefficient (a_r) is indeterminate when $c = c_1$, the complete general solution is given by using this value of c. Using $c = c_1 + n$ gives a series which is a simple multiple of one of the series in the first solution.

Make a note of these two points in your record book. Then move on

68

There is a third category to be added to (a) and (b) above.

(c) If the roots $c = c_1$ and $c = c_1 + n$ of the indicial equation differ by an integer and one coefficient (a_r) becomes infinite when $c = c_1$, the series is rewritten with a_0 replaced by $k(c - c_1)$.

Putting $c = c_1$ in the rewritten series and that of its derivative with respect to c gives two independent solutions.

Add this to the previous two. Then we will see how it works in practice

69

Example 4

Find the series solution of the equation

$$xy'' + (2 + x)y' - 2y = 0.$$

Using $y = x^c(a_0 + a_1x + a_2x^2 + a_3x^3 + \ldots + a_rx^r + \ldots)$ and its first two derivatives, the expansions for

$$xy'' = \ldots\ldots\ldots$$
$$2y' = \ldots\ldots\ldots$$
$$xy' = \ldots\ldots\ldots$$
$$-2y = \ldots\ldots\ldots$$

Method as before.

70

$$xy'' = a_0c(c-1)x^{c-1} + a_1(c+1)cx^c + a_2(c+2)(c+1)x^{c+1} + \ldots$$
$$+ a_r(c+r)(c+r-1)x^{c+r-1} + \ldots$$
$$2y' = 2a_0cx^{c-1} + 2a_1(c+1)x^c + 2a_2(c+2)x^{c+1} + 2a_3(c+3)x^{c+2}$$
$$+ \ldots + 2a_r(c+r)x^{c+r-1} + \ldots$$
$$xy' = a_0cx^c + a_1(c+1)x^{c+1} + a_2(c+2)c^{c+2} + \ldots$$
$$+ a_r(c+r)x^{c+r} + \ldots$$
$$-2y = -2a_0x^c - 2a_1x^{c+1} - 2a_2x^{c+2} - 2a_3x^{c+3} - \ldots$$
$$- 2a_rx^{c+r} - \ldots$$

From which, the indicial equation is $\ldots\ldots\ldots$

71

$$\boxed{a_0(c^2 + c) = 0}$$

i.e. equating the coefficient of the lowest power of x, (x^{c-1}), to zero.

$a_0 \neq 0 \quad \therefore c = 0$ or -1

Also, from the expansions, the total coefficient of x^c gives

$$a_1 = \ldots\ldots\ldots$$

72

$$\boxed{a_1 = \frac{-a_0(c-2)}{(c+1)(c+2)}}$$

From the terms in x^c, all four expansions are involved, so we can form the recurrence relation from the coefficient of x^{c+r}.

$$a_{r+1} = \ldots\ldots\ldots$$

$$a_{r+1} = \frac{-a_r(c+r-2)}{(c+r+1)(c+r+2)}$$

Because

$$a_{r+1}(c+r+1)(c+r) + 2a_{r+1}(c+r+1) + a_r(c+r) - 2a_r = 0$$
$$a_{r+1}(c+r+1)(c+r+2) + a_r(c+r-2) = 0$$

$$\therefore a_{r+1} = \frac{-a_r(c+r-2)}{(c+r+1)(c+r+2)} \qquad r \geq 0$$

$$\therefore a_2 = \ldots\ldots\ldots\ldots$$

$$a_2 = \frac{a_0(c-1)(c-2)}{(c+1)(c+2)^2(c+3)}$$

and, from the recurrence relation, when $r = 2$

$$a_3 = \ldots\ldots\ldots\ldots$$

$$a_3 = \frac{-a_0 c(c-1)(c-2)}{(c+1)(c+2)^2(c+3)^2(c+4)}$$

$$\therefore y = a_0 x^c \left\{ 1 - \frac{c-2}{(c+1)(c+2)}x + \frac{(c-1)(c-2)}{(c+1)(c+2)^2(c+3)}x^2 \right.$$
$$\left. - \frac{c(c-1)(c-2)}{(c+1)(c+2)^2(c+3)^2(c+4)}x^3 + \ldots \right\}$$

From the indicial equation above, the values of c are 0 and -1.

Putting c = 0, we have one solution

$$y = u = \ldots\ldots\ldots\ldots$$

$$y = u = a_0 \left\{ 1 + x + \frac{x^2}{6} \right\}$$

Note that coefficients after the x^2 term are zero, because of the factor c in the numerator.

Putting c = −1, we soon find that $\ldots\ldots\ldots\ldots$

77

coefficients become infinite, because of
the factor $(c+1)$ in the denominator.

Therefore, we substitute $a_0 = k(c - c_1) = k(c - [-1]) = k(c + 1)$.

$$\therefore\ y = k(c+1)x^c\left\{1 - \frac{c-2}{(c+1)(c+2)}x + \frac{(c-1)(c-2)}{(c+1)(c+2)^2(c+3)}x^2\right.$$
$$\left. - \frac{c(c-1)(c-2)}{(c+1)(c+2)^2(c+3)^2(c+4)}x^3 + \ldots\right\}$$
$$= kx^c\left\{(c+1) - \frac{c-2}{c+2}x + \frac{(c-1)(c-2)}{(c+2)^2(c+3)}x^2\right.$$
$$\left. - \frac{c(c-1)(c-2)}{(c+2)^2(c+3)^2(c+4)}x^3 + \ldots\right\}$$

Now, putting $c = -1$:

$$y = \ldots\ldots\ldots$$

78

$$y = kx^{-1}\left\{3x + 3x^2 + \frac{x^3}{2}\right\}$$

All subsequent terms are zero, since the numerators all contain a factor $(c+1)$.

$$\therefore\ y = v = \left\{3 + 3x + \frac{x^2}{2}\right\}$$

is a solution.

A solution is also given by $\dfrac{\partial y}{\partial c} = 0$.

So, starting from

$$y = kx^c\left\{(c+1) - \frac{c-2}{c+2}x + \frac{(c-1)(c-2)}{(c+2)^2(c+3)}x^2\right.$$
$$\left. - \frac{c(c-1)(c-2)}{(c+2)^2(c+3)^2(c+4)}x^3 + \ldots\right\}$$
$$\frac{\partial y}{\partial c} = kx^c \ln x\left\{(c+1) - \frac{c-2}{c+2}x + \frac{(c-1)(c-2)}{(c+2)^2(c+3)}x^2\right.$$
$$\left. - \frac{c(c-1)(c-2)}{(c+2)^2(c+3)^2(c+4)}x^3 + \ldots\right\}$$
$$+ kx^c\frac{\partial}{\partial c}\left\{(c+1) - \frac{c-2}{c+2}x + \frac{(c-1)(c-2)}{(c+2)^2(c+3)}x^2 - \ldots\right\}$$

We now have to determine the partial derivative of each term.

▶

$$\frac{\partial}{\partial c}(c+1) = 1$$

$$\frac{\partial}{\partial c}\left\{\frac{c-2}{c+2}\right\} = \ldots\ldots\ldots\ldots$$

$$\boxed{\frac{\partial}{\partial c}\left\{\frac{c-2}{c+2}\right\} = \frac{4}{(c+2)^2}}$$

Now we have to differentiate $\dfrac{(c-1)(c-2)}{(c+2)^2(c+3)}$

Let $t = \dfrac{(c-1)(c-2)}{(c+2)^2(c+3)}$

$$\therefore \ \ln t = \ln(c-1) + \ln(c-2) - 2\ln(c+2) - \ln(c+3)$$

$$\therefore \ \frac{1}{t}\frac{\partial t}{\partial c} = \frac{1}{c-1} + \frac{1}{c-2} - \frac{2}{c+2} - \frac{1}{c+3}$$

$$\therefore \ \frac{\partial t}{\partial c} = \frac{(c-1)(c-2)}{(c+2)^2(c+3)}\left\{\frac{1}{c-1} + \frac{1}{c-2} - \frac{2}{c+2} - \frac{1}{c+3}\right\}$$

$$\therefore \ \text{when } c=-1, \ \frac{\partial}{\partial c}(c+1) = 1$$

$$\frac{\partial}{\partial c}\left\{\frac{c-2}{c+2}\right\} = 4$$

$$\frac{\partial}{\partial c}\left\{\frac{(c-1)(c-2)}{(c+2)^2(c+3)}\right\} = \ldots\ldots\ldots\ldots$$

$$\boxed{-10}$$

Therefore, when $c = -1$:

$$\frac{\partial y}{\partial c} = kx^{-1}\ln x\left\{0 + 3x + 3x^2 + \frac{x^3}{2} + \ldots\right\}$$
$$+ kx^{-1}\{1 - 4x - 10x^2 + \ldots\}$$

$\therefore$ Another solution is

$$y = w = C\left\{\ln x\left(3 + 3x + \frac{x^2}{2} + \ldots\right) + x^{-1}(1 - 4x - 10x^2 + \ldots)\right\}$$

▶

Now we have a problem, for we seem to have three separate series solutions for a second-order differential equation.

(a) $y = u = A\left(1 + x + \dfrac{x^2}{6}\right)$

(b) $y = v = B\left(3 + 3x + \dfrac{x^2}{2}\right)$

(c) $y = w = C\left\{\ln x\left(3 + 3x + \dfrac{x^2}{2} + \ldots\right) + x^{-1}(1 - 4x - 10x^3 + \ldots)\right\}$

But (b) is clearly a simple multiple of (a) and thus not a distinct solution. So finally, we have just (a) and (c).

i.e. $\quad y = u = A\left(1 + x + \dfrac{x^2}{6}\right)$

and $\quad y = w = B\left\{\ln x\left(3 + 3x + \dfrac{x^2}{2} + \ldots\right) + x^{-1}(1 - 4x - 10x^3 + \ldots)\right\}$

The complete solution is then $\quad y = u + w$

In general if $c_1 - c_2 = n$ where n is a non-zero integer the solution is of the form:

$$y = (1 + k\ln x)x^{c_1}\{a_0 + a_1 x + a_2 x^2 + \ldots\} + x^{c_2}\{b_0 + b_1 x + b_2 x^2 + \ldots\}$$

Finally we have just one more variation to the list in Frames 67 and 68,
so move on

81

Example 5

Solve the equation $\quad xy'' + y' - xy = 0$.

Start off as before and build up expansions for the terms in the left-hand side of the equation.

$$xy'' = \ldots\ldots\ldots\ldots$$
$$y' = \ldots\ldots\ldots\ldots$$
$$-xy = \ldots\ldots\ldots\ldots$$

82

$$
\begin{aligned}
xy'' &= a_0 c(c-1)x^{c-1} + a_1(c+1)cx^c + a_2(c+2)(c+1)x^{c+1} + \ldots \\
&\quad + a_r(c+r)(c+r-1)x^{c+r-1} + \ldots \\
y' &= a_0 c x^{c-1} + a_1(c+1)x^c + a_2(c+2)x^{c+1} + \ldots \\
&\quad + a_r(c+r)x^{c+r-1} + \ldots \\
-xy &= -a_0 x^{c+1} - a_1 x^{c+2} - \ldots \\
&\quad - a_r x^{c+r+1} - \ldots
\end{aligned}
$$

The indicial equation, therefore, gives $c = \ldots\ldots\ldots\ldots$

83

$$\boxed{c = 0 \quad \text{(twice)}}$$

Because $a_0 \{c(c-1) + c\} = 0$ $a_0 \neq 0$ $\therefore$ $c^2 = 0$ $\therefore$ $c = 0$ (twice)

Coefficient of x^c gives ....

84

$$\boxed{a_1 = 0}$$

$[x^c]$: $a_1(c^2 + c + c + 1) = 0$ $\therefore$ $a_1(c+1)^2 = 0$ $\therefore$ $a_1 = 0$

$[x^{c+1}]$:   This involves all three expansions and from this point, we can use the general recurrence relation.

$[x^{c+r-1}]$: $a_r\{(c+r)(c+r-1) + (c+r)\} - a_{r-2} = 0$

$$\therefore \ a_r(c+r)^2 = a_{r-2} \quad \therefore \ a_r = \frac{a_{r-2}}{(c+r)^2}$$

$$\therefore \ y = \ldots\ldots\ldots$$

85

$$\boxed{y = x^c\left\{a_0 + \frac{a_0}{(c+2)^2}x^2 + \frac{a_0}{(c+2)^2(c+4)^2}x^4 + \ldots\right\}}$$

i.e. $y = a_0 x^c\left\{1 + \dfrac{x^2}{(c+2)^2} + \dfrac{x^4}{(c+2)^2(c+4)^2} + \ldots\right\}$

$\therefore$ When $c = 0$

$$y = u = A\left\{1 + \frac{x^2}{2^2} + \frac{x^4}{2^2 \times 4^2} + \ldots\right\} \tag{1}$$

▶

This is one solution. Another is given by $v = \dfrac{\partial y}{\partial c}$

$$\frac{\partial y}{\partial c} = a_0 x^c \ln x \left\{ 1 + \frac{x^2}{(c+2)^2} + \frac{x^4}{(c+2)^2(c+4)^2} + \ldots \right\}$$

$$+ a_0 x^c \frac{\partial}{\partial c} \left\{ 1 + \frac{x^2}{(c+2)^2} + \frac{x^4}{(c+2)^2(c+4)^2} + \ldots \right\}$$

Now $\dfrac{\partial}{\partial c}(1) = 0; \quad \dfrac{\partial}{\partial c} \left\{ \dfrac{1}{(c+2)^2} \right\} = \dfrac{-2}{(c+2)^3}$

Let $t = \dfrac{1}{(c+2)^2(c+4)^2} \qquad \therefore \quad \ln t = -2\ln(c+2) - 2\ln(c+4)$

$$\therefore \quad \frac{1}{t}\frac{\partial t}{\partial c} = \frac{-2}{c+2} - \frac{2}{c+4} \qquad \therefore \quad \frac{\partial t}{\partial c} = \frac{-2}{(c+2)^2(c+4)^2}\left\{ \frac{1}{c+2} + \frac{1}{c+4} \right\}$$

$$\therefore \quad \frac{\partial y}{\partial c} = a_0 x^c \ln x \left\{ 1 + \frac{x^2}{(c+2)^2} + \frac{x^4}{(c+2)^2(c+4)^2} + \ldots \right\}$$

$$+ a_0 x^c \left\{ 0 - \frac{2x^2}{(c+2)^3} - \frac{4x^4(c+3)}{(c+2)^3(c+4)^3} + \ldots \right\}$$

$\therefore$ When $c = 0$

$$y = v = \ldots\ldots\ldots\ldots$$

86

$$\boxed{ y = v = B\left\{ \ln x\left(1 + \frac{x^2}{2^2} + \frac{x^4}{2^2 \times 4^2} + \ldots \right) - \frac{x^2}{2^2} - \frac{3x^4}{2^3 \times 4^2} + \ldots \right\} } \qquad (2)$$

So our two solutions are $y = u$ (at 1) and $y = v$ (at 2). The complete solution is therefore $y = u + v$.

In general if $c_1 = c_2 = c$ the solution is of the form

$$y = (1 + k\ln x)x^c\{a_0 + a_1 x + a_2 x^2 + \ldots\} + x^c\{b_1 x + b_2 x^2 + \ldots\}$$

Summary

Let us now summarise the four types of procedures in the method of Frobenius that we have covered.

(a) Assume a series of the form

$$y = x^c(a_0 + a_1 x + a_2 x^2 + \ldots + a_r x^r + \ldots)$$

(b) Indicial equation gives $c = c_1$ and $c = c_2$.

(c) *Case 1.* c_1 and c_2 differ by a quantity *not an integer*. Substitute $c = c_1$ and $c = c_2$ in the series for y.

(d) *Case 2.* c_1 and c_2 differ by *an integer* and make a coefficient *indeterminate* with $c = c_1$. Substitution of $c = c_1$ gives the complete solution.

(e) *Case 3.* c_1 and c_2 $(c_1 < c_2)$ differ by *an integer* and make a coefficient *infinite* for $c = c_1$. Replace a_0 by $k\,(c - c_1)$. Put $c = c_1$ in the new series for y and for $\dfrac{\partial y}{\partial c}$.

In general if $c_1 - c_2 = n$ where n is a non-zero integer, the solution is of the form

$$y = (1 + k \ln x)x^{c_1}\{a_0 + a_1 x + a_2 x^2 + \ldots\} + x^{c_2}\{b_0 + b_1 x + b_2 x^2 + \ldots\}$$

(f) *Case 4.* c_1 and c_2 *equal*. Substitute $c = c_1$ in the series for y and for $\dfrac{\partial y}{\partial c}$. Make the substitution after differentiating.

In general if $c_1 = c_2 = c$, the solution is of the form

$$y = (1 + k \ln x)x^c\{a_0 + a_1 x + a_2 x^2 + \ldots\} + x^c\{b_1 x + b_2 x^2 + \ldots\}$$

Make a note of this summary for future reference

Bessel's equation

A second-order differential equation that occurs frequently in branches of technology is of the form

$$x^2 y'' + xy' + (x^2 - v^2)y = 0$$

where v is a real constant.

Starting with $y = x^c(a_0 + a_1 x + a_2 x^2 + a_3 x^3 + \ldots + a_r x^r + \ldots)$ and proceeding as before, we obtain

$$c = \pm v \quad \text{and} \quad a_1 = 0$$

The recurrence relation is $\quad a_r = \dfrac{a_{r-2}}{v^2 - (c + r)^2} \quad$ for $r \geq 2$.

It follows that $a_1 = a_3 = a_5 = a_7 = \ldots = 0$

and that $\quad a_2 = \ldots\ldots\ldots\ldots; \quad a_4 = \ldots\ldots\ldots\ldots; \quad a_6 = \ldots\ldots\ldots\ldots$

89

$$a_2 = \frac{a_0}{v^2 - (c+2)^2}; \quad a_4 = \frac{a_0}{\left[v^2 - (c+2)^2\right]\left[v^2 - (c+4)^2\right]};$$

$$a_6 = \frac{a_0}{\left[v^2 - (c+2)^2\right]\left[v^2 - (c+4)^2\right]\left[v^2 - (c+6)^2\right]}$$

$\therefore$ When $c = +v$ $\quad a_2 = \ldots\ldots\ldots\ldots;$ $\quad a_4 = \ldots\ldots\ldots\ldots$
$\qquad\qquad\qquad a_6 = \ldots\ldots\ldots\ldots;$ $\quad a_r = \ldots\ldots\ldots\ldots$

90

$$a_2 = \frac{-a_0}{2^2(v+1)}; \qquad a_4 = \frac{a_0}{2^4 \times 2(v+1)(v+2)}$$

$$a_6 = \frac{-a_0}{2^6 \times 3!(v+1)(v+2)(v+3)}$$

$$a_r = \frac{(-1)^{r/2}a_0}{2^r \times (r/2)!(v+1)(v+2)\ldots(v+r/2)} \text{ for } r \text{ even}$$

The resulting series solution is therefore

$$y = u = \ldots\ldots\ldots\ldots$$

91

$$y = u = Ax^v\left\{1 - \frac{x^2}{2^2(v+1)} + \frac{x^4}{2^4 \times 2!(v+1)(v+2)}\right.$$
$$\left. - \frac{x^6}{2^6 \times 3!(v+1)(v+2)(v+3)} + \ldots\right\}$$

This is valid provided v is not a negative integer.

Similarly, *when $c = -v$*

$$y = w = Bx^{-v}\left\{1 + \frac{x^2}{2^2(v-1)} + \frac{x^4}{2^4 \times 2!(v-1)(v-2)}\right.$$
$$\left. + \frac{x^6}{2^6 \times 3!(v-1)(v-2)(v-3)} + \ldots\right\}$$

This is valid provided v is not a positive integer.

Except for these two restrictions, the complete solution of Bessel's equation is therefore $y = u + w$ with the two arbitrary constants A and B.

Bessel functions

92

It is convenient to present the two results obtained above in terms of gamma functions, remembering that for $x > 0$

$$\Gamma(x+1) = x\Gamma(x)$$
$$\Gamma(x+2) = (x+1)\Gamma(x+1) = (x+1)x\Gamma(x)$$
$$\Gamma(x+3) = (x+2)\Gamma(x+2) = (x+2)(x+1)x\Gamma(x), \quad \text{etc.}$$

If, at the same time, we assign to the arbitrary constant a_0 the value $\dfrac{1}{2^v\Gamma(v+1)}$, then we have, for $c = v$

$$a_2 = \frac{a_0}{v^2 - (c+2)^2} = \frac{a_0}{(v-c-2)(v+c+2)} = \frac{a_0}{-2(2v+2)}$$
$$= \frac{-1}{2^2(v+1)} \cdot \frac{1}{2^v\Gamma(v+1)} = \frac{-1}{2^{v+2}(1!)\Gamma(v+2)}$$

Similarly $\qquad a_4 = \ldots\ldots\ldots$

$$\boxed{a_4 = \frac{1}{2^{v+4}(2!)\Gamma(v+3)}}$$

Because

$$a_4 = \frac{a_2}{v^2 - (c+4)^2} = \frac{a_2}{(v-c-4)(v+c+4)} = \frac{a_2}{-4(2v+4)}$$
$$= \frac{-1}{2^3(v+2)} \cdot \frac{-1}{2^{v+2}(1!)\Gamma(v+2)} = \frac{1}{2^{v+4}(2!)\Gamma(v+3)}$$

and $\quad a_6 = \ldots\ldots\ldots$

$$\boxed{a_6 = \frac{-1}{2^{v+6}(3!)\Gamma(v+4)}}$$

We can see the pattern taking shape.

$$a_r = \frac{(-1)^{r/2}}{2^{v+r}\left(\frac{r}{2}!\right)\Gamma\left(v+\frac{r}{2}+1\right)} \quad \text{for } r \text{ even.} \quad \therefore \text{ Put } r = 2k$$

The result then becomes

$$a_{2k} = \ldots\ldots\ldots$$

95

$$a_{2k} = \frac{(-1)^k}{2^{v+2k}(k!)\Gamma(v+k+1)} \quad k = 1, 2, 3, \ldots$$

Therefore, we can write the new form of the series for y as

$$y = x^v \left\{ \frac{1}{2^v\Gamma(v+1)} - \frac{x^2}{2^{v+2}(1!)\Gamma(v+2)} + \frac{x^4}{2^{v+4}(2!)\Gamma(v+3)} - \cdots \right\}$$

This is called the *Bessel function of the first kind of order v* and is denoted by $J_v(x)$.

$$\therefore\ J_v(x) = \left(\frac{x}{2}\right)^v \left\{ \frac{1}{\Gamma(v+1)} - \frac{x^2}{2^2(1!)\Gamma(v+2)} + \frac{x^4}{2^4(2!)\Gamma(v+3)} - \cdots \right\}$$

This is valid provided v is not ....

96

a negative integer

– otherwise some of the terms would become infinite.
If we take the other value for c, i.e. $c = -v$, the corresponding result becomes

$$J_{-v}(x) = \ldots\ldots\ldots$$

97

$$J_{-v}(x) = \left(\frac{x}{2}\right)^{-v} \left\{ \frac{1}{\Gamma(1-v)} - \frac{x^2}{2(1!)\Gamma(2-v)} + \frac{x^4}{2^2(2!)\Gamma(3-v)} - \cdots \right\}$$

provided that v is not a positive integer.
In general terms

$$J_v(x) = \left(\frac{x}{2}\right)^v \sum_{k=0}^{\infty} \frac{(-1)^k x^{2k}}{2^{2k}(k!)\Gamma(v+k+1)}$$

$$J_{-v}(x) = \left(\frac{x}{2}\right)^{-v} \sum_{k=0}^{\infty} \frac{(-1)^k x^{2k}}{2^{2k}(k!)\Gamma(k-v+1)}$$

The convergence of the series for all values of x can be established by the normal ratio test.

$J_v(x)$ and $J_{-v}(x)$ are two independent solutions of the original equation. Hence, the complete solution is

$$y = AJ_v(x) + BJ_{-v}(x)$$

where A and B are constants.

Make a note of the expressions for $J_v(x)$ and $J_{-v}(x)$.
Then on to the next frame

Some Bessel functions are commonly used and are worthy of special mention. **98**
This arises when v is a positive integer, denoted by n.

$$\therefore J_n(x) = \left(\frac{x}{2}\right)^n \sum_{k=0}^{\infty} \frac{(-1)^k x^{2k}}{2^{2k}(k!)\Gamma(n+k+1)}$$

From our work on gamma functions, $\Gamma(k+1) = k!$ for $k = 0, 1, 2, \ldots$

$$\therefore \Gamma(n+k+1) = (n+k)!$$

and the result above then becomes

$$J_n(x) = \ldots\ldots\ldots\ldots$$

99

$$\boxed{J_n(x) = \left(\frac{x}{2}\right)^n \sum_{k=0}^{\infty} \frac{(-1)^k x^{2k}}{2^{2k}(k!)(n+k)!}}$$

We have seen that $J_v(x)$ and $J_{-v}(x)$ are two solutions of Bessel's equation. When v and $-v$ are not integers, the two solutions are independent of each other. Then $y = AJ_v(x) + BJ_{-v}(x)$.

When, however, $v = n$ (integer), then $J_n(x)$ and $J_{-n}(x)$ are not independent, but are related by $J_{-n}(x) = (-1)^n J_n(x)$. This can be shown by referring once again to our knowledge of gamma functions.

$$\Gamma(x+1) = x\Gamma(x) \quad \therefore \quad \Gamma(x) = \frac{\Gamma(x+1)}{x}$$

and for negative integral values of x, or zero, $\Gamma(x)$ is infinite.

From the previous result:

$$J_{-v}(x) = \left(\frac{x}{2}\right)^{-v} \sum_{k=0}^{\infty} \frac{(-1)^k x^{2k}}{2^{2k}(k!)\Gamma(k-v+1)} \qquad k = 0, 1, 2, \ldots$$

Let us consider the gamma function $\Gamma(k-v+1)$ in the denominator and let v approach closely to a positive integer n.

Then $\quad \Gamma(k-v+1) \to \Gamma(k-n+1)$.

When $k-n+1 \leq 0$, i.e. when $k \leq (n-1)$, then $\Gamma(k-n+1)$ is infinite. The first finite value of $\Gamma(k-n+1)$ occurs for $k = n$.

When values of $\Gamma(k-v+1)$ are infinite the coefficients of $J_{-v}(x)$ are

$$\ldots\ldots\ldots\ldots$$

100

zero

The series, therefore, starts at $k = n$

$$\therefore \; J_{-n}(x) = \left(\frac{x}{2}\right)^{-n} \sum_{k=n}^{\infty} \frac{(-1)^k x^{2k}}{2^{2k}(k!)\Gamma(k - n + 1)}$$

$$= \sum_{k=n}^{\infty} \frac{(-1)^k x^{2k-n}}{2^{2k-n}(k!)\Gamma(k - n + 1)} \qquad \text{Put } k = p + n$$

$$= \sum_{p=0}^{\infty} \frac{(-1)^{p+n} x^{2p+n}}{2^{2p+n}(k!)(k - n)!}$$

$$= (-1)^n \sum_{p=0}^{\infty} \frac{(-1)^p x^{2p+n}}{2^{2p+n}(p!)(p + n)!}$$

$$= (-1)^n \left(\frac{x}{2}\right)^n \sum_{p=0}^{\infty} \frac{(-1)^p x^{2p}}{2^{2p}(p!)(p + n)!}$$

$$= (-1)^n \left(\frac{x}{2}\right)^n \sum_{k=0}^{\infty} \frac{(-1)^k x^{2k}}{2^{2k}(k!)(k + n)!}$$

$$\therefore \; J_{-n}(x) = (-1)^n J_n(x)$$

So, after all that, the series for $J_n(x) = \ldots\ldots\ldots\ldots$

101

$$\boxed{J_n(x) = \left(\frac{x}{2}\right)^n \left\{ \frac{1}{n!} - \frac{1}{(n + 1)!}\left(\frac{x}{2}\right)^2 + \frac{1}{(2!)(n + 2)!}\left(\frac{x}{2}\right)^4 - \ldots\ldots\ldots\ldots \right\}}$$

From this we obtain two commonly used functions

$$J_0(x) = \ldots\ldots\ldots\ldots$$

102

$$\boxed{J_0(x) = 1 - \frac{1}{(1!)^2}\left(\frac{x}{2}\right)^2 + \frac{1}{(2!)^2}\left(\frac{x}{2}\right)^4 - \frac{1}{(3!)^2}\left(\frac{x}{2}\right)^6 + \ldots}$$

and $\qquad\qquad\qquad\qquad J_1(x) = \ldots\ldots\ldots\ldots$

103

$$\boxed{J_1(x) = \frac{x}{2}\left\{ 1 - \frac{1}{(1!)(2!)}\left(\frac{x}{2}\right)^2 + \frac{1}{(2!)(3!)}\left(\frac{x}{2}\right)^4 + \ldots \right\}}$$

Bessel functions for a range of values of n and x are tabulated in published lists of mathematical data. Of these, $J_0(x)$ and $J_1(x)$ are most commonly used.

Graphs of Bessel functions $J_0(x)$ and $J_1(x)$

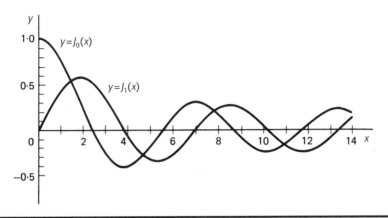

Legendre's equation

Another equation of special interest in engineering applications is Legendre's [105] equation of the form

$$(1 - x^2)y'' - 2xy' + k(k + 1)y = 0$$

where k is a real constant.

This may be solved by the Frobenius method as before. In this case, the indicial equation gives $c = 0$ and $c = 1$, and the two corresponding solutions are

(a) $c = 0$: $\quad y = a_0 \left\{ 1 - \dfrac{k(k + 1)}{2!} x^2 + \dfrac{k(k - 2)(k + 1)(k + 3)}{4!} x^4 - \cdots \right\}$

(b) $c = 1$: $\quad y = a_1 \left\{ x - \dfrac{(k - 1)(k + 2)}{3!} x^3 \right.$

$$\left. + \dfrac{(k - 1)(k - 3)(k + 2)(k + 4)}{5!} x^5 - \cdots \right\}$$

where a_0 and a_1 are the usual arbitrary constants

Legendre polynomials

When k is an integer (n), one of the solution series terminates after a finite number of terms. The resulting polynomial in x, denoted by $P_n(x)$, is called a *Legendre polynomial*, with a_0 or a_1 being chosen so that the polynomial has unit value when $x = 1$.

For example $\qquad\qquad P_2(x) = \ldots\ldots\ldots$

106

$$P_2(x) = \frac{1}{2}(3x^2 - 1)$$

Because, in $P_2(x)$, $n = k = 2$

$$\therefore \ y = a_0 \left\{ 1 - \frac{2 \times 3}{2!} x^2 + 0 + 0 + \ldots \right\}$$

$$= a_0 \{ 1 - 3x^2 \}$$

The constant a_0 is then chosen to make $y = 1$ when $x = 1$

i.e. $1 = a_0(1 - 3)$ $\therefore \ a_0 = -\frac{1}{2}$

$\therefore \ P_2(x) = -\frac{1}{2}(1 - 3x^2) = \frac{1}{2}(3x^2 - 1)$

Similarly $P_3(x) = \ldots\ldots\ldots$

107

$$P_3(x) = \frac{1}{2}(5x^3 - 3x)$$

Here $n = k = 3$

$$\therefore \ y = a_1 \left\{ x - \frac{2 \times 5}{3!} x^3 + 0 + 0 + \ldots \right\}$$

$$= a_1 \left\{ x - \frac{5x^3}{3} \right\}$$

$y = 1$ when $x = 1$ $\therefore \ a_1 \left(1 - \frac{5}{3} \right) = 1$ $\therefore \ a_1 = -\frac{3}{2}$

$$\therefore \ P_3(x) = -\frac{3}{2} \left(x - \frac{5x^3}{3} \right) = \frac{1}{2}(5x^3 - 3x)$$

108   **Rodrigue's formula and the generating function**

Legendre polynomials can be derived by using *Rodrigue's formula*

$$P_n(x) = \frac{1}{2^n n!} \frac{d^n}{dx^n} (x^2 - 1)^n$$

so using this formula

$$P_4(x) = \ldots\ldots\ldots$$

$$P_4(x) = \frac{1}{8}\left(35x^4 - 30x^2 + 3\right)$$

Because

$$P_4(x) = \frac{1}{2^4 4!}\frac{d^4}{dx^4}\left(x^2 - 1\right)^4$$

$$= \frac{1}{384}\frac{d^4}{dx^4}\left(x^8 - 4x^6 + 6x^4 - 4x^2 + 1\right)$$

$$= \frac{1}{384}\frac{d^3}{dx^3}\left(8x^7 - 24x^5 + 24x^3 - 8x\right)$$

$$= \frac{1}{384}\frac{d^2}{dx^2}\left(56x^6 - 120x^4 + 72x^2 - 8\right)$$

$$= \frac{1}{384}\frac{d}{dx}\left(336x^5 - 480x^3 + 144x\right)$$

$$= \frac{1}{384}\left(1680x^4 - 1440x^2 + 144\right)$$

$$= \frac{1}{8}\left(35x^4 - 30x^2 + 3\right)$$

In addition to Rodrigue's formula, the function

$$\frac{1}{\sqrt{1 - 2xt + t^2}} = \sum_{n=0}^{\infty} P_n(x)t^n, \qquad |t| < 1$$

is called the *generating function* for Legendre polynomials and can be used to obtain some of their properties. For example using this generating function we find that

$$P_n(1) = \ldots\ldots\ldots$$

110

$$\boxed{P_n(1) = 1}$$

Because

When $x = 1$ the generating function becomes

$$\frac{1}{\sqrt{1 - 2t + t^2}} = \sum_{n=0}^{\infty} P_n(1)t^n, \qquad |t| < 1$$

Noting that $\dfrac{1}{\sqrt{1 - 2t + t^2}} = \dfrac{1}{\sqrt{(1-t)^2}} = \dfrac{1}{1-t} = (1-t)^{-1}$, the left-hand side

is expanded by the binomial theorem to give

$$(1-t)^{-1} = 1 + t + t^2 + t^3 + \dots = \sum_{n=0}^{\infty} t^n.$$

Therefore $\displaystyle\sum_{n=0}^{\infty} t^n = \sum_{n=0}^{\infty} P_n(1)t^n$ and so $P_n(1) = 1$

By a similar reasoning

$$P_n(-1) = \dots\dots\dots$$

111

$$\boxed{P_n(-1) = (-1)^n}$$

Because

When $x = -1$ the generating function becomes

$$\frac{1}{\sqrt{1 + 2t + t^2}} = \sum_{n=0}^{\infty} P_n(-1)t^n$$

Noting that $\dfrac{1}{\sqrt{1 + 2t + t^2}} = \dfrac{1}{\sqrt{(1+t)^2}} = \dfrac{1}{1+t} = (1+t)^{-1}$, the left-hand side

is expanded by the binomial theorem to give

$$(1+t)^{-1} = 1 - t + t^2 - t^3 + \dots = \sum_{n=0}^{\infty} (-1)^n t^n. \text{ Therefore}$$

$$\sum_{n=0}^{\infty} (-1)^n t^n = \sum_{n=0}^{\infty} P_n(-1)t^n \text{ and so } P_n(-1) = (-1)^n$$

Legendre's equation, whose solutions are expressed in terms of Legendre polynomials, is an example of a particular class of differential equations referred to as Sturm–Liouville systems. In the following frames we shall look at such systems more closely.

So on to the next frame

Sturm–Liouville systems

112

A boundary value problem that is described by a differential equation of the general form

$$(p(x)y')' + (q(x) + \lambda r(x))y = 0 \quad \text{for} \quad a \le x \le b \text{ and } r(x) > 0$$

where the boundary conditions can be written in the form

$$\alpha_1 y(a) + \alpha_2 y'(a) = 0 \quad \text{and} \quad \beta_1 y(b) + \beta_2 y'(b) = 0$$

is called a **Sturm–Liouville** system. Solutions of such a system are in the form of an infinite sequence of *eigenfunctions* y_n, each corresponding to an *eigenvalue* λ_n of the system for $n = 0, 1, 2, \ldots$.

For example, consider the differential equation

$$y'' + \lambda y = 0 \quad \text{for} \quad 0 \le x \le 5$$

where here, $a = 0$ and $b = 5$. Also

$$y(0) = 0 \quad \text{and} \quad y(5) = 0$$

By comparing this equation with the general form given above we can see that

$$p(x) = \ldots\ldots\ldots\ldots; \quad q(x) = \ldots\ldots\ldots\ldots; \quad r(x) = \ldots\ldots\ldots\ldots;$$

$$\alpha_2 = \ldots\ldots\ldots\ldots; \quad \beta_2 = \ldots\ldots\ldots\ldots$$

113

$$\boxed{p(x) = 1; \quad q(x) = 0; \quad r(x) = 1; \quad \alpha_2 = 0; \quad \beta_2 = 0}$$

Because

By performing the differentiation on the left-hand term of $(p(x)y')' + (q(x) + \lambda r(x))y = 0$ we find that the differential equation can be written as

$$p(x)y'' + p'(x)y' + (q(x) + \lambda r(x))y = 0$$

By inspection, comparing this form with the differential equation $y'' + \lambda y = 0$ it is easily seen that $p(x) = 1$, $q(x) = 0$, $r(x) = 1$ and comparing boundary conditions gives $\alpha_2 = 0$ and $\beta_2 = 0$.

To solve the equation $y'' + \lambda y = 0$ we use the auxiliary equation $m^2 + \lambda = 0$ which has solutions $m = \pm j\sqrt{\lambda}$ (refer to *Engineering Mathematics (Fifth Edition)*, page 1077). This means that the solution can be written in the form

$$y = A\sin\ldots\ldots\ldots\ldots + B\cos\ldots\ldots\ldots\ldots$$

114

$$\boxed{y = A\sin\sqrt{\lambda}x + B\cos\sqrt{\lambda}x}$$

Because

When the solutions to the auxiliary equation are of the form $m = \alpha \pm j\beta$ the solution to the differential equation is of the form

$$y = e^{\alpha x}(A\sin\beta x + B\cos\beta x) \quad \text{and here} \quad \alpha = 0 \quad \text{and} \quad \beta = \sqrt{\lambda}$$

Applying the boundary condition $y(0) = 0$ then $B = \ldots\ldots\ldots\ldots$

115

$$B = 0$$

Because

$y = A \sin \sqrt{\lambda} x + B \cos \sqrt{\lambda} x$ and so $y(0) = A \sin 0 + B \cos 0 = B = 0$. Therefore
$y = A \sin \sqrt{\lambda} x$

Applying the boundary condition $y(5) = 0$ then

$$\lambda = \dots\dots\dots$$

116

$$\lambda = \frac{n^2 \pi^2}{25}$$

Because

$y = A \sin \sqrt{\lambda} x$ therefore $y(5) = A \sin \sqrt{\lambda} 5 = 0$. If $A = 0$ the solution reduces to the trivial solution $y = 0$. For a non-trivial solution $\sin \sqrt{\lambda} 5 = 0$ and so $\sqrt{\lambda} 5 = n\pi$, $n = 0, 1, 2, 3, \dots$. This means that

$$\sqrt{\lambda} = \frac{n\pi}{5} \quad \text{and so} \quad \lambda = \frac{n^2 \pi^2}{25}$$

There is an infinity of eigenvalues, the nth eigenvalue being denoted by λ_n where $\lambda_n = \frac{n^2 \pi^2}{25}$ and to each eigenvalue there is an eigenvector solution $y_n = A_n \sin \frac{n\pi x}{5}$.

117 **Orthogonality**

If two different functions $f(x)$ and $g(x)$ are defined on the interval $a \le x \le b$ and

$$\int_a^b f(x)g(x)\,dx = 0$$

then we say that the two functions are mutually **orthogonal**. If, on the other hand, a third function $w(x) > 0$ exists such that

$$\int_a^b f(x)g(x)w(x)\,dx = 0$$

then we say that $f(x)$ and $g(x)$ are mutually orthogonal *with respect to the weight function $w(x)$*.

One important property of the solutions to a Sturm–Liouville system is that the solutions are all mutually orthogonal with respect to the weight function $r(x)$. For instance, in the previous example the individual solutions were given as

$$y_n = A_n \sin \frac{n\pi x}{5} \quad \text{where} \quad r(x) = 1$$

and so if $m \ne n$

$$\int_0^5 y_m(x)y_n(x)r(x)\,dx = \dots\dots\dots$$

$$\int_0^5 y_m(x)y_n(x)r(x)\,\mathrm{d}x = 0$$

Because

$$\int_0^5 y_m(x)y_n(x)r(x)\,\mathrm{d}x = \int_0^5 A_m \sin\frac{m\pi x}{5} A_n \sin\frac{n\pi x}{5}\,\mathrm{d}x \qquad \text{where } r(x) = 1$$

$$= A_m A_n \int_0^5 \sin\frac{m\pi x}{5}\sin\frac{n\pi x}{5}\,\mathrm{d}x$$

$$= \frac{A_m A_n}{2}\int_0^5 \left(\cos\frac{(m-n)\pi x}{5} - \cos\frac{(m+n)\pi x}{5}\right)\mathrm{d}x$$

$$= \frac{A_m A_n}{2}\left[-\frac{5}{(m-n)\pi}\sin\frac{(m-n)\pi x}{5}\right.$$

$$\left. + \frac{5}{(m+n)\pi}\sin\frac{(m+n)\pi x}{5}\right]_0^5 \qquad \text{provided } m \neq n$$

$$= 0$$

Summary

1 A Sturm–Liouville system is a differential equation of the form

$$p(x)y'' + p'(x)y' + (q(x) + \lambda r(x))y = 0 \quad \text{for}$$
$$a \le x \le b \text{ and } r(x) > 0$$

where the boundary conditions can be written in the form

$$\alpha_1 y(a) + \alpha_2 y'(a) = 0 \ . \ \text{and} \quad \beta_1 y(b) + \beta_2 y'(b) = 0$$

2 Solutions y_n to a Sturm–Liouville system are called eigenvectors, each corresponding to an eigenvalue λ_n for $n = 0, 1, 2, \ldots$.
3 The solutions y_n are mutually orthogonal with respect to the weighting $r(x)$. That is

$$\int_a^b y_m(x)y_n(x)r(x)\,\mathrm{d}x = 0 \qquad (m \neq n)$$

Keep going

Legendre's equation revisited

The equation $(1-x^2)y'' - 2xy' + n(n+1)y = 0$ is Legendre's equation and has Legendre polynomials as solutions. That is

$$y_n = P_n(x) \qquad \text{where } P_n(1) = 1 \text{ and } P_n(-1) = (-1)^n$$

This equation is an example of a Sturm–Liouville system

$$p(x)y'' + p'(x)y' + (q(x) + \lambda r(x))y = 0$$

with boundary conditions

$$\alpha_1 y(a) + \alpha_2 y'(a) = 0 \quad \text{and} \quad \beta_1 y(b) + \beta_2 y'(b) = 0 \quad \text{where}$$
$$p(x) = \ldots\ldots\ldots; \quad q(x) = \ldots\ldots\ldots; \quad r(x) = \ldots\ldots\ldots;$$
$$\alpha_1, \alpha_2 = \ldots\ldots\ldots; \quad \beta_1, \beta_2 = \ldots\ldots\ldots$$

121

$$p(x) = 1 - x^2; \quad q(x) = 0; \quad r(x) = 1; \quad \alpha_1, \alpha_2 = 1, 0; \quad \beta_1, \beta_2 = 1, 0$$

Consequently, Legendre polynomials are mutually orthogonal. That is, if $m \neq n$

$$\int_{-1}^{1} P_m(x) P_n(x)\, \mathrm{d}x = \dots\dots\dots$$

122

$$\int_{-1}^{1} P_m(x) P_n(x)\, \mathrm{d}x = 0$$

Polynomials as a finite series of Legendre polynomials

Many differential equations cannot be solved by the normal analytical means and solution by power series provides a powerful tool in many situations. Furthermore, any polynomial can be written as a finite series of Legendre polynomials.

Example 1

Show that $f(x) = x^2$ can be written as a series of Legendre polynomials. Assume that

$$f(x) = x^2 = \sum_{n=0}^{\infty} a_n P_n(x), \text{ then}$$

$$x^2 = a_0 P_0(x) + a_1 P_1(x) + a_2 P_2(x) + \dots$$

$$= a_0(1) + a_1(x) + a_2 \frac{3x^2 - 1}{2} + a_3 \frac{5x^3 - 3x}{2} + \dots$$

Since the left-hand side is a polynomial of degree 2 then any Legendre polynomial on the right-hand side containing powers of x greater than 2 must be excluded. Therefore $a_3 = a_4 = \dots = 0$, so that

$$x^2 = a_0 - \frac{a_2}{2} + a_1 x + \frac{3}{2} a_2 x^2 \quad \text{giving} \quad a_2 = \frac{2}{3}, \, a_1 = 0, \, a_0 - \frac{a_2}{2} = 0$$

therefore $a_0 = \frac{1}{3}$, and

$$x^2 = \frac{1}{3} P_0(x) + \frac{2}{3} P_2(x)$$

Now you try one

123

Example 2

The polynomial $1 + x + x^3$ can be written as a series of Legendre polynomials in the form

$$1 + x + x^3 = \dots\dots\dots$$

$$1 + x + x^3 = P_0(x) + \frac{8}{5}P_1(x) + \frac{2}{5}P_3(x)$$

Because

$$1 + x + x^3 = a_0P_0(x) + a_1P_1(x) + a_2P_2(x) + \cdots$$

$$= a_0(1) + a_1(x) + a_2\frac{3x^2 - 1}{2} + a_3\frac{5x^3 - 3x}{2} + \cdots$$

Since the left-hand side is a polynomial of degree 3 then any Legendre polynomial on the right-hand side containing powers of x greater than 3 must be excluded. Therefore $a_4 = a_5 = \ldots = 0$, so that

$$1 + x + x^3 = a_0 - \frac{a_2}{2} + \left(a_1 - \frac{3}{2}a_3\right)x + \frac{3}{2}a_2x^2 + \frac{5}{2}a_3x^3$$

This gives $a_3 = \frac{2}{5}$, $a_2 = 0$, $a_1 - \frac{3}{2}a_3 = 1$, $a_0 - \frac{a_2}{2} = 1$ therefore $a_0 = 1$,

and $a_1 = \frac{8}{5}$ so

$$1 + x + x^3 = P_0(x) + \frac{8}{5}P_1(x) + \frac{2}{5}P_3(x)$$

As usual, the main points that we have covered in this Program are listed in the **Review summary** that follows. Read this in conjunction with the **Can You?** checklist and note any sections that may need further attention: refer back to the relevant parts of the Program, if necessary. There will then be no trouble with the **Test exercise**. The set of **Further problems** provides an opportunity for further practice.

Review summary

1 *Higher derivatives*

y	$y^{(n)}$
x^a	$\dfrac{a!}{(a-n)!}x^{a-n}$
e^{ax}	$a^n e^{ax}$
$\sin ax$	$a^n \sin\left(ax + \dfrac{n\pi}{2}\right)$
$\cos ax$	$a^n \cos\left(ax + \dfrac{n\pi}{2}\right)$
$\sinh ax$	$\dfrac{a^n}{2}\{[1 + (-1)^n]\sinh ax + [1 - (-1)^n]\cosh ax\}$
$\cosh ax$	$\dfrac{a^n}{2}\{[1 - (-1)^n]\sinh ax + [1 + (-1)^n]\cosh ax\}$

▶

2   *Leibnitz theorem* — *n*th derivative of a product of functions.
If $y = uv$

$$y^{(n)} = u^{(n)}v + nu^{(n-1)}v^{(1)} + \frac{n(n-1)}{2!}u^{(n-2)}v^{(2)}$$
$$+ \frac{n(n-1)(n-2)}{3!}u^{(n-3)}v^{(3)} + \ldots$$
$$\ldots + \frac{n(n-1)(n-2)\ldots(n-r+1)}{r!}u^{(n-r)}v^{(r)} + \ldots$$

i.e. $y^{(n)} = \sum_{r=0}^{\infty} {}^nC_r u^{(n-r)}v^{(r)}$.

$(uv)^{(n)}$ can be obtained by expanding $(u+v)^{(n)}$ using the binomial theorem where the 'powers' are interpreted as derivatives.

3   *Power series solution of second-order differential equations*

(a) *Leibnitz–Maclaurin method*
   (1) Differentiate the equation n times by the Leibnitz theorem.
   (2) Put $x = 0$ to establish a recurrence relation.
   (3) Substitute $n = 1, 2, 3, \ldots$ to obtain $y', y'', y''', \ldots$ at $x = 0$.
   (4) Substitute in Maclaurin's series and simplify where possible.

(b) *Frobenius' method*

Assume a series solution of the form
$$y = x^c\{a_0 + a_1 x + a_2 x^2 + \ldots + a_r x^r + \ldots\} \quad a_0 \neq 0$$
   (1) Differentiate the assumed series to find y' and y''.
   (2) Substitute in the equation.
   (3) Equate coefficients of corresponding powers of x on each side of the equation – usually written with zero on the right-hand side.
   (4) Coefficient of the lowest power of x gives the *indicial equation* from which values of c are obtained, $c = c_1$ and $c = c_2$.

Case 1: c_1 and c_2 differ by a quantity *not an integer*. Substitute $c = c_1$ and $c = c_2$ in the series for y.

Case 2: c_1 and c_2 differ by an *integer* and make a coefficient *indeterminate* when $c = c_1$. Substitute $c = c_1$ to obtain the complete solution.

Case 3: c_1 and c_2 ($c_1 < c_2$) differ by an *integer* and make a coefficient *infinite* when $c = c_1$. Replace a_0 by $k(c - c_1)$. Two independent solutions then obtained by putting $c = c_1$ in the new series for y and for $\frac{\partial y}{\partial c}$.

In general if $c_1 - c_2 = n$ where n is a non-zero integer, the solution is of the form

$$y = (1 + k \ln x)x^{c_1}\{a_0 + a_1 x + a_2 x^2 + \ldots\}$$
$$+ x^{c_2}\{b_0 + b_1 x + b_2 x^2 + \ldots\}$$

Case 4: c_1 and c_2 are *equal*. Substitute $c = c_1$ in the series for y and for $\dfrac{\partial y}{\partial c}$. Make the substitution after differentiating. The second solution will consist of the product of the first solution and $\ln x$, together with a further series.

In general if $c_1 = c_2 = c$, the solution is of the form

$$y = (1 + k \ln x)x^{c}\{a_0 + a_1 x + a_2 x^2 + \ldots\}$$
$$+ x^{c}\{b_1 x + b_2 x^2 + \ldots\}$$

4 *Bessel's equation*

$$x^2 y'' + xy' + (x^2 - v^2) = 0$$

where v is a real constant.

Bessel functions: Express the two solutions obtained in terms of gamma functions.

$$J_v(x) = \left(\frac{x}{2}\right)^v \left\{\frac{1}{\Gamma(v+1)} - \frac{x^2}{2^2(1!)\Gamma(v+2)} + \frac{x^4}{2^4(2!)\Gamma(v+3)} - \ldots\right\}$$

This is the *Bessel function of the first kind of order v* – valid for v not a negative integer.

Also $J_{-v}(x) = \left(\frac{x}{2}\right)^{-v} \left\{\frac{1}{\Gamma(1-v)} - \frac{x^2}{2(1!)\Gamma(2-v)} + \frac{x^4}{2^2(2!)\Gamma(3-v)} - \ldots\right\}$

provided that v is not a positive integer.

Complete solution is therefore $y = AJ_v(x) + BJ_{-v}(x)$.

When $v = n$ (an integer) $\qquad J_{-n}(x) = (-1)^n J_n(x)$

$$J_n(x) = \left(\frac{x}{2}\right)^n \left\{\frac{1}{n!} - \frac{1}{(n+1)!}\left(\frac{x}{2}\right)^2 + \frac{1}{(2!)(n+2)!}\left(\frac{x}{2}\right)^4\right.$$
$$\left. - \frac{1}{(3!)(n+3)!}\left(\frac{x}{2}\right)^6 + \ldots\right\}$$

In particular

$$J_0(x) = 1 - \frac{1}{(1!)^2}\left(\frac{x}{2}\right)^2 + \frac{1}{(2!)^2}\left(\frac{x}{2}\right)^4 - \frac{1}{(3!)^2}\left(\frac{x}{2}\right)^6 + \ldots$$

and

$$J_1(x) = \frac{x}{2}\left\{1 - \frac{1}{(1!)(2!)}\left(\frac{x}{2}\right)^2 + \frac{1}{(2!)(3!)}\left(\frac{x}{2}\right)^4 - \frac{1}{(3!)(4!)}\left(\frac{x}{2}\right)^6 + \ldots\right\}$$

▶

5   *Legendre's equation*

$$(1 - x^2)y'' - 2xy' + k(k+1)y = 0$$

where k is a real constant.

Solution by Frobenius gives

$$c = 0: \quad y = a_0\left\{1 - \frac{k(k+1)}{2!}x^2 + \frac{k(k-2)(k+1)(k+3)}{4!}x^4 - \dots\right\}$$

$$c = 1: \quad y = a_1\left\{x - \frac{(k-1)(k+2)}{3!}x^3 \right.$$

$$\left. + \frac{(k-1)(k-3)(k+2)(k+4)}{5!}x^5 - \dots\right\}$$

When *k is an integer*, one series terminates. The resulting polynomial in x, $P_n(x)$, is a *Legendre polynomial*, with a_0 or a_1 being chosen so that the polynomial has unit value when $x = 1$.

6   *Rodrigue's formula*

$$P_n(x) = \frac{1}{2^n n!}\frac{\mathrm{d}^n}{\mathrm{d}x^n}\left(x^2 - 1\right)^n$$

Generating function

$$\frac{1}{\sqrt{1 - 2xt + t^2}} = \sum_{n=0}^{\infty} P_n(x)t^n$$

7   *Sturm–Liouville systems*

$(p(x)y')' + (q(x) + \lambda r(x))y = 0$ for $a \leq x \leq b$ and $r(x) > 0$ with boundary conditions $\alpha_1 y(a) + \alpha_2 y'(a) = 0$ and $\beta_1 y(b) + \beta_2 y(b) = 0$

Solutions y_n to a Sturm–Liouville system are called eigenvectors, each corresponding to an eigenvalue λ_n for $n = 0, 1, 2, \dots$

8   *Orthogonality*

If two different functions $f(x)$ and $g(x)$ are defined on the interval $a \leq x \leq b$ and

$$\int_a^b f(x)g(x)\,\mathrm{d}x = 0$$

then the two functions are **orthogonal** to each other. If a function $w(x) > 0$ exists such that

$$\int_a^b f(x)g(x)w(x)\,\mathrm{d}x = 0$$

then $f(x)$ and $g(x)$ are orthogonal to each other *with respect to the weight function $w(x)$*.

The solutions of a Sturm–Liouville system y_n are mutually orthogonal with respect to the weighting $r(x)$. That is

$$\int_a^b y_m(x)y_n(x)r(x)\,\mathrm{d}x = 0 \qquad (m \neq n)$$

▶

9　*Legendre polynomials are mutually orthogonal*
　If $m \neq n$ then

$$\int_{-1}^{1} P_m(x)P_n(x)\, dx = 0$$

The orthogonality of the Legendre polynomials permits any polynomial to be written as a finite series of Legendre polynomials.

☑ Can You?

Checklist 5

126

Check this list before and after you try the end of Program test.

On a scale of 1 to 5 how confident are you that you can:　　Frames

- Obtain the nth derivative of the exponential and circular and hyperbolic functions?　　　1 to 9
  Yes ☐ ☐ ☐ ☐ ☐ *No*

- Apply the Leibnitz theorem to derive the nth derivative of a product of expressions?　　　10 to 17
  Yes ☐ ☐ ☐ ☐ ☐ *No*

- Apply the Leibnitz–Maclaurin method of obtaining a series solution to a second-order homogeneous differential equation with constant coefficients?　　　18 to 36
  Yes ☐ ☐ ☐ ☐ ☐ *No*

- Apply Frobenius' method of obtaining a series solution to a second-order homogeneous differential equation for different cases of the indicial equation?　　　37 to 87
  Yes ☐ ☐ ☐ ☐ ☐ *No*

- Apply Frobenius' method to Bessel's equation to derive Bessel functions of the first kind?　　　88 to 104
  Yes ☐ ☐ ☐ ☐ ☐ *No*

- Apply Frobenius' method to Legendre's equation to derive Legendre polynomials?　　　104 to 107
  Yes ☐ ☐ ☐ ☐ ☐ *No*

- Use Rodrigue's formula to derive Legendre polynomials and the generating function to obtain some of their properties?　　　108 to 111
  Yes ☐ ☐ ☐ ☐ ☐ *No*

▶

- Recognise a Sturm–Liouville system and the orthogonality properties of its eigenfunctions?

  Yes ☐ ☐ ☐ ☐ ☐ *No* $\boxed{112}$ to $\boxed{121}$

- Write a polynomial in x as a finite series of Legendre polynomials?

  Yes ☐ ☐ ☐ ☐ ☐ *No* $\boxed{122}$ to $\boxed{124}$

🚲 Test exercise 5

$\boxed{127}$

1 If $y = e^{x^2+x}$, show that $y'' = y'(2x+1) + 2y$ and hence prove that $y^{(n+2)} = (2x+1)y^{(n+1)} + 2(n+1)y^{(n)}$.

2 Obtain a power series solution of the equation
$(1 + x^2)y'' - 3xy' - 5y = 0$
up to and including the term in x^6.

3 Determine a series solution for each of the following.
(a) $3xy'' + 2y' + y = 0$
(b) $y'' + x^2y = 0$
(c) $xy'' + 3y' - y = 0$.

4 Use Rodrigue's formula $P_n(x) = \dfrac{1}{2^n n!} \dfrac{d^n}{dx^n} (x^2 - 1)^n$ to derive the Legendre polynomials $P_2(x)$ and $P_3(x)$, and show that $P_2(x)$ and $P_3(x)$ are orthogonal on $(-1, 1)$.

5 Write $f(x) = 1 - 2x^2$ as a series of Legendre polynomials.

🚵 Further problems 5

$\boxed{128}$

(a) *Use the Leibnitz theorem* for the following.
1 If $y = x^3 e^{4x}$, determine $y^{(5)}$.
2 Find the nth derivative of $y = x^3 e^{-x}$ for $n > 3$.
3 If $y = x^3(2x + 1)^2$, find $y^{(4)}$.
4 Find the 6th derivative of $y = x^4 \cos x$.
5 If $y = e^{-x} \sin x$, obtain an expression for $y^{(4)}$.
6 Determine $y^{(3)}$ when $y = x^4 \ln x$.

▶

7 If $x^2y'' + xy' + y = 0$, show that

$$x^2y^{(n+2)} + (2n+1)xy^{(n+1)} + (n^2+1)y^{(n)} = 0.$$

8 If $y = (2x - \pi)^4 \sin\left(\dfrac{x}{2}\right)$, evaluate $y^{(6)}$ when $x = \pi/2$.

9 If $y = e^{-x}\cos x$, show that $y^{(4)} + 4y = 0$.

10 Find the $(2n)$th derivative of (a) $y = x^2 \sinh x$
(b) $y = x^3 \cosh x$.

11 If $y = (x^3 + 3x^2)e^{2x}$, determine an expression for $y^{(6)}$.

12 Find the nth derivative of $y = e^{-ax}\cos ax$ and hence determine $y^{(3)}$.

13 If $y = \dfrac{\sin x}{1 - x^2}$, show that

(a) $(1 - x^2)y'' - 4xy' - (1 + x^2)y = 0$
(b) $y^{(n+2)} - (n^2 + 3n + 1)y^{(n)} - n(n-1)y^{(n-2)} = 0$ at $x = 0$.

(b) *Use the Leibnitz–Maclaurin method* to determine series solutions for the following.

14 $(1 + x^2)y'' + xy' - 9y = 0$.

15 $(x + 1)y'' + (x - 1)y' - 2y = 0$.

16 $(1 - x^2)y'' - 7xy' - 9y = 0$.

17 $(1 - x^2)y'' - 2xy' + 2y = 0$.

18 $xy'' + y' + 2xy = 0$.

(c) *Use the method of Frobenius* to obtain series solutions of the following.

19 $3xy'' + y' - y = 0$.

20 $y'' + y = 0$.

21 $y'' - xy = 0$.

22 $3xy'' + 4y' + y = 0$.

23 $y'' - xy' + y = 0$.

24 $xy'' - 3y' + y = 0$.

25 $xy'' + y' - 3y = 0$.

▶

26 Verify that $y'' + \lambda y = 0$ where $y'(0) = 0$ and $y(2) = 0$ is a Sturm–Liouville system. Find the eigenvalues and eigenfunctions of the system and prove that they are orthogonal in $(0, 2)$.

27 Series solutions of the equation $y'' - 2xy' + 2ny = 0$ are known as Hermite polynomials, $H_n(x)$, where

$$H_n(x) = (-1)^n e^{x^2} \frac{d^n}{dx^n}\left(e^{-x^2}\right)$$

Derive the first four Hermite polynomials and show that they are orthogonal with respect to the weight e^{-x^2} in $(-\infty, \infty)$.

28 Series solutions of the equation $xy'' + (1 - x)y' + ny = 0$ are known as Laguerre polynomials, $L_n(x)$, where

$$L_n(x) = e^x \frac{d^n}{dx^n}(x^n e^{-x})$$

Derive the first four Laguerre polynomials and show that they are orthogonal with respect to the weight e^{-x} in $(0, \infty)$.

29 Given the generating function for Laguerre polynomials $L_n(x)$ as

$$\frac{e^{-xt/(1-t)}}{1-t} = \sum_{n=0}^{\infty} \frac{L_n(x)}{n!} t^n$$

show that $L_n(0) = n!$

30 Given the generating function for Hermite polynomials $H_n(x)$ as

$$e^{2tx-t^2} = \sum_{n=0}^{\infty} \frac{H_n(x)}{n!} t^n$$

show that $H_{2n+1}(0) = 0$.

31 Given the generating function for Legendre polynomials $P_n(x)$ as

$$\frac{1}{\sqrt{1 - 2xt + t^2}} = \sum_{n=0}^{\infty} P_n(x)t^n$$

show that $P_{2n+1}(0) = 0$.

Introduction to Laplace transforms

Learning outcomes

When you have completed this Program you will be able to:

- Derive the Laplace transform of an expression by using the integral definition
- Obtain inverse Laplace transforms with the help of a Table of Laplace transforms
- Derive the Laplace transform of the derivative of an expression
- Solve first-order, constant-coefficient, inhomogeneous differential equations using the Laplace transform
- Derive further Laplace transforms from known transforms
- Use the Laplace transform to obtain the solution to linear, constant-coefficient, inhomogeneous differential equations of second and higher order

The Laplace transform

1

All the differential equations you have looked at so far have had solutions containing a number of unknown integration constants A, B, C etc. The values of these constants have then been found by applying boundary conditions to the solution, a procedure that can often prove to be tedious. Fortunately, for a certain type of differential equation there is a method of obtaining the solution where these unknown integration constants are evaluated *during the process of solution*. Furthermore, rather than employing integration as the way of unravelling the differential equation, you use straightforward algebra.

The method hinges on what is called the *Laplace transform*. If $f(x)$ represents some expression in x defined for $x \geq 0$, the *Laplace transform* of $f(x)$, denoted by $L\{f(x)\}$, is defined to be:

$$L\{f(x)\} = \int_{x=0}^{\infty} e^{-sx} f(x)\, dx$$

where s is a variable whose values are chosen so as to ensure that the semi-infinite integral converges. More will be said about the variable s in Frame 3. For now, what would you say is the Laplace transform $f(x) = 2$ for $x \geq 0$?

Substitute for $f(x)$ in the integral above and then perform the integration.
The answer is in the next frame

2

$$L\{2\} = \frac{2}{s} \text{ provided } s > 0$$

Because:

$$L\{f(x)\} = \int_{x=0}^{\infty} e^{-sx} f(x)\, dx$$

so

$$L\{2\} = \int_{x=0}^{\infty} e^{-sx} 2\, dx$$
$$= 2\left[\frac{e^{-sx}}{-s}\right]_{x=0}^{\infty}$$
$$= 2(0 - (-1/s))$$
$$= \frac{2}{s}$$

Notice that $s > 0$ is demanded because if $s < 0$ then $e^{-sx} \to \infty$ as $x \to \infty$ and if $s = 0$ then $L\{2\}$ is not defined (in both of these two cases the integral diverges), so that

$$L\{2\} = \frac{2}{s} \text{ provided } s > 0$$

▶

By the same reasoning, if k is some constant then

$$L\{k\} = \frac{k}{s} \text{ provided } s > 0$$

Now, how about the Laplace transform of $f(x) = e^{-kx}$, $x \geq 0$ where k is a constant?

Go back to the integral definition and work it out.
Again, the answer is in the next frame

3

$$L\{e^{-kx}\} = \frac{1}{s+k} \text{ provided } s > -k$$

Because

$$L\{e^{-kx}\} = \int_{x=0}^{\infty} e^{-sx} e^{-kx} \, dx$$

$$= \int_{x=0}^{\infty} e^{-(s+k)x} \, dx$$

$$= \left[\frac{e^{-(s+k)x}}{-(s+k)} \right]_{x=0}^{\infty}$$

$$= \left(0 - \left(-\frac{1}{(s+k)} \right) \right) \quad \begin{array}{l} s+k > 0 \text{ is demanded to ensure that the} \\ \text{integral converges at both limits} \end{array}$$

$$= \frac{1}{(s+k)} \quad \text{provided } s+k > 0, \text{ that is provided } s > -k$$

These two examples have demonstrated that you need to be careful about the finite existence of the Laplace transform and not just take the integral definition without some thought. For the Laplace transform to exist the integrand

$$e^{-sx} f(x)$$

must converge to zero as $x \to \infty$ and this will impose some conditions on the values of s for which the integral does converge and, hence, the Laplace transform exists. In this Program you can be assured that there are no problems concerning the existence of any of the Laplace transforms that you will meet.

Move on to the next frame

The inverse Laplace transform

4

The Laplace transform is an expression in the variable s which is denoted by $F(s)$. It is said that $f(x)$ and $F(s) = L\{f(x)\}$ form a *transform pair*. This means that if $F(s)$ is the *Laplace transform* of $f(x)$ then $f(x)$ is the *inverse Laplace transform* of $F(s)$. We write:

$$f(x) = L^{-1}\{F(s)\}$$

▶

There is no simple integral definition of the inverse transform so you have to find it by working backwards. For example:

if $f(x) = 4$ then the Laplace transform $L\{f(x)\} = F(s) = \dfrac{4}{s}$

so

if $F(s) = \dfrac{4}{s}$ then the inverse Laplace transform $L^{-1}\{F(s)\} = f(x) = 4$

It is this ability to find the Laplace transform of an expression and then reverse it that makes the Laplace transform so useful in the solution of differential equations, as you will soon see.

For now, what is the inverse Laplace transform of $F(s) = \dfrac{1}{s-1}$?

To answer this, look at the Laplace transforms you now know.
The answer is in the next frame

5

$$L^{-1}\{F(s)\} = f(x) = e^x$$

Because you know that:

$L\{e^{-kx}\} = \dfrac{1}{s+k}$ you can say that $L^{-1}\left\{\dfrac{1}{s+k}\right\} = e^{-kx}$

so when $k = -1$, $L^{-1}\left\{\dfrac{1}{s-1}\right\} = e^{-(-1)x} = e^x$

To assist in the process of finding Laplace transforms and their inverses a table is used. In the next frame is a short table containing what you know to date.

6   **Table of Laplace transforms**

$f(x) = L^{-1}\{F(s)\}$	$F(s) = L\{f(x)\}$	
k	$\dfrac{k}{s}$	$s > 0$
e^{-kx}	$\dfrac{1}{s+k}$	$s > -k$

Reading the table from left to right gives the Laplace transform and reading the table from right to left gives the inverse Laplace transform.

*Use these, where possible, to answer the questions in the **Review exercise** that follows. Otherwise use the basic definition in Frame 1.*

Review summary

7

1   The *Laplace transform* of $f(x)$, denoted by $L\{f(x)\}$, is defined to be:

$$L\{f(x)\} = \int_{x=0}^{\infty} e^{-sx} f(x)\,dx$$

where s is a variable whose values are chosen so as to ensure that the semi-infinite integral converges.

2   If $F(s)$ is the *Laplace transform* of $f(x)$ then $f(x)$ is the *inverse Laplace transform* of $F(s)$. We write:

$$f(x) = L^{-1}\{F(s)\}$$

There is no simple integral definition of the inverse transform so you have to find it by working backwards using a *Table of Laplace transforms*.

Review exercise

8

1   Find the Laplace transform of each of the following. In each case $f(x)$ is defined for $x \geq 0$:

   (a) $f(x) = -3$ (b) $f(x) = e$ (c) $f(x) = e^{2x}$

   (d) $f(x) = -5e^{-3x}$ (e) $f(x) = 2e^{7x-2}$

2   Find the inverse Laplace transform of each of the following:

   (a) $F(s) = -\dfrac{1}{s}$ (b) $F(s) = \dfrac{1}{s-5}$ (c) $F(s) = \dfrac{3}{s+2}$

   (d) $F(s) = -\dfrac{3}{4s}$ (e) $F(s) = \dfrac{1}{2s-3}$

Solutions in next frame

9

1   (a) $f(x) = -3$

   Because $L\{k\} = \dfrac{k}{s}$ provided $s > 0$, $L\{-3\} = -\dfrac{3}{s}$ provided $s > 0$

   (b) $f(x) = e$

   Because $L\{k\} = \dfrac{k}{s}$ provided $s > 0$, $L\{e\} = \dfrac{e}{s}$ provided $s > 0$

   (c) $f(x) = e^{2x}$

   Because $L\{e^{-kx}\} = \dfrac{1}{s+k}$ provided $s > -k$, $L\{e^{2x}\} = \dfrac{1}{s-2}$ provided $s > 2$

   (d) $f(x) = -5e^{-3x}$

   $$L\{-5e^{-3x}\} = \int_{x=0}^{\infty} e^{-sx}(-5e^{-3x})\,dx = -5\int_{x=0}^{\infty} e^{-sx}e^{-3x}\,dx = -5L\{e^{-3x}\}$$

   $$L\{-5e^{-3x}\} = -\dfrac{5}{s+3} \text{ provided } s > -3$$

   (e) $f(x) = 2e^{7x-2}$

   $$L\{2e^{7x-2}\} = \int_{x=0}^{\infty} e^{-sx}(2e^{7x-2})\,dx = 2e^{-2}\int_{x=0}^{\infty} e^{-sx}e^{7x}\,dx = 2e^{-2}L\{e^{7x}\}$$

   $$L\{2e^{7x-2}\} = \dfrac{2e^{-2}}{s-7} \text{ provided } s > 7$$

▶

2 (a) $F(s) = -\dfrac{1}{s}$

Because $L^{-1}\left\{\dfrac{k}{s}\right\} = k$, $L^{-1}\left\{-\dfrac{1}{s}\right\} = L^{-1}\left\{\dfrac{-1}{s}\right\} = -1$

(b) $F(s) = \dfrac{1}{s-5}$

Because $L^{-1}\left\{\dfrac{1}{s+k}\right\} = e^{-kx}$, $L^{-1}\left\{\dfrac{1}{s-5}\right\} = e^{-(-5)x} = e^{5x}$

(c) $F(s) = \dfrac{3}{s+2}$

Because $L^{-1}\left\{\dfrac{1}{s+2}\right\} = e^{-2x}$ and $L\{3e^{-2x}\} = 3L\{e^{-2x}\} = \dfrac{3}{s+2}$ so

$L^{-1}\left\{\dfrac{3}{s+2}\right\} = 3e^{-2x}$

(d) $F(s) = -\dfrac{3}{4s}$

$F(s) = -\dfrac{3}{4s} = \dfrac{(-3/4)}{s}$ so that $L^{-1}\left\{-\dfrac{3}{4s}\right\} = L^{-1}\left\{\dfrac{-3/4}{s}\right\} = -3/4$

(e) $F(s) = \dfrac{1}{2s-3}$

$F(s) = \dfrac{1}{2s-3} = \dfrac{\frac{1}{2}}{s-\frac{3}{2}}$ so that $f(x) = L^{-1}\left\{\dfrac{1}{2s-3}\right\} = L^{-1}\left\{\dfrac{\frac{1}{2}}{s-\frac{3}{2}}\right\} = \dfrac{1}{2}e^{\frac{3}{2}x}$

Next frame

10 Laplace transform of a derivative

Before you can use the Laplace transform to solve a differential equation you need to know the Laplace transform of a derivative. Given some expression $f(x)$ with Laplace transform $L\{f(x)\} = F(s)$, the Laplace transform of the derivative $f'(x)$ is:

$$L\{f'(x)\} = \int_{x=0}^{\infty} e^{-sx}f'(x)\,dx$$

This can be integrated by parts as follows:

$$L\{f'(x)\} = \int_{x=0}^{\infty} e^{-sx}f'(x)\,dx$$
$$= \int_{x=0}^{\infty} u(x)dv(x)$$
$$= \left[u(x)v(x)\right]_{x=0}^{\infty} - \int_{x=0}^{\infty} v(x)du(x) \quad \text{(This is called the Parts formula)}$$

where $u(x) = e^{-sx}$ so $du(x) = -se^{-sx}dx$ and where $dv(x) = f'(x)dx$ so $v(x) = f(x)$.

Therefore, substitution in the Parts formula gives:

$$L\{f'(x)\} = \left[e^{-sx}f(x)\right]_{x=0}^{\infty} + s\int_{x=0}^{\infty} e^{-sx}f(x)\mathrm{d}x$$
$$= (0 - f(0)) + sF(s) \text{ assuming } e^{-sx}f(x) \to 0 \text{ as } x \to \infty$$

That is:

$$L\{f'(x)\} = sF(s) - f(0)$$

So the Laplace transform of the derivative of $f(x)$ is given in terms of the Laplace transform of $f(x)$ itself and the value of $f(x)$ when $x = 0$. Before you use this fact just consider two properties of the Laplace transform in the next frame.

Two properties of Laplace transforms

11

Both the Laplace transform and its inverse are *linear transforms*, by which is meant that:

(1) *The transform of a sum (or difference) of expressions is the sum (or difference) of the individual transforms. That is:*

$$L\{f(x) \pm g(x)\} = L\{f(x)\} \pm L\{g(x)\}$$
$$\text{and} \quad L^{-1}\{F(s) \pm G(s)\} = L^{-1}\{F(s)\} \pm L^{-1}\{G(s)\}$$

(2) *The transform of an expression that is multiplied by a constant is the constant multiplied by the transform of the expression. That is:*

$$L\{kf(x)\} = kL\{f(x)\} \text{ and } L^{-1}\{kF(s)\} = kL^{-1}\{F(s)\} \text{ where } k \text{ is a constant}$$

These are easily proved using the basic definition of the Laplace transform in Frame 1.

Armed with this information let's try a simple differential equation. By using

$$L\{f'(x)\} = sF(s) - f(0)$$

take the Laplace transform of both sides of the equation

$$f'(x) + f(x) = 1 \text{ where } f(0) = 0$$

and find an expression for the Laplace transform $F(s)$.

Work through this steadily using what you know;
you will find the answer in Frame 12

12

$$F(s) = \frac{1}{s(s+1)}$$

Because, taking Laplace transforms of both sides of the equation you have that:

$$L\{f'(x) + f(x)\} = L\{1\}$$ The Laplace transform of the left-hand side equals the Laplace transform of the right-hand side

That is:

$$L\{f'(x)\} + L\{f(x)\} = L\{1\}$$ The transform of a sum is the sum of the transforms.

From what you know about the Laplace transform of $f(x)$ and its derivative $f'(x)$ this gives:

$$[sF(s) - f(0)] + F(s) = \frac{1}{s}$$

That is:

$$(s+1)F(s) - f(0) = \frac{1}{s} \quad \text{and you are given that } f(0) = 0 \text{ so}$$

$$(s+1)F(s) = \frac{1}{s}, \text{ that is } F(s) = \frac{1}{s(s+1)}$$

Well done. Now, separate the right-hand side into partial fractions.

The answer is in Frame 13

13

$$F(s) = \frac{1}{s} - \frac{1}{s+1}$$

Because

Assume that $\dfrac{1}{s(s+1)} = \dfrac{A}{s} + \dfrac{B}{s+1}$ then, $1 = A(s+1) + Bs$ from which you

find that $A = 1$ and $B = -1$ so that $F(s) = \dfrac{1}{s} - \dfrac{1}{s+1}$

That was straightforward enough. Now take the inverse Laplace transform and find the solution to the differential equation.

The answer is in Frame 14

14

$$\boxed{f(x) = 1 - e^{-x}}$$

Because

$$f(x) = L^{-1}\{F(s)\}$$

$$= L^{-1}\left\{\frac{1}{s} - \frac{1}{s+1}\right\}$$

$$= L^{-1}\left\{\frac{1}{s}\right\} - L^{-1}\left\{\frac{1}{s+1}\right\} \quad \text{The inverse Laplace transform of a difference is the difference of the inverse transforms}$$

$$= 1 - e^{-x} \quad \text{Using the Table of Laplace transforms in Frame 6}$$

You now have a method for solving a differential equation of the form:

$$af'(x) + bf(x) = g(x) \text{ given that } f(0) = k$$

where a, b and k are known constants and $g(x)$ is a known expression in x:

(a) Take the Laplace transform of both sides of the differential equation
(b) Find the expression $F(s) = L\{f(x)\}$ in the form of an algebraic fraction
(c) Separate $F(s)$ into its partial fractions
(d) Find the inverse Laplace transform $L^{-1}\{F(s)\}$ to find the solution $f(x)$ to the differential equation.

Now you try some but before you do just look at the Table of Laplace transforms in the next frame. You will need them to solve the equations in the **Review exercise** *that follows.*

Table of Laplace transforms

15

$f(x) = L^{-1}\{F(s)\}$	$F(s) = L\{f(x)\}$	
k	$\dfrac{k}{s}$	$s > 0$
e^{-kx}	$\dfrac{1}{s+k}$	$s > -k$
xe^{-kx}	$\dfrac{1}{(s+k)^2}$	$s > -k$

We will derive this third transform later in the Program. For now, use these to answer the questions that follow the **Review summary** *in the next frame*

16 **Review summary**

1 If $F(s)$ is the Laplace transform of $f(x)$ then the Laplace transform of $f'(x)$ is:

$$L\{f'(x)\} = sF(s) - f(0)$$

2 (a) The Laplace transform of a sum (or difference) of expressions is the sum (or difference) of the individual transforms. That is:

$$L\{f(x) \pm g(x)\} = L\{f(x)\} \pm L\{(g(x)\}$$
$$\text{and}\quad L^{-1}\{F(s) \pm G(s)\} = L^{-1}\{F(s)\} \pm L^{-1}\{G(s)\}$$

(b) The transform of an expression multiplied by a constant is the constant multiplied by the transform of the expression. That is:

$$L\{kf(x)\} = kL\{f(x)\} \text{ and } L^{-1}\{kF(s)\} = kL^{-1}\{F(s)\}$$

where k is a constant.

3 To solve a differential equation of the form:

$$af'(x) + bf(x) = g(x) \text{ given that } f(0) = k$$

where a, b and k are known constants and $g(x)$ is a known expression in x:

(a) Take the Laplace transform of both sides of the differential equation
(b) Find the expression $F(s) = L\{f(x)\}$ in the form of an algebraic fraction
(c) Separate $F(s)$ into its partial fractions
(d) Find the inverse Laplace transform $L^{-1}\{F(s)\}$ to find the solution $f(x)$ to the differential equation.

17 **Review exercise**

Solve each of the following differential equations:
(a) $f'(x) - f(x) = 2$ where $f(0) = 0$
(b) $f'(x) + f(x) = e^{-x}$ where $f(0) = 0$
(c) $f'(x) + f(x) = 3$ where $f(0) = -2$
(d) $f'(x) - f(x) = e^{2x}$ where $f(0) = 1$
(e) $3f'(x) - 2f(x) = 4e^{-x} + 2$ where $f(0) = 0$

Solutions in next frame

18

(a) $f'(x) - f(x) = 2$ where $f(0) = 0$

Taking Laplace transforms of both sides of this equation gives:

$$sF(s) - f(0) - F(s) = \frac{2}{s} \text{ so that } F(s) = \frac{2}{s(s-1)} = -\frac{2}{s} + \frac{2}{s-1}$$

The inverse transform then gives the solution as

$$f(x) = -2 + 2e^x = 2(e^x - 1)$$

▶

(b) $f'(x) + f(x) = e^{-x}$ where $f(0) = 0$

Taking Laplace transforms of both sides of this equation gives:

$$sF(s) - f(0) + F(s) = \frac{1}{s+1} \text{ so that } F(s) = \frac{1}{(s+1)^2}$$

The Table of inverse transforms then gives the solution as $f(x) = xe^{-x}$

(c) $f'(x) + f(x) = 3$ where $f(0) = -2$

Taking Laplace transforms of both sides of this equation gives:

$$sF(s) - f(0) + F(s) = \frac{3}{s} \text{ so that}$$

$$F(s) = -\frac{2}{s+1} + \frac{3}{s(s+1)} = \frac{3 - 2s}{s(s+1)} = \frac{3}{s} - \frac{5}{s+1}$$

The inverse transform then gives the solution as $f(x) = 3 - 5e^{-x}$

(d) $f'(x) - f(x) = e^{2x}$ where $f(0) = 1$

Taking Laplace transforms of both sides of this equation gives:

$$sF(s) - f(0) - F(s) = \frac{1}{s-2} \text{ giving } (s-1)F(s) - 1 = \frac{1}{s-2}$$

so that $F(s) = \dfrac{1}{s-1} + \dfrac{1}{(s-1)(s-2)} = \dfrac{1}{s-2}$

The inverse transform then gives the solution as $f(x) = e^{2x}$

(e) $3f'(x) - 2f(x) = 4e^{-x} + 2$ where $f(0) = 0$

Taking Laplace transforms of both sides of this equation gives:

$$3[sF(s) - f(0)] - 2F(s) = \frac{4}{s+1} + \frac{2}{s} = \frac{6s+2}{s(s+1)} \text{ so that}$$

$$F(s) = \frac{6s+2}{s(s+1)(3s-2)} = \frac{27}{5}\left(\frac{1}{3s-2}\right) - \frac{1}{s} - \frac{4}{5}\left(\frac{1}{s+1}\right)$$

$$= \frac{27}{15}\left(\frac{1}{s-\dfrac{2}{3}}\right) - \frac{1}{s} - \frac{4}{5}\left(\frac{1}{s+1}\right)$$

The inverse transform then gives the solution as:

$$f(x) = \frac{9}{5}e^{2x/3} - \frac{4}{5}e^{-x} - 1$$

On now to Frame 19

19   **Generating new transforms**

Deriving the Laplace transform of $f(x)$ often requires you to integrate by parts, sometimes repeatedly. However, because $L\{f'(x)\} = sL\{f(x)\} - f(0)$ you can sometimes avoid this involved process when you know the transform of the derivative $f'(x)$. Take as an example the problem of finding the Laplace transform of the expression $f(x) = x$. Now $f'(x) = 1$ and $f(0) = 0$ so that substituting in the equation:

$$L\{f'(x)\} = sL\{f(x)\} - f(0)$$

gives

$$L\{1\} = sL\{x\} - 0$$

that is

$$\frac{1}{s} = sL\{x\}$$

therefore

$$L\{x\} = \frac{1}{s^2}$$

That was easy enough, so what is the Laplace transform of $f(x) = x^2$?

The answer is in the next frame

20

$$\boxed{\dfrac{2}{s^3}}$$

Because

$$f(x) = x^2, \ f'(x) = 2x \text{ and } f(0) = 0$$

Substituting in

$$L\{f'(x)\} = sL\{f(x)\} - f(0)$$

gives

$$L\{2x\} = sL\{x^2\} - 0$$

that is

$$2L\{x\} = sL\{x^2\} \text{ so } \frac{2}{s^2} = sL\{x^2\}$$

therefore

$$L\{x^2\} = \frac{2}{s^3}$$

Just try another one. Verify the third entry in the Table of Laplace transforms in Frame 15, that is:

$$L\{xe^{-x}\} = \frac{1}{(s+1)^2}$$

This is a littler harder but just follow the procedure laid out in the previous two frames and try it. The explanation is in the next frame

Because

$$f(x) = xe^{-x}, \ f'(x) = e^{-x} - xe^{-x} \text{ and } f(0) = 0$$

Substituting in

$$L\{f'(x)\} = sL\{f(x)\} - f(0)$$

gives

$$L\{e^{-x} - xe^{-x}\} = sL\{xe^{-x}\} - 0$$

that is

$$L\{e^{-x}\} - L\{xe^{-x}\} = sL\{xe^{-x}\}$$

therefore

$$L\{e^{-x}\} = (s+1)L\{xe^{-x}\}$$

giving

$$\frac{1}{s+1} = (s+1)L\{xe^{-x}\} \text{ and so } L\{xe^{-x}\} = \frac{1}{(s+1)^2}$$

On now to Frame 22

Laplace transforms of higher derivatives

The Laplace transforms of derivatives higher than the first are readily derived. Let $F(s)$ and $G(s)$ be the respective Laplace transforms of $f(x)$ and $g(x)$. That is

$$L\{f(x)\} = F(s) \text{ so that } L\{f'(x)\} = sF(s) - f(0)$$

and

$$L\{g(x)\} = G(s) \text{ and } L\{g'(x)\} = sG(s) - g(0)$$

Now let $g(x) = f'(x)$ so that $L\{g(x)\} = L\{f'(x)\}$ where

$$g(0) = f'(0) \text{ and } G(s) = sF(s) - f(0)$$

Now, because $g(x) = f'(x)$

$$g'(x) = f''(x)$$

This means that

$$L\{g'(x)\} = L\{f''(x)\} = sG(s) - g(0) = s[sF(s) - f(0)] - f'(0)$$

so

$$L\{f''(x)\} = s^2 F(s) - sf(0) - f'(0)$$

By a similar argument it can be shown that

$$L\{f'''(x)\} = s^3 F(s) - s^2 f(0) - sf'(0) - f''(0)$$

and so on. Can you see the pattern developing here?

The Laplace transform of $f^{iv}(x)$ is ....

Next frame

23

$$L\{f^{iv}(x)\} = s^4 F(s) - s^3 f(0) - s^2 f'(0) - s f''(0) - f'''(0)$$

Now, using $L\{f''(x)\} = s^2 F(s) - s f(0) - f'(0)$ the Laplace transform of $f(x) = \sin kx$ where k is a constant is . . . .

Differentiate $f(x)$ twice and follow the procedure that you used in Frames 19 to 21. Take it carefully, the answer and working are in the following frame

24

$$L\{\sin kx\} = \frac{k}{s^2 + k^2}$$

Because

$f(x) = \sin kx,\ f'(x) = k \cos kx$ and $f''(x) = -k^2 \sin kx$.
Also $f(0) = 0$ and $f'(0) = k$.

Substituting in

$L\{f''(x)\} = s^2 F(s) - s f(0) - f'(0)$ where $F(s) = L\{f(x)\}$

gives

$L\{-k^2 \sin kx\} = s^2 L\{\sin kx\} - s \cdot 0 - k$

that is

$-k^2 L\{\sin kx\} = s^2 L\{\sin kx\} - k$

so

$(s^2 + k^2) L\{\sin kx\} = k$ and $L\{\sin kx\} = \dfrac{k}{s^2 + k^2}$

And $L\{\cos kx\} = \ldots \ldots \ldots$

It's just the same method

25

$$L\{\cos kx\} = \frac{s}{s^2 + k^2}$$

Because

$f(x) = \cos kx,\ f'(x) = -k \sin kx$ and $f''(x) = -k^2 \cos kx$.
Also $f(0) = 1$ and $f'(0) = 0$.

Substituting in

$L\{f''(x)\} = s^2 F(s) - s f(0) - f'(0)$ where $F(s) = L\{f(x)\}$

gives

$L\{-k^2 \cos kx\} = s^2 L\{\cos kx\} - s \cdot 1 - 0$

that is

$-k^2 L\{\cos kx\} = s^2 L\{\cos kx\} - s$

so

$(s^2 + k^2) L\{\cos kx\} = s$ and $L\{\cos kx\} = \dfrac{s}{s^2 + k^2}$

The Table of transforms is now extended in the next frame.

Table of Laplace transforms

$f(x) = L^{-1}\{F(s)\}$	$F(s) = L\{f(x)\}$	
k	$\dfrac{k}{s}$	$s > 0$
e^{-kx}	$\dfrac{1}{s+k}$	$s > -k$
xe^{-kx}	$\dfrac{1}{(s+k)^2}$	$s > -k$
x	$\dfrac{1}{s^2}$	$s > 0$
x^2	$\dfrac{2}{s^3}$	$s > 0$
$\sin kx$	$\dfrac{k}{s^2+k^2}$	$s^2+k^2 > 0$
$\cos kx$	$\dfrac{s}{s^2+k^2}$	$s^2+k^2 > 0$

Linear, constant-coefficient, inhomogeneous differential equations

The Laplace transform can be used to solve equations of the form:

$$a_n f^{(n)}(x) + a_{n-1} f^{(n-1)}(x) + \cdots + a_2 f''(x) + a_1 f'(x) + a_0 f(x) = g(x)$$

where $a_n, a_{n-1}, \ldots, a_2, a_1, a_0$ are known constants, $g(x)$ is a known expression in x and the values of $f(x)$ and its derivatives are known at $x = 0$. This type of equation is called a *linear, constant-coefficient, inhomogeneous differential equation* and the values of $f(x)$ and its derivatives at $x = 0$ are called *boundary conditions*. The method of obtaining the solution follows the procedure laid down in Frame 14. For example:

To find the solution of:

$$f''(x) + 3f'(x) + 2f(x) = 4x \text{ where } f(0) = f'(0) = 0$$

(a) *Take the Laplace transform of both sides of the equation*

$$L\{f''(x)\} + 3L\{f'(x)\} + 2L\{f(x)\} = 4L\{x\}$$

to give $\quad [s^2 F(s) - sf(0) - f'(0)] + 3[sF(s) - f(0)] + 2F(s) = \dfrac{4}{s^2}$

▶

(b) *Find the expression $F(s) = L\{f(x)\}$ in the form of an algebraic fraction*

Substituting the values for $f(0)$ and $f'(0)$ and then rearranging gives

$$(s^2 + 3s + 2)F(s) = \frac{4}{s^2}$$

so that

$$F(s) = \frac{4}{s^2(s+1)(s+2)}$$

(c) *Separate $F(s)$ into its partial fractions*

$$\frac{4}{s^2(s+1)(s+2)} = \frac{A}{s} + \frac{B}{s^2} + \frac{C}{s+1} + \frac{D}{s+2}$$

Adding the right-hand side partial fractions together and then equating the left-hand side numerator with the right-hand side numerator gives

$$4 = As(s+1)(s+2) + B(s+1)(s+2) + Cs^2(s+2) + Ds^2(s+1)$$

Let $\quad s = 0 \qquad 4 = 2B$ therefore $B = 2$

$\qquad\quad s = -1 \qquad 4 = C(-1)^2(-1+2) = C$

$\qquad\quad s = -2 \qquad 4 = D(-2)^2(-2+1) = -4D$ therefore $D = -1$

Equate the coefficients of s:

$0 = 2A + 3B = 2A + 6$ therefore $A = -3$

Consequently:

$$F(s) = -\frac{3}{s} + \frac{2}{s^2} + \frac{4}{s+1} - \frac{1}{s+2}$$

(d) *Use the Tables to find the inverse Laplace transform $L^{-1}\{F(s)\}$ and so find the solution $f(x)$ to the differential equation*

$$f(x) = -3 + 2x + 4e^{-x} - e^{-2x}$$

So that was all very straightforward even if it was involved. Now try your hand at the differential equations in the next frame

---

28   **Review summary**

1 If $F(s)$ is the Laplace transform of $f(x)$ then:

$$L\{f''(x)\} = s^2F(s) - sf(0) - f'(0)$$
$$\text{and}\quad L\{f'''(x)\} = s^3F(s) - s^2f(0) - sf'(0) - f''(0)$$

2 Equations of the form:

$$a_nf^{(n)}(x) + a_{n-1}f^{(n-1)}(x) + \cdots + a_2f''(x) + a_1f'(x) + a_0f(x) = g(x)$$

where $a_n, a_{n-1}, \ldots, a_2, a_1, a_0$ are constants are called linear, constant-coefficient, inhomogeneous differential equations.

3 The Laplace transform can be used to solve constant-coefficient, inhomogeneous differential equations provided a_n, $a_{n-1}, \ldots, a_2$, a_1, a_0 are known constants, $g(x)$ is a known expression in x, and the values of $f(x)$ and its derivatives are known at $x = 0$.

4 The procedure for solving these equations of second and higher order is the same as that for solving the equations of first order. Namely:

(a) Take the Laplace transform of both sides of the differential equation

(b) Find the expression $F(s) = L\{f(x)\}$ in the form of an algebraic fraction

(c) Separate $F(s)$ into its partial fractions

(d) Find the inverse Laplace transform $L^{-1}\{F(s)\}$ to find the solution $f(x)$ to the differential equation.

 Review exercise   **29**

Use the Laplace transform to solve each of the following equations:
(a) $f'(x) + f(x) = 3$ where $f(0) = 0$
(b) $3f'(x) + 2f(x) = x$ where $f(0) = -2$
(c) $f''(x) + 5f'(x) + 6f(x) = 2e^{-x}$ where $f(0) = 0$ and $f'(0) = 0$
(d) $f''(x) - 4f(x) = \sin 2x$ where $f(0) = 1$ and $f'(0) = -2$

Answers in next frame

30

(a) $f'(x) + f(x) = 3$ where $f(0) = 0$

Taking Laplace transforms of both sides of the equation gives

$$L\{f'(x)\} + L\{f(x)\} = L\{3\} \text{ so that } [sF(s) - f(0)] + F(s) = \frac{3}{s}$$

That is $(s+1)F(s) = \dfrac{3}{s}$ so $F(s) = \dfrac{3}{s(s+1)} = \dfrac{3}{s} - \dfrac{3}{s+1}$

giving the solution as $f(x) = 3 - 3e^{-x} = 3(1 - e^{-x})$

(b) $3f'(x) + 2f(x) = x$ where $f(0) = -2$

Taking Laplace transforms of both sides of the equation gives

$$L\{3f'(x)\} + L\{2f(x)\} = L\{x\} \text{ so that } 3[sF(s) - f(0)] + 2F(s) = \frac{1}{s^2}$$

That is $(3s+2)F(s) - (-6) = \dfrac{1}{s^2}$ so $F(s) = \dfrac{1 - 6s^2}{s^2(3s+2)}$

The partial fraction breakdown gives

$$F(s) = -\frac{3}{4} \cdot \frac{1}{s} + \frac{1}{2} \cdot \frac{1}{s^2} - \frac{15}{4} \cdot \frac{1}{(3s+2)} = -\frac{3}{4} \cdot \frac{1}{s} + \frac{1}{2} \cdot \frac{1}{s^2} - \frac{5}{4} \cdot \frac{1}{(s+\frac{2}{3})}$$

giving the solution as

$$f(x) = -\frac{3}{4} + \frac{x}{2} - \frac{5e^{-2x/3}}{4}$$

▶

(c) $f''(x) + 5f'(x) + 6f(x) = 2e^{-x}$ where $f(0) = 0$ and $f'(0) = 0$

Taking Laplace transforms of both sides of the equation gives

$$L\{f''(x)\} + L\{5f'(x)\} + L\{6f(x)\} = L\{2e^{-x}\}$$

so that $[s^2F(s) - sf(0) - f'(0)] + 5[sF(s) - f(0)] + 6F(s) = \dfrac{2}{s+1}$

That is $(s^2 + 5s + 6)F(s) = \dfrac{2}{s+1}$

so $F(s) = \dfrac{2}{(s+1)(s+2)(s+3)} = \dfrac{1}{s+1} - \dfrac{2}{s+2} + \dfrac{1}{s+3}$

giving the solution as

$$f(x) = e^{-x} - 2e^{-2x} + e^{-3x}$$

(d) $f''(x) - 4f(x) = \sin 2x$ where $f(0) = 1$ and $f'(0) = -2$

Taking Laplace transforms of both sides of the equation gives

$$L\{f''(x)\} - L\{4f(x)\} = L\{\sin 2x\}$$

so that $[s^2F(s) - sf(0) - f'(0)] - 4F(s) = \dfrac{2}{s^2 + 2^2}$

That is $(s^2 - 4)F(s) - s \cdot 1 - (-2) = \dfrac{2}{s^2 + 2^2}$

so $F(s) = \dfrac{2}{(s^2 - 4)(s^2 + 2^2)} + \dfrac{s - 2}{s^2 - 4}$

$\qquad = \dfrac{15}{16} \cdot \dfrac{1}{s+2} + \dfrac{1}{16} \cdot \dfrac{1}{s-2} - \dfrac{1}{8} \cdot \dfrac{2}{s^2 + 2^2}$

giving the solution as

$$f(x) = \dfrac{15}{16}e^{-2x} + \dfrac{1}{16}e^{2x} - \dfrac{\sin 2x}{8}$$

So, finally, the **Can You?** *checklist followed by the* **Test exercise** *and* **Further exercises**

☑ **Can You?**

Checklist 6

31

Check this list before and after you try the end of Program test.

On a scale of 1 to 5 how confident are you that you can: Frames

- Derive the Laplace transform of an expression by using the integral definition?   **1** to **3**
  Yes ☐ ☐ ☐ ☐ ☐ *No*

- Obtain inverse Laplace transforms with the help of a Table of Laplace transforms?   **4** to **9**
  Yes ☐ ☐ ☐ ☐ ☐ *No*

- Derive the Laplace transform of the derivative of an expression?   **10**
  Yes ☐ ☐ ☐ ☐ ☐ *No*

- Solve first-order, constant-coefficient, inhomogeneous differential equations using the Laplace transform?   **11** to **18**
  Yes ☐ ☐ ☐ ☐ ☐ *No*

- Derive further Laplace transforms from known transforms?   **19** to **26**
  Yes ☐ ☐ ☐ ☐ ☐ *No*

- Use the Laplace transform to obtain the solution to linear, constant-coefficient, inhomogeneous differential equations of higher order than the first?   **27** to **30**
  Yes ☐ ☐ ☐ ☐ ☐ *No*

🚲 **Test exercise 6**

32

1 Using the integral definition, find the Laplace transforms for each of the following:

  (a) $f(x) = 8$ (b) $f(x) = e^{5x}$ (c) $f(x) = -4e^{2x+3}$

2 Using the Table of Laplace transforms, find the inverse Laplace transforms of each of the following:

  (a) $F(s) = -\dfrac{5}{(s-2)^2}$ (b) $F(s) = \dfrac{2e^3}{s^3}$

  (c) $F(s) = \dfrac{3}{s^2+9}$ (d) $F(s) = -\dfrac{2s-5}{s^2+3}$

▶

3 Given that the Laplace transform of xe^{-kx} is $F(s) = \dfrac{1}{(s+k)^2}$ derive the Laplace transform of $x^2 e^{3x}$ without using the integral definition.

4 Use the Laplace transform to solve each of the following equations:
  (a) $f'(x) + 2f(x) = x$ where $f(0) = 0$
  (b) $f'(x) - f(x) = e^{-x}$ where $f(0) = -1$
  (c) $f''(x) + 4f'(x) + 4f(x) = e^{-2x}$ where $f(0) = 0$ and $f'(0) = 0$
  (d) $4f''(x) - 9f(x) = -18$ where $f(0) = 0$ and $f'(0) = 0$

🚴 Further problems 6

33

1 Find the Laplace transform of each of the following expressions (each being defined for $x \geq 0$):
  (a) $f(x) = a^{kx}$, $a > 0$ (b) $f(x) = \sinh kx$ (c) $f(x) = \cosh kx$
  (d) $f(x) = \begin{cases} k & \text{for } 0 \leq x \leq a \\ 0 & \text{for } x > a \end{cases}$

2 Find the inverse Laplace transform of each of the following:
  (a) $F(s) = -\dfrac{2}{3s - 4}$ (b) $F(s) = \dfrac{1}{s^2 - 8}$ (c) $F(s) = \dfrac{3s - 4}{s^2 + 16}$
  (d) $F(s) = \dfrac{7s^2 + 27}{s^3 + 9s}$ (e) $F(s) = \dfrac{4s}{(s^2 - 1)^2}$ (f) $F(s) = -\dfrac{s^2 - 6s + 14}{s^3 - s^2 + 4s - 4}$

3 Show that if $F(s) = L\{f(x)\} = \displaystyle\int_{x=0}^{\infty} e^{-sx} f(x)\,dx$ then:

  (a) (i) $F'(s) = -L\{xf(x)\}$ (ii) $F''(s) = L\{x^2 f(x)\}$
  Use part (a) to find
  (b) (i) $L\{x \sin 2x\}$ (ii) $L\{x^2 \cos 3x\}$
  (c) What would you say the nth derivative of $F(s)$ is equal to?

4 Show that if $L\{f(x)\} = F(s)$ then $L\{e^{kx} f(x)\} = F(s - k)$ where k is a constant. Hence find:
  (a) $L\{e^{ax} \sin bx\}$
  (b) $L\{e^{ax} \cos bx\}$ where a and b are constants in both cases.

▶

5 Solve each of the following differential equations:

(a) $f''(x) - 5f'(x) + 6f(x) = 0$ where $f(0) = 0$ and $f'(0) = 1$

(b) $f''(x) - 5f'(x) + 6f(x) = 1$ where $f(0) = 0$ and $f'(0) = 0$

(c) $f''(x) - 5f'(x) + 6f(x) = e^{2x}$ where $f(0) = 0$ and $f'(0) = 0$

(d) $2f''(x) - f'(x) - f(x) = e^{-3x}$ where $f(0) = 2$ and $f'(0) = 1$

(e) $f(x) + f'(x) - 2f''(x) = xe^{-x}$ where $f(0) = 0$ and $f'(0) = 1$

(f) $f''(x) + 16f(x) = 0$ where $f(0) = 1$ and $f'(0) = 4$

(g) $2f''(x) - f'(x) - f(x) = \sin x - \cos x$ where $f(0) = 0$ and $f'(0) = 0$

Laplace transforms 2

Learning outcomes

When you have completed this Program you will be able to:

- Use the Heaviside unit step function to 'switch' expressions on and off
- Obtain the Laplace transform of expressions involving the Heaviside unit step function

Introduction

1

In the previous Program, we introduced the Laplace transforms of a function $f(t)$ and we tacitly assumed that $f(t)$ was continuous. In practical applications, however, it is convenient to have a function which, in effect, 'switches on' or 'switches off' a given term at pre-described values of t. This we can do with the *Heaviside unit step function*.

Heaviside unit step function

Consider a function that maintains a zero value for all values of t up to $t = c$ and a unit value for $t = c$ and all values of $t \geq c$.

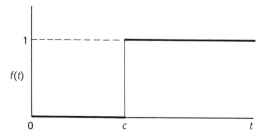

$$f(t) = 0 \quad \text{for } t < c$$
$$f(t) = 1 \quad \text{for } t \geq c$$

This function is the *Heaviside unit step function* and is denoted by

$$f(t) = u(t - c)$$

where the c indicates the value of t at which the function changes from a value of 0 to a value of 1.

Thus, the function

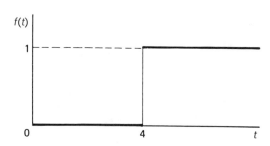

is denoted by $f(t) = \ldots\ldots\ldots\ldots$

$$f(t) = u(t - 4)$$

Similarly, the graph of $f(t) = 2u(t - 3)$ is

$\ldots\ldots\ldots\ldots$

3

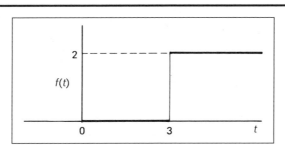

So $u(t-c)$ has just two values

for $t < c$, $u(t-c) = \ldots\ldots\ldots$

for $t \geq c$, $u(t-c) = \ldots\ldots\ldots$

4

$$t < c, \ u(t-c) = 0; \quad t \geq c, \ u(t-c) = 1$$

Unit step at the origin

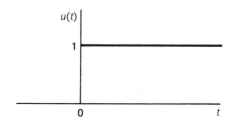

If the unit step occurs at the origin, then $c = 0$ and $f(t) = u(t-c)$ becomes

$$f(t) = u(t)$$

i.e. $u(t) = 0$ for $t < 0$

$u(t) = 1$ for $t \geq 0$.

Effect of the unit step function

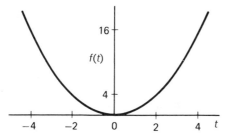

The graph of $f(t) = t^2$ is, of course, as shown.

Remembering the definition of $u(t-c)$, the graph of

$$f(t) = u(t-2) \cdot t^2 \text{ is}$$

$\ldots\ldots\ldots$

5

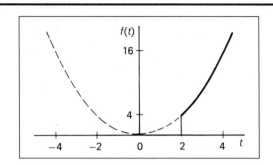

For $t < 2$, $u(t-2) = 0$ $\therefore u(t-2) \cdot t^2 = 0 \cdot t^2 = 0$

$t \geq 2$, $u(t-2) = 1$ $\therefore u(t-2) \cdot t^2 = 1 \cdot t^2 = t^2$

So the function $u(t-2)$ suppresses the function t^2 for all values of t up to $t = 2$ and 'switches on' the function t^2 at $t = 2$.

Now we can sketch the graphs of the following functions.

(a) $f(t) = \sin t$ for $0 < t < 2\pi$

(b) $f(t) = u(t - \pi/4) \cdot \sin t$ for $0 < t < 2\pi$.

These give ....

and ....

6

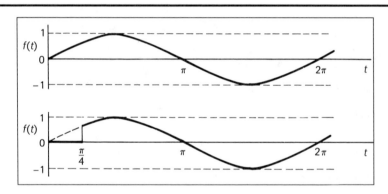

That is, the graph of $f(t) = u(t - \pi/4) \cdot \sin t$ is the graph of $f(t) = \sin t$ but suppressed for all values prior to $t = \pi/4$.

If we sketch the graph of $f(t) = \sin(t - \pi/4)$ we have

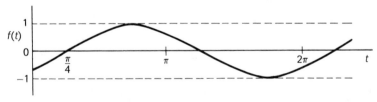

Since $u(t - c)$ has the effect of suppressing a function for $t < c$, then the graph of $f(t) = u(t - \pi/4) \cdot \sin(t - \pi/4)$ is

....

7

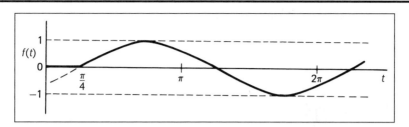

That is, the graph of $f(t) = u(t - \pi/4) \cdot \sin(t - \pi/4)$ is the graph of $f(t) = \sin t$ $(t > 0)$, shifted $\pi/4$ units along the t-axis.

In general, the graph of $f(t) = u(t - c) \cdot \sin(t - c)$ is the graph of $f(t) = \sin t$ $(t > 0)$, shifted along the t-axis through an interval of c units.

Similarly, for $t > 0$, sketch the graphs of

(a) $f(t) = e^{-t}$

(b) $f(t) = u(t - c) \cdot e^{-t}$

(c) $f(t) = u(t - c) \cdot e^{-(t-c)}$

(d) $f(t) = e^{-t}\{u(t - 1) - u(t - 2)\}$.

Arrange the graphs under each other to show the important differences.

8

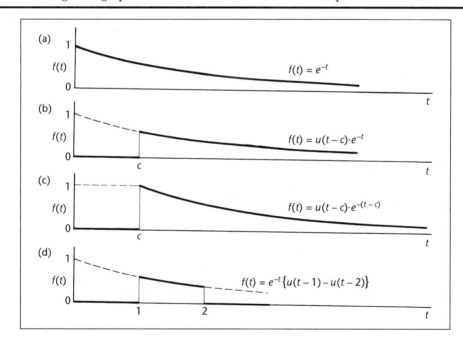

In (a), we have the graph of $f(t) = e^{-t}$
In (b), the same graph is suppressed prior to $t = c$
In (c), the graph of $f(t) = e^{-t}$ is shifted c units along the t-axis
In (d), the graph of $f(t) = e^{-t}$ is turned on at $t = 1$ and off at $t = 2$.

▶

Laplace transform of $u(t - c)$

$$L\{u(t - c)\} = \frac{e^{-cs}}{s}$$

Because

$$L\{u(t - c)\} = \int_0^\infty e^{-st} u(t - c)\, dt$$

but

$$e^{-st} u(t - c) = \begin{cases} 0 & \text{for } 0 < t < c \\ e^{-st} & \text{for } t \ge c \end{cases}$$

so that

$$L\{u(t - c)\} = \int_0^\infty e^{-st} u(t - c)\, dt = \int_c^\infty e^{-st}\, dt$$

$$= \left[\frac{e^{-st}}{-s}\right]_c^\infty = \frac{e^{-sc}}{s}$$

Therefore, the Laplace transform of the unit step at the origin is

$$L\{u(t)\} = \ldots\ldots\ldots\ldots$$

9

$$\boxed{\dfrac{1}{s}}$$

Because $c = 0$.

So

$$L\{u(t - c)\} = \frac{e^{-cs}}{s}$$

and

$$L\{u(t)\} = \frac{1}{s}.$$

Also from the definition of $u(t)$:

$$L(1) = L\{1 \cdot u(t)\}$$

$$L(t) = L\{t \cdot u(t)\}$$

$$L\{f(t)\} = L\{f(t) \cdot u(t)\}$$

Make a note of these results: we shall be using them

As we have seen, the unit step function $u(t-c)$ is often combined with other functions of t, so we now consider the Laplace transform of $u(t-c) \cdot f(t-c)$.

10

Laplace transform of $u(t-c) \cdot f(t-c)$ (the second shift theorem)

$$L\{u(t-c) \cdot (f(t-c)\} = e^{-cs}L\{f(t)\} = e^{-cs}F(s)$$

Because

$$L\{u(t-c) \cdot f(t-c)\} = \int_0^\infty e^{-st}u(t-c) \cdot f(t-c)\,dt$$

$$\text{but } e^{-st}u(t-c) = \begin{cases} 0 & \text{for } 0 < t < c \\ e^{-st} & \text{for } t \geq c \end{cases}$$

so that

$$L\{u(t-c) \cdot f(t-c)\} = \int_c^\infty e^{-st}f(t-c)\,dt$$

We now make the substitution $t - c = v$ so that $t = c + v$ and $dt = dv$. Also for the limits, when $t = c$, $v = 0$ and when $t \to \infty$, $v \to \infty$. Therefore

$$L\{u(t-c) \cdot f(t-c)\} = \int_0^\infty e^{-s(c+v)}f(v)\,dv$$

$$= e^{-cs}\int_0^\infty e^{-sv}f(v)\,dv$$

Now $\int_0^\infty e^{-sv}f(v)\,dv$ has exactly the same value as $\int_0^\infty e^{-st}f(t)\,dt$ which is, of course, the Laplace transform of $f(t)$. Therefore

$$L\{u(t-c) \cdot f(t-c)\} = e^{-cs}L\{f(t)\} = e^{-cs}F(s)$$

11

$$\boxed{L\{u(t-c) \cdot f(t-c)\} = e^{-cs} \cdot F(s) \quad \text{where } F(s) = L\{f(t)\}}$$

So $L\left\{u(t-4) \cdot (t-4)^2\right\} = e^{-4s} \cdot F(s) \quad$ where $F(s) = L\{t^2\}$

$$= e^{-4s}\left(\frac{2!}{s^3}\right) = \frac{2e^{-4s}}{s^3}$$

Note that $F(s)$ is the transform of t^2 and *not* of $(t-4)^2$.

In the same way:

$$L\{u(t-3) \cdot \sin(t-3)\} = \ldots\ldots\ldots\ldots$$

12

$$\boxed{\dfrac{e^{-3s}}{s^2 + 1}}$$

Because $L\{u(t-3) \cdot \sin(t-3)\} = e^{-3s} \cdot F(s)$　where $F(s) = L\{\sin t\}$

$$= \dfrac{1}{s^2 + 1}$$

$\therefore\ L\{u(t-3) \cdot \sin(t-3)\} = e^{-3s}\left(\dfrac{1}{s^2 + 1}\right)$

So now do these in the same way.

(a) $L\left\{u(t-2) \cdot (t-2)^3\right\}$　　$= \ldots\ldots\ldots$

(b) $L\{u(t-1) \cdot \sin 3(t-1)\}$　　$= \ldots\ldots\ldots$

(c) $L\left\{u(t-5) \cdot e^{(t-5)}\right\}$　　$= \ldots\ldots\ldots$

(d) $L\{u(t - \pi/2) \cdot \cos 2(t - \pi/2)\} = \ldots\ldots\ldots$

13

Here they are

(a) $L\left\{u(t-2) \cdot (t-2)^3\right\} = e^{-2s} \cdot F(s)$　where $F(s) = L\{t^3\}$

$$= e^{-2s}\left(\dfrac{3!}{s^4}\right) = \dfrac{6e^{-2s}}{s^4}$$

(b) $L\{u(t-1) \cdot \sin 3(t-1)\} = e^{-s} \cdot F(s)$　where $F(s) = L\{\sin 3t\}$

$$= e^{-s}\left(\dfrac{3}{s^2 + 9}\right) = \dfrac{3e^{-s}}{s^2 + 9}$$

(c) $L\left\{u(t-5) \cdot e^{(t-5)}\right\} = e^{-5s} \cdot F(s)$　where $F(s) = L\{e^t\}$

$$= e^{-5s}\left(\dfrac{1}{s-1}\right) = \dfrac{e^{-5s}}{s-1}$$

(d) $L\{u(t-\pi/2) \cdot \cos 2(t-\pi/2)\} = e^{-\pi s/2} \cdot F(s)$　where $F(s) = L\{\cos 2t\}$

$$= e^{-\pi s/2}\left(\dfrac{s}{s^2 + 4}\right) = \dfrac{s \cdot e^{-\pi s/2}}{s^2 + 4}$$

So $L\{u(t-c) \cdot f(t-c)\} = e^{-cs} \cdot F(s)$　where $F(s) = L\{f(t)\}$.

Written in reverse, this becomes

If $F(s) = L\{f(t)\}$, then $e^{-cs} \cdot F(s) = L\{u(t-c) \cdot f(t-c)\}$

where c is real and positive.

This is known as the *second shift theorem*.

Make a note of it: then we will use it

> If $F(s) = L\{f(t)\}$, then $e^{-cs} \cdot F(s) = L\{u(t - c) \cdot f(t - c)\}$

14

This is useful in finding inverse transforms, as we shall now see.

Example 1

Find the function whose transform is $\dfrac{e^{-4s}}{s^2}$.

The numerator corresponds to e^{-cs} where $c = 4$ and therefore indicates $u(t - 4)$.

Then $\dfrac{1}{s^2} = F(s) = L\{t\}$ $\therefore f(t) = t$.

$\therefore L^{-1}\left\{\dfrac{e^{-4s}}{s^2}\right\} = u(t - 4) \cdot (t - 4)$

Remember that in writing the final result, $f(t)$ is replaced by

. . . .

15

> $f(t - c)$

Example 2

Determine $L^{-1}\left\{\dfrac{6e^{-2s}}{s^2 + 4}\right\}$.

The numerator contains e^{-2s} and therefore indicates . . . .

16

> $u(t - 2)$

The remainder of the transform, i.e. $\dfrac{6}{s^2 + 4}$, can be written as $3\left(\dfrac{2}{s^2 + 4}\right)$

$\therefore \dfrac{6}{s^2 + 4} = F(s) = L\{\ldots\ldots\ldots\ldots\}$

17

> $L\{3 \sin 2t\}$

$\therefore L^{-1}\left\{\dfrac{6e^{-2s}}{s^2 + 4}\right\} = \ldots\ldots\ldots\ldots$

18

$$3u(t-2)\cdot\sin 2(t-2)$$

Because

$$L^{-1}\left\{\frac{6e^{-2s}}{s^2+4}\right\} = u(t-2)\cdot f(t-2) \quad \text{where } f(t)=L^{-1}\left\{\frac{6}{s^2+4}\right\}$$

$$= u(t-2)\cdot 3\sin 2(t-2)$$

Example 3

Determine $L^{-1}\left\{\dfrac{s\cdot e^{-s}}{s^2+9}\right\}$.

This, in similar manner, is ....

19

$$u(t-1)\cdot\cos 3(t-1)$$

Because the numerator contains e^{-s} which indicates $u(t-1)$.
Also $\dfrac{s}{s^2+9}=F(s)=L\{\cos 3t\}$

$$\therefore \ f(t)=\cos 3t \quad \therefore \ f(t-1)=\cos 3(t-1).$$

$$\therefore L^{-1}\left\{\frac{s\cdot e^{-s}}{s^2+9}\right\} = u(t-1)\cdot\cos 3(t-1)$$

Remember that, having obtained $f(t)$, the result contains $f(t-c)$.
Here is a short exercise by way of practice.

Exercise

Determine the inverse transforms of the following.

(a) $\dfrac{2e^{-5s}}{s^3}$

(b) $\dfrac{3e^{-2s}}{s^2-1}$

(c) $\dfrac{8e^{-4s}}{s^2+4}$

(d) $\dfrac{2s\cdot e^{-3s}}{s^2-16}$

(e) $\dfrac{5e^{-s}}{s}$

(f) $\dfrac{s\cdot e^{-s/2}}{s^2+2}$

20

Results – all very straightforward.

(a) $u(t-5)\cdot(t-5)^2$
(b) $3u(t-2)\cdot\sinh(t-2)$
(c) $4u(t-4)\cdot\sin 2(t-4)$
(d) $2u(t-3)\cdot\cosh 4(t-3)$
(e) $5u(t-1)$
(f) $u(t-1/2)\cdot\cos\sqrt{2}(t-1/2)$.

Before looking at a more interesting example, let us collect our results together as far as we have gone.

21

The main points are

(a) $u(t - c) = 0 \qquad 0 < t < c$
$\qquad\qquad = 1 \qquad t \geq c$ } (1)

(b) $L\{u(t - c)\} = \dfrac{e^{-cs}}{s}$

$\qquad L\{u(t)\} = \dfrac{1}{s}$ } (2)

(c) $L\{u(t - c) \cdot f(t - c)\} = e^{-cs} \cdot F(s)$ where $F(s) = L\{f(t)\}$ (3)

(d) If $F(s) = L\{f(t)\}$, then $e^{-cs} \cdot F(s) = L\{u(t - c)\} \cdot f(t - c)\}$ (4)

Now let us apply these to some further examples.

Example 1

Determine the expression $f(t)$ for which

$$L\{f(t)\} = \frac{3}{s} - \frac{4e^{-s}}{s^2} + \frac{5e^{-2s}}{s^2}$$

We take each term in turn and find its inverse transform.

(a) $L^{-1}\left\{\dfrac{3}{s}\right\} = 3L^{-1}\left\{\dfrac{1}{s}\right\} = 3$ i.e. $3u(t)$

(b) $L^{-1}\left\{\dfrac{4e^{-s}}{s^2}\right\} = u(t - 1) \cdot 4(t - 1)$

(c) $L^{-1}\left\{\dfrac{5e^{-2s}}{s^2}\right\} = \ldots\ldots\ldots$

$$\boxed{u(t - 2) \cdot 5(t - 2)}$$

So we have $L^{-1}\left\{\dfrac{3}{s}\right\} = 3u(t)$

$\qquad L^{-1}\left\{\dfrac{4e^{-s}}{s^2}\right\} = u(t - 1) \cdot 4(t - 1)$

$\qquad L^{-1}\left\{\dfrac{5e^{-2s}}{s^2}\right\} = u(t - 2) \cdot 5(t - 2)$

$\qquad \therefore F(t) = 3u(t) - u(t - 1) \cdot 4(t - 1) + u(t - 2) \cdot 5(t - 2)$

To sketch the graph of $f(t)$ we consider the values of the function within the three sections $0 < t < 1$, $1 < t < 2$, and $2 < t$.

$\qquad\qquad$ Between $t = 0$ and $t = 1$, $f(t) = \ldots\ldots\ldots$

$$\boxed{f(t) = 3}$$

Because in this interval, $u(t) = 1$, but $u(t - 1) = 0$ and $u(t - 2) = 0$. In the same way, between $t = 1$ and $t = 2$, $f(t) = \ldots\ldots\ldots$

24

$$f(t) = 7 - 4t$$

Because between $t = 1$ and $t = 2$, $u(t) = 1$, $u(t - 1) = 1$, but $u(t - 2) = 0$.

$\therefore\ f(t) = 3 - 4(t - 1) + 0 = 3 - 4t + 4 = 7 - 4t$

Similarly, for $t > 2$, $f(t) = \dots\dots\dots$

25

$$f(t) = t - 3$$

Because for $t > 2$, $u(t) = 1$, $u(t - 1) = 1$ and $u(t - 2) = 1$

$\therefore\ f(t) = 3 - 4(t - 1) + 5(t - 2)$
$\quad\quad\quad = 3 - 4t + 4 + 5t - 10 = t - 3$

So, collecting the results together, we have

for $\quad 0 < t < 1,\quad f(t) = 3$
$\quad\quad 1 < t < 2,\quad f(t) = 7 - 4t\quad (t = 1, f(t) = 3; t = 2, f(t) = -1)$
$\quad\quad 2 < t,\quad\quad\ f(t) = t - 3\quad (t = 2, f(t) = -1; t = 3, f(t) = 0)$

Using these facts we can sketch the graph of $f(t)$, which is

$\dots\dots\dots$

26

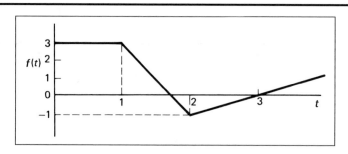

Here is another.

Example 2

Determine the expression $f(t) = L^{-1}\left\{\dfrac{2}{s} + \dfrac{3e^{-s}}{s^2} - \dfrac{3e^{-3s}}{s^2}\right\}$ and sketch the graph of $f(t)$.

First we express the inverse transform of each term in terms of the unit step function.

$\quad\quad\quad\quad\quad\quad\quad$ This gives $\dots\dots\dots$

27

$$L^{-1}\left\{\frac{2}{s}\right\} = 2u(t); \quad L^{-1}\left\{\frac{3e^{-s}}{s^2}\right\} = u(t-1) \cdot 3(t-1)$$
$$L^{-1}\left\{\frac{3e^{-3s}}{s^2}\right\} = u(t-3) \cdot 3(t-3)$$

$$\therefore \ f(t) = 2u(t) + u(t-1) \cdot 3(t-1) - u(t-3) \cdot 3(t-3)$$

So there are 'break points', i.e. changes of function, at $t = 1$ and $t = 3$, and we investigate $f(t)$ within the three intervals.

$$0 < t < 1 \qquad f(t) = \dots\dots\dots$$
$$1 < t < 3 \qquad f(t) = \dots\dots\dots$$
$$3 < t \qquad\quad f(t) = \dots\dots\dots$$

28

$$0 < t < 1, f(t) = 2; \quad 1 < t < 3, f(t) = 3t - 1; \quad 3 < t, f(t) = 8$$

Because with

$$0 < t < 1, \quad u(t) = 1, \text{ but } u(t-1) = u(t-3) = 0 \qquad \therefore \ f(t) = 2$$
$$1 < t < 3, \quad u(t) = 1, u(t-1) = 1, \text{ but } u(t-3) = 0$$
$$\therefore \ f(t) = 2 + 3(t-1) = 3t - 1 \qquad\qquad \therefore \ f(t) = 3t - 1$$
$$3 < t, \qquad u(t) = 1, u(t-1) = 1, u(t-3) = 1$$
$$\therefore \ f(t) = 2 + 3t - 3 - 3t + 9 \qquad\qquad \therefore \ f(t) = 8$$

Therefore, the graph of $f(t)$ is $\dots\dots\dots$

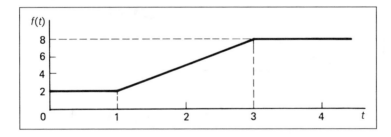

Between the break points, $f(t) = 3t - 1$ $\quad \begin{cases} t = 1, f(t) = 2 \\ t = 3, f(t) = 8 \end{cases}$

Now move on for the next example

Example 3

If $f(t) = L^{-1}\left\{\dfrac{\left(1 - e^{-2s}\right)\left(1 + e^{-4s}\right)}{s^2}\right\}$, determine $f(t)$ and sketch the graph of the function.

Although at first sight this looks more complicated, we simply multiply out the numerator and proceed as before. ▶

$$f(t) = L^{-1}\left\{\frac{1 - e^{-2s} + e^{-4s} - e^{-6s}}{s^2}\right\}$$

$$= L^{-1}\left\{\frac{1}{s^2} - \frac{e^{-2s}}{s^2} + \frac{e^{-4s}}{s^2} - \frac{e^{-6s}}{s^2}\right\}$$

We now write down the inverse transform of each term in terms of the unit function, so that

$$f(t) = \ldots\ldots\ldots\ldots$$

31

$$\boxed{f(t) = u(t) \cdot t - u(t - 2) \cdot (t - 2) + u(t - 4) \cdot (t - 4) - u(t - 6) \cdot (t - 6)}$$

and we can see there are break points at $t = 2$, $t = 4$, $t = 6$.

For		
$0 < t < 2$,	$f(t) = t - 0 + 0 - 0$	$f(t) = t$
$2 < t < 4$,	$f(t) = t - (t - 2) + 0 - 0$	$f(t) = 2$
$4 < t < 6$,	$f(t) = t - (t - 2) + (t - 4) - 0$	$f(t) = t - 2$
$6 < t$,	$f(t) = t - (t - 2) + (t - 4) - (t - 6)$	$f(t) = 4$

The second and fourth components are constant, but before sketching the graph of the function, we check the values of $f(t) = t$ and $f(t) = t - 2$ at the relevant break points.

$f(t) = t$.   At $t = 0$, $f(t) = 0$;   at $t = 2$, $f(t) = 2$

$f(t) = t - 2$.   At $t = 4$, $f(t) = 2$;   at $t = 6$, $f(t) = 4$.

So the graph of the function is $\ldots\ldots\ldots\ldots$

32

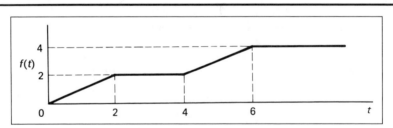

It is always wise to calculate the function values at break points, since discontinuities, or jumps, sometimes occur.

On to the next frame

33

Now for one in reverse.

Example 4

A function $f(t)$ is defined by

$$f(t) = 4 \qquad \text{for } 0 < t < 2$$
$$= 2t - 3 \quad \text{for } 2 < t.$$

Sketch the graph of the function and determine its Laplace transform.

We see that for $t = 0$ to $t = 2$, $f(t) = 4$.

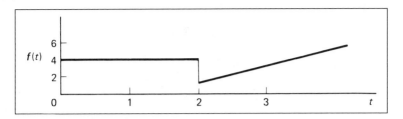

34

Notice the discontinuity at $t = 2$.

Expressing the function in unit step form:

$$f(t) = 4u(t) - 4u(t-2) + u(t-2) \cdot (2t - 3)$$

Note that the second term cancels $f(t) = 4$ at $t = 2$ and that the third switches on $f(t) = 2t - 3$ at $t = 2$.

Before we can express this in Laplace transforms, $(2t - 3)$ in the third term must be written as a function of $(t - 2)$ to correspond to $u(t - 2)$. Therefore, we write $2t - 3$ as $2(t - 2) + 1$.

Then
$$\begin{aligned} f(t) &= 4u(t) - 4u(t-2) + u(t-2) \cdot \{2(t-2) + 1\} \\ &= 4u(t) - 4u(t-2) + u(t-2) \cdot 2(t-2) + u(t-2) \\ &= 4u(t) - 3u(t-2) + u(t-2) \cdot 2(t-2) \end{aligned}$$

$$\therefore \ L\{f(t)\} = \ldots\ldots\ldots\ldots$$

35

$$\boxed{L\{f(t)\} = \frac{4}{s} - \frac{3e^{-2s}}{s} + \frac{2e^{-2s}}{s^2}}$$

Here is one for you to work through in much the same way.

Example 5

A function is defined by
$$\begin{aligned} f(t) &= 6 & 0 < t < 1 \\ &= 8 - 2t & 1 < t < 3 \\ &= 4 & 3 < t. \end{aligned}$$

Sketch the graph and find the Laplace transform of the function.

36

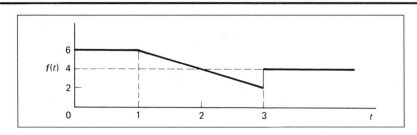

Expressing this in unit step form we have

$$f(t) = 6u(t) - 6u(t-1) + u(t-1) \cdot (8-2t)$$
$$- u(t-3) \cdot (8-2t) + u(t-3) \cdot 4$$

where the second term switches off the first function $f(t) = 6$ at $t = 1$ and the third term switches on the second function $f(t) = 8 - 2t$, which in turn is switched off by the fourth term at $t = 3$ and replaced by $f(t) = 4$ in the fifth term.

Before we can write down the transforms of the third and fourth terms, we must express $f(t) = 8 - 2t$ in terms of $(t-1)$ and $(t-3)$ respectively.

$$8 - 2t = 6 + 2 - 2t = 6 - 2(t-1)$$
$$8 - 2t = 2 + 6 - 2t = 2 - 2(t-3)$$
$$\therefore\ f(t) = 6u(t) - 6u(t-1) + u(t-1) \cdot \{6 - 2(t-1)\}$$
$$- u(t-3) \cdot \{2 - 2(t-3)\} + 4u(t-3)$$
$$= 6u(t) - 6u(t-1) + 6u(t-1)$$
$$- u(t-1) \cdot 2(t-1) - 2u(t-3)$$
$$+ u(t-3) \cdot 2(t-3) + 4u(t-3)$$

which simplifies finally to $f(t) = \ldots\ldots\ldots\ldots$

37

$$\boxed{f(t) = 6u(t) - u(t-1) \cdot 2(t-1) + u(t-3) \cdot 2(t-3) + 2u(t-3)}$$

from which $L\{f(t)\} = \ldots\ldots\ldots\ldots$

38

$$\boxed{L\{f(t)\} = \frac{6}{s} - \frac{2e^{-s}}{s^2} + \frac{2e^{-3s}}{s^2} + \frac{2e^{-3s}}{s}}$$

Note that, in building up the function in unit step form
(a) to 'switch on' a function $f(t)$ at $t = c$, we add the term $u(t-c) \cdot f(t-c)$
(b) to 'switch off' a function $f(t)$ at $t = c$, we subtract $u(t-c) \cdot f(t-c)$.

You have now reached the end of this Program and this brings you to the **Review summary** and the **Can You?** checklist. Following that is the **Test exercise**. Work through this *at your own pace*. A set of **Further problems** provides additional valuable practice.

Review summary

39

1 *Heaviside unit step function:* $u(t - c)$

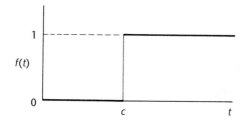

$$f(t) = 0 \quad 0 < t < c$$
$$\quad = 1 \quad c < t$$

2 *Suppression and shift*

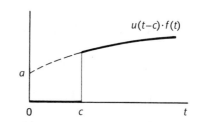

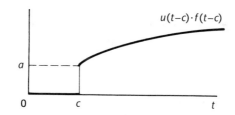

3 *Laplace transform of* $u(t - c)$

$$L\{u(t - c)\} = \frac{e^{-cs}}{s}; \quad L\{u(t)\} = \frac{1}{s}.$$

4 *Laplace transform of* $u(t - c) \cdot f(t - c)$

$$L\{u(t - c) \cdot f(t - c)\} = e^{-cs} \cdot F(s) \quad \text{where } F(s) = L\{f(t)\}.$$

5 *Second shift theorem*
If $F(s) = L\{f(t)\}$, then $e^{-cs} \cdot F(s) = L\{u(t - c) \cdot f(t - c)\}$ where c is real and positive.

✓ **Can You?**

Checklist 7

Check this list before and after you try the end of Program test.

On a scale of 1 to 5 how confident are you that you can: Frames

- Use the Heaviside unit step function to 'switch' expressions on and off?   **1** to **8**
  Yes ☐ ☐ ☐ ☐ ☐ *No*

- Obtain the Laplace transform of expressions involving the Heaviside unit step function?   **8** to **38**
  Yes ☐ ☐ ☐ ☐ ☐ *No*

🚲 **Test exercise 7**

41

1 In each of the following cases, sketch the graph of the function and find its Laplace transform.

(a) $f(t) = 3t \quad 0 < t < 2$
  $= 6 \quad 2 < t$

(b) $f(t) = e^{-2t} \quad 0 < t < 3$
  $= 0 \quad 3 < t$

(c) $f(t) = t^2 \quad 0 < t < 2$
  $= 2 \quad 2 < t < 3$
  $= 4 \quad 3 < t$

(d) $f(t) = \sin 2t \quad 0 < t < \pi$
  $= 0 \quad \pi < t.$

2 Determine the function $f(t)$ whose transform $F(s)$ is

$$F(s) = \frac{1}{s}\left\{2 - 5e^{-s} + 8e^{-3s}\right\}.$$

Sketch the graph of the function between $t = 0$ and $t = 4$.

3 If $f(t) = L^{-1}\left\{\frac{(1 + 3e^{-2s})(1 - e^{-3s})}{s^2}\right\}$, determine $f(t)$ and sketch the graph of the function.

4 Determine the function $f(t)$ for which

$$f(t) = L^{-1}\left\{\frac{2(1 - e^{-s})}{s(1 - e^{-3s})}\right\}.$$

Sketch the waveform and express the function in analytical form.

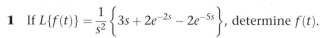
Further problems 7

42

1 If $L\{f(t)\} = \dfrac{1}{s^2}\left\{3s + 2e^{-2s} - 2e^{-5s}\right\}$, determine $f(t)$.

2 If $f(t) = L^{-1}\left\{\dfrac{(1 - e^{-s})(1 + e^{-2s})}{s^2}\right\}$, find $f(t)$ in terms of the unit step function.

3 A function $f(t)$ is defined by

$$\begin{aligned}f(t) &= 4 && 0 < t < 3 \\ &= 2t + 1 && 3 < t.\end{aligned}$$

Sketch the graph of the function and determine its Laplace transform.

4 Express in terms of the Heaviside unit step function
(a) $\begin{aligned}f(t) &= t^2 && 0 < t < 3 \\ &= 5t && 3 < t.\end{aligned}$

(b) $\begin{aligned}f(t) &= \cos t && 0 < t < \pi \\ &= \cos 2t && \pi < t < 2\pi \\ &= \cos 3t && 2\pi < t.\end{aligned}$

5 A function $f(t)$ is defined by

$$\begin{aligned}f(t) &= 0 && 0 < t < 2 \\ &= t + 1 && 2 < t < 3 \\ &= 0 && 3 < t.\end{aligned}$$

Determine $L\{f(t)\}$.

6 A function $f(t)$ is defined by

$$\begin{aligned}f(t) &= t^2 && 0 < t < 2 \\ &= 4 && 2 < t < 5 \\ &= 0 && 5 < t.\end{aligned}$$

Determine (a) the function in terms of the unit step function
(b) the Laplace transform of $f(t)$.

Laplace transforms 3

Frames
1 to 70

Learning outcomes

When you have completed this Program you will be able to:

- Find the Laplace transforms of periodic functions
- Obtain the inverse Laplace transforms of transforms of periodic functions
- Describe and use the unit impulse to evaluate integrals
- Obtain the Laplace transform of the unit impulse
- Use the Laplace transform to solve differential equations involving the unit impulse
- Solve the equation and describe the behaviour of an harmonic oscillator

Laplace transforms of periodic functions

1 Periodic functions

Let $f(t)$ represent a periodic function with period T so that $f(t+nT) = f(t)$ with a graph of the following form

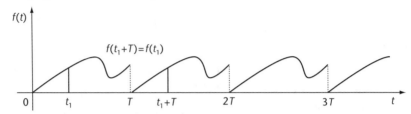

If we describe the first cycle by $\bar{f}(t)$ then

$$\bar{f}(t) = \begin{cases} f(t) & \text{for } 0 \leq t < T \\ 0 & \text{otherwise} \end{cases}$$

The second cycle is identical to the first cycle except that it is shifted by T units of time along the t-axis. Therefore the second cycle can be described in terms of the Heaviside unit step function as $\bar{f}(t-T)u(t-T)$. That is

$$\bar{f}(t-T)u(t-T) = \begin{cases} f(t) & \text{for } T \leq t < 2T \\ 0 & \text{otherwise} \end{cases}$$

By this reasoning the periodic function $f(t)$ is represented by

$$f(t) = \bar{f}(t)u(t) + \dots\dots\dots$$

2

$$\boxed{f(t) = \bar{f}(t)u(t) + \bar{f}(t-T)u(t-T) + \bar{f}(t-2T)u(t-2T) + \cdots}$$

Because

$u(t)$ switches on $\bar{f}(t)$ at time $t = 0$, $u(t-T)$ switches on $\bar{f}(t-T)$ at time $t = T$ and $u(t-2T)$ switches on $\bar{f}(t-2T)$ at time $t = 2T$, etc.

Consider now the Laplace transform of $\bar{f}(t)$. By definition

$$L\{\bar{f}(t)\} = \int_0^\infty e^{-st}\bar{f}(t)\,\mathrm{d}t = \int_0^T e^{-st}f(t)\,\mathrm{d}t = \bar{F}(s)$$

because for $t > T$, $\bar{f}(t) = 0$ and so the semi-infinite integral becomes an integral just over the period of $f(t)$. Using the second shift theorem (see Frame 10 of Program 7), the Laplace transform of $f(t)$ is

$$L\{f(t)\} = L\{\bar{f}(t)u(t)\} + L\{\bar{f}(t-T)u(t-T)\}$$
$$+ L\{\bar{f}(t-2T)u(t-2T)\} + \cdots$$

That is

$$L\{f(t)\} = \dots\dots\dots$$

3

$$L\{f(t)\} = \bar{F}(s) + e^{-sT}\bar{F}(s) + e^{-2sT}\bar{F}(s) + \cdots$$

Because

$L\{\bar{f}(t)u(t-c)\} = e^{-sc}L\{\bar{f}(t)\}$ by the second shift theorem.

We can factor out $\bar{F}(s)$ and write $L\{f(t)\}$ as

$$L\{f(t)\} = (1 + e^{-sT} + e^{-2sT} + \ldots)\bar{F}(s)$$

Now, do you remember the series $1 + x + x^2 + x^3 + \ldots$? This can be written in closed form as

$$1 + x + x^2 + x^3 + \ldots = \ldots\ldots\ldots\ldots$$

$$1 + x + x^2 + x^3 + \ldots = \frac{1}{1-x}$$

Because

$$\frac{1}{1-x} = (1-x)^{-1} = 1 + x + x^2 + x^3 + \ldots$$

either by the binomial theorem or by performing the long division.

So, if we let $x = e^{-sT}$ then

$$1 + e^{-sT} + e^{-2sT} + \ldots = \ldots\ldots\ldots\ldots$$

$$1 + e^{-sT} + e^{-2sT} + \ldots = \frac{1}{1 - e^{-sT}}$$

And so the Laplace transform of $f(t)$ is given as

$$L\{f(t)\} = (1 + e^{-sT} + e^{-2sT} + \ldots)\bar{F}(s) = \ldots\ldots\ldots\ldots \text{ where } \bar{F}(s) = \ldots\ldots\ldots\ldots$$

$$L\{f(t)\} = \frac{1}{(1 - e^{-sT})}\bar{F}(s) \text{ where } \bar{F}(s) = \int_0^T e^{-st}f(t)\,\mathrm{d}t$$

Note that we integrate $e^{-st}f(t)$ over one cycle, that is from $t = 0$ to $t = T$, and not from $t = 0$ to $t = \infty$ as we did previously.

This is an important result. Make a note of it – then we shall apply it

▶

Example 1

Find the Laplace transform of the function $f(t)$ defined by

$$f(t) = 3 \quad 0 < t < 2 \atop = 0 \quad 2 < t < 4 \Big\} \quad f(t+4) = f(t)$$

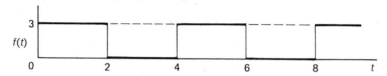

The expression for $L\{f(t)\}$ is

. . . .   (do not evaluate it yet)

7

$$L\{f(t)\} = \frac{1}{1 - e^{-4s}} \int_0^4 e^{-st} \cdot f(t)\, dt$$

Because the period $= 4$, i.e. $T = 4$.

The function $f(t) = 3$ for $0 < t < 2$ and $f(t) = 0$ for $2 < t < 4$.

$$\therefore \ L\{f(t)\} = \frac{1}{1 - e^{-4s}} \int_0^2 e^{-st} \cdot 3\, dt = \ldots \ldots \ldots$$

8

$$L\{f(t)\} = \frac{3}{s(1 + e^{-2s})}$$

Because

$$L\{f(t)\} = \frac{3}{1 - e^{-4s}} \left[\frac{e^{-st}}{-s}\right]_0^2 = \frac{3}{1 - e^{-4s}} \left\{ \left(\frac{e^{-2s}}{-s}\right) - \left(\frac{1}{-s}\right) \right\}$$

$$= \frac{3}{1 - e^{-4s}} \left\{ \frac{1 - e^{-2s}}{s} \right\} = \frac{3}{s(1 + e^{-2s})}$$

That is all there is to it. Now for another, so move on

Example 2

9

Find the Laplace transform of the periodic function defined by

$$f(t) = t/2 \qquad 0 < t < 3$$
$$f(t+3) = f(t)$$

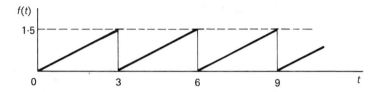

Because in this case, period $= 3$, i.e. $T = 3$.

$$\therefore \ L\{f(t)\} = \frac{1}{1 - e^{-Ts}} \int_0^T e^{-st} \cdot f(t)\,dt$$

$$= \frac{1}{1 - e^{-3s}} \int_0^3 e^{-st} \cdot \left(\frac{t}{2}\right) dt$$

$$\therefore \ 2(1 - e^{-3s})L\{f(t)\} = \int_0^3 t \cdot e^{-st}\,dt$$

Integrating by parts and simplifying the result gives

$$L\{f(t)\} = \dots\dots\dots$$

10

$$L\{f(t)\} = \frac{1}{2s^2}\left\{1 - \frac{3s}{e^{3s} - 1}\right\}$$

Because

$$2(1 - e^{-3s})L\{f(t)\} = \int_0^3 te^{-st}\,dt$$

$$= \left[t\left(\frac{e^{-st}}{-s}\right)\right]_0^3 + \frac{1}{s}\int_0^3 e^{-st}\,dt$$

$$= -\frac{3e^{-3s}}{s} + \frac{1}{s}\left[\frac{e^{-st}}{-s}\right]_0^3$$

$$= -\frac{3e^{-3s}}{s} - \frac{e^{-3s}}{s^2} + \frac{1}{s^2}$$

$$\therefore \ L\{f(t)\} = \frac{1}{2s^2}\left\{1 - \frac{3se^{-3s}}{1 - e^{-3s}}\right\}$$

$$= \frac{1}{2s^2}\left\{1 - \frac{3s}{e^{3s} - 1}\right\}$$

▶

Example 3

Sketch the graph of the function

$$f(t) = e^t \qquad 0 < t < 5$$
$$f(t+5) = f(t)$$

and determine its Laplace transform.

First we sketch the graph of $f(t)$, which is ....

11

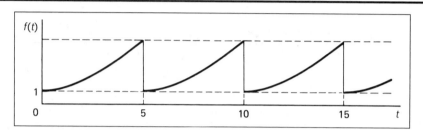

Clearly, period $= 5$ $\therefore$ $T = 5$

$$L\{f(t)\} = \frac{1}{1 - e^{-Ts}} \int_0^T e^{-st} \cdot f(t)\, dt \quad \text{gives}$$

$$L\{f(t)\} = ...$$

Complete the working

12

$$L\{f(t)\} = \frac{1 - e^{-5(s-1)}}{(s-1)(1 - e^{-5s})}$$

Because

$$L\{f(t)\} = \frac{1}{1 - e^{-5s}} \int_0^5 e^{-st} \cdot e^t\, dt$$

$$\therefore \ (1 - e^{-5s})L\{f(t)\} = \int_0^5 e^{-(s-1)t}\, dt$$

$$= \left[\frac{e^{-(s-1)t}}{-(s-1)} \right]_0^5 = \frac{1}{s-1}\left\{ 1 - e^{-5(s-1)} \right\}$$

$$\therefore \ L\{f(t)\} = \frac{1 - e^{-5(s-1)}}{(s-1)(1 - e^{-5s})}$$

All very straightforward.

Example 4

Determine the Laplace transform of the half-wave rectifier output waveform defined by

$$f(t) = 8\sin t \quad 0 < t < \pi \atop = 0 \qquad \pi < t < 2\pi \Bigg\} \quad f(t + 2\pi) = f(t)$$

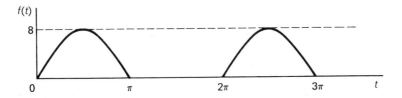

Here the period is 2π i.e. $T = 2\pi$.

In general, for a periodic function of period T

$$L\{f(t)\} = \dots\dots\dots$$

13

$$\boxed{L\{f(t)\} = \frac{1}{1 - e^{-Ts}} \int_0^T e^{-st} \cdot f(t)\,\mathrm{d}t}$$

So, for this example

$$L\{f(t)\} = \frac{1}{1 - e^{-2\pi s}} \int_0^{2\pi} e^{-st} \cdot f(t)\,\mathrm{d}t$$

$$\therefore \ (1 - e^{-2\pi s})L\{f(t)\} = \int_0^{\pi} e^{-st} \cdot 8\sin t\,\mathrm{d}t$$

Writing $\sin t$ as the imaginary part of e^{it}, i.e. $\sin t \equiv \mathscr{I}e^{it}$,

$$(1 - e^{-2\pi s})L\{f(t)\} = 8\mathscr{I} \int_0^{\pi} e^{-st} \cdot e^{it}\,\mathrm{d}t$$

$$= 8\mathscr{I} \int_0^{\pi} e^{-(s-i)t}\,\mathrm{d}t$$

and this you can finish off in the usual manner, giving

$$L\{f(t)\} = \dots\dots\dots$$

14

$$L\{f(t)\} = \frac{8}{(s^2+1)(1-e^{-\pi s})}$$

Because

$$(1-e^{-2\pi s})L\{f(t)\} = 8 \cdot \mathscr{I}\int_0^{\pi} e^{-(s-i)t}\, dt$$

$$= 8 \cdot \mathscr{I}\left[\frac{e^{-(s-i)t}}{-(s-i)}\right]_0^{\pi}$$

$$= \mathscr{I}\left\{\frac{-8}{s-i}\left[e^{-(s-i)\pi}-1\right]\right\}$$

$$= 8 \cdot \mathscr{I}\left\{\frac{1}{s-i}\left[1-e^{-s\pi}e^{i\pi}\right]\right\}$$

But $e^{i\pi} = \cos\pi + i\sin\pi = -1$.

$$\therefore \ (1-e^{-2\pi s})L\{f(t)\} = 8 \cdot \mathscr{I}\left\{\frac{1}{s-i}(1+e^{-s\pi})\right\}$$

$$= 8 \cdot \mathscr{I}\left\{\frac{s+i}{s^2+1}(1+e^{-\pi s})\right\} = 8\left\{\frac{1+e^{-\pi s}}{s^2+1}\right\}$$

$$\therefore \ L\{f(t)\} = \frac{1}{1-e^{-2\pi s}} \times 8\left\{\frac{1+e^{-\pi s}}{s^2+1}\right\}$$

$$= \frac{8}{(1-e^{-\pi s})(s^2+1)}$$

Now let us consider the corresponding inverse transforms when periodic functions are involved.

15   ## Inverse transforms

Finding inverse transforms of functions of s which are transforms of periodic functions is not as straightforward as in earlier examples, for the transforms result from integration over one cycle and not from $t=0$ to $t=\infty$. Hence we have no simple table of inverse transforms upon which to draw.

However, all difficulties can be surmounted and an example will show how we deal with this particular problem.

Example 1

Determine the inverse transform

$$L^{-1}\left\{\frac{2+e^{-2s}-3e^{-s}}{s(1-e^{-2s})}\right\}$$

The first thing we see is the factor $(1-e^{-2s})$ in the denominator, which suggests a periodic function of period 2 units, i.e. $\frac{1}{1-e^{-Ts}}$ where $T=2$.

The key to the solution is to write $(1-e^{-2s})$ in the denominator as $(1-e^{-2s})^{-1}$ in the numerator and to expand this as a binomial series.

We remember that $(1-x)^{-1} = \ldots\ldots\ldots$

$$\boxed{(1-x)^{-1} = 1 + x + x^2 + x^3 + \dots}$$

$$\therefore \ \left(1 - e^{-2s}\right)^{-1} = 1 + \left(e^{-2s}\right) + \left(e^{-2s}\right)^2 + \left(e^{-2s}\right)^3 + \dots$$

$$= 1 + e^{-2s} + e^{-4s} + e^{-6s} + \dots$$

$$\therefore \ L\{f(t)\} = \frac{2 + e^{-2s} - 3e^{-s}}{s(1 - e^{-2s})} = \frac{1}{s}\left(2 + e^{-2s} - 3e^{-s}\right)\left(1 - e^{-2s}\right)^{-1}$$

$$= \frac{1}{s}\left(2 + e^{-2s} - 3e^{-s}\right)\left(1 + e^{-2s} + e^{-4s} + e^{-6s} + e^{-8s} + \dots\right)$$

We now multiply the second series by each term of the first in turn and collect up like terms, giving

$$L\{f(t)\} = \frac{1}{s}\left\{\begin{array}{llll} 2 & +2e^{-2s} & +2e^{-4s} & +2e^{-6s} \ \dots \\ & +\ e^{-2s} & +\ e^{-4s} & +\ e^{-6s} \ \dots \\ -3e^{-s} & -3e^{-3s} & -3e^{-5s} & \dots \end{array}\right\}$$

$$= \dots\dots\dots\dots$$

$$\boxed{L\{f(t)\} = \frac{1}{s}\left\{2 - 3e^{-s} + 3e^{-2s} - 3e^{-3s} + 3e^{-4s} - 3e^{-5s} + \dots\right\}}$$

Each term is of the form $\dfrac{e^{-cs}}{s}$, so, expressing $f(t)$ in unit step form, we have

$$f(t) = \dots\dots\dots\dots$$

$$\boxed{f(t) = 2u(t) - 3u(t-1) + 3u(t-2) - 3u(t-3) + 3u(t-4)\dots}$$

and from this we can sketch the waveform, which is therefore

$$\dots\dots\dots\dots$$

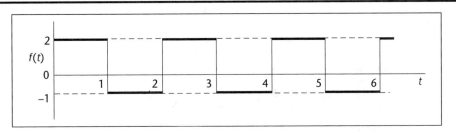

19

We can finally define this periodic function in analytical terms.

$$f(t) = \dots\dots\dots\dots$$

20

$$\boxed{\begin{aligned} f(t) &= 2 \quad 0 < t < 1 \\ &= -1 \quad 1 < t < 2 \end{aligned} \Bigg\} \; f(t+2) = f(t)}$$

The key to the whole process is thus to ....

21

$$\boxed{\begin{aligned} &\text{express } (1 - e^{-Ts}) \text{ in the denominator} \\ &\text{as } (1 - e^{-Ts})^{-1} \text{ in the numerator and} \\ &\text{to expand this as a binomial series.} \end{aligned}}$$

We do this by making use of the basic series

$$(1 - x)^{-1} = \ldots\ldots\ldots$$

22

$$\boxed{(1 - x)^{-1} = 1 + x + x^2 + x^3 + x^4 + \ldots}$$

Example 2

Determine $L^{-1}\left\{\dfrac{3(1 - e^{-s})}{s(1 - e^{-3s})}\right\}$ and sketch the resulting waveform of $f(t)$.

$$L\{f(t)\} = \frac{3}{s}(1 - e^{-s})(1 - e^{-3s})^{-1}$$

$$= \ldots\ldots\ldots \qquad \text{(next step)}$$

23

$$\boxed{L\{f(t)\} = \frac{3}{s}(1 - e^{-s})(1 + e^{-3s} + e^{-6s} + e^{-9s} + \ldots)}$$

which multiplied out gives

$$L\{f(t)\} = \frac{3}{s}(1 - e^{-s} + e^{-3s} - e^{-4s} + e^{-6s} - e^{-7s} + \ldots)$$

$$= \frac{3}{s} - \frac{3e^{-s}}{s} + \frac{3e^{-3s}}{s} - \frac{3e^{-4s}}{s} + \frac{3e^{-6s}}{s} - \ldots$$

And in unit step form, this gives

$$f(t) = \ldots\ldots\ldots$$

24

$$\boxed{f(t) = 3u(t) - 3u(t - 1) + 3u(t - 3) - 3u(t - 4) + \ldots}$$

The waveform is thus ....

25

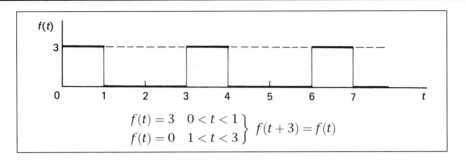

$$f(t) = 3 \quad 0 < t < 1$$
$$f(t) = 0 \quad 1 < t < 3 \Big\} \; f(t+3) = f(t)$$

And now, one more. They are all done in the same way

Example 3

26

If $L\{f(t)\} = \dfrac{1}{2s^2} - \dfrac{2e^{-4s}}{s(1 - e^{-4s})}$, determine $f(t)$ and sketch the waveform.

The first term is easy enough. In unit step form $L^{-1}\left\{\dfrac{1}{2s^2}\right\} = \dfrac{t}{2} \cdot u(t)$

From the second term

$$\frac{2e^{-4s}}{s(1 - e^{-4s})} = \frac{2}{s}\left\{e^{-4s}\left(1 - e^{-4s}\right)^{-1}\right\}$$

$$= \frac{2}{s}\left\{e^{-4s}\left(1 + e^{-4s} + e^{-8s} + e^{-12s} + \ldots\right)\right\}$$

$$= \frac{2e^{-4s}}{s} + \frac{2e^{-8s}}{s} + \frac{2e^{-12s}}{s} + \frac{2e^{-16s}}{s} + \ldots$$

$$\therefore \; f(t) = \ldots\ldots\ldots\ldots \quad \text{(in unit step form)}$$

27

$$f(t) = \frac{t}{2} \cdot u(t) - 2u(t - 4) - 2u(t - 8) - 2u(t - 12) - \ldots$$

Now we have to draw the waveform. Consider the function terms up to each break point in turn.

$$0 < t < 4 \quad f(t) = \frac{t}{2} \qquad f(0) = 0; \;\; f(4) = 2$$

$$4 < t < 8 \quad f(t) = \frac{t}{2} - 2 \qquad f(4) = 0; \;\; f(8) = 2$$

$$8 < t < 12 \quad f(t) = \frac{t}{2} - 2 - 2 \quad f(8) = 0; \;\; f(12) = 2 \text{ etc.}$$

So the waveform is $\ldots\ldots\ldots\ldots$

28

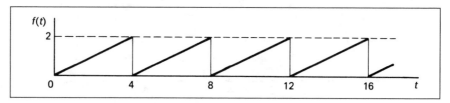

Expressed analytically, we finally have

$$f(t) = \frac{t}{2} \quad 0 < t < 4, \quad f(t+4) = f(t)$$

The Dirac delta – the unit impulse

29

So far we have dealt with a number of standard Laplace transforms and then the Heaviside unit step function with some of its applications. We now come to consider an entity that is different from any of the functions we have used before because it is not a proper function. Rather than being defined by its inputs and corresponding outputs it is defined by its effect on other functions. If $f(t)$ represents a function then the Dirac delta $\delta(t)$ is defined by the integral

$$\int_{-\infty}^{\infty} f(t)\delta(t-a)\,\mathrm{d}t = f(a)$$

$\delta(t)$ is often referred to as the **Dirac delta function** even though it is not a function in the conventional sense of being completely defined in terms of its outputs for the corresponding inputs. The nearest that can be achieved in defining it in function terms is

$$\delta(t) = \begin{cases} 0 & t \neq 0 \\ \text{undefined} & t = 0 \end{cases}$$

From the definition, if $f(t) = 1$ then

$$\int_{-\infty}^{\infty} \delta(t-a)\,\mathrm{d}t = \ldots\ldots\ldots\ldots$$

30

$$\boxed{\int_{-\infty}^{\infty} \delta(t-a)\,\mathrm{d}t = 1}$$

Because

$$\int_{-\infty}^{\infty} f(t)\delta(t-a)\,\mathrm{d}t = f(a) \text{ and } f(t) = 1 \text{ so } f(a) = 1, \text{ therefore}$$

$$\int_{-\infty}^{\infty} \delta(t-a)\,\mathrm{d}t = 1 \text{ hence the name } \textit{unit impulse}.$$

Also, if $p < a < q$ then

$$\int_{p}^{q} \delta(t-a)\,\mathrm{d}t = \ldots\ldots\ldots\ldots$$

$$\int_p^q \delta(t-a)\,\mathrm{d}t = 1$$

Because

$$\int_{-\infty}^{\infty} \delta(t-a)\,\mathrm{d}t = \int_{-\infty}^{p} \delta(t-a)\,\mathrm{d}t + \int_p^q \delta(t-a)\,\mathrm{d}t + \int_q^{\infty} \delta(t-a)\,\mathrm{d}t$$

$$= 0 + \int_p^q \delta(t-a)\,\mathrm{d}t + 0 \quad \begin{matrix} \text{since } \delta(t-a) = 0 \\ \text{for } -\infty < t \le p \\ \text{and } q \le t < \infty \end{matrix}$$

$$= 1$$

So that $\displaystyle\int_p^q \delta(t-a)\,\mathrm{d}t = 1$

Graphical representation

Graphically the Dirac delta or unit impulse $\delta(t-a)$ is represented by the horizontal axis with a vertical line of infinite length at $t = a$.

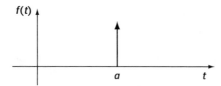

So far, then, we have

(a) $\displaystyle\int_p^q \delta(t-a)\,\mathrm{d}t = 1$

(b) $\displaystyle\int_p^q f(t)\cdot\delta(t-a)\,\mathrm{d}t = f(a)$

provided, in each case, that $p < a < q$.

Example 1

To evaluate $\displaystyle\int_1^3 (t^2 + 4)\cdot\delta(t-2)\,\mathrm{d}t$.

The factor $\delta(t-2)$ shows that the impulse occurs at $t = 2$, i.e. $a = 2$.

$$f(t) = t^2 + 4 \qquad \therefore\ f(a) = f(2) = 4 + 4 = 8$$

$$\therefore\ \int_1^3 (t^2 + 4)\cdot\delta(t-2)\,\mathrm{d}t = f(2) = 8$$

▶

Example 2

To evaluate $\int_0^\pi \cos 6t \cdot \delta(t - \pi/2)\, dt$.

$$\int_0^\pi \cos 6t \cdot \delta(t - \pi/2)\, dt = f(\pi/2) = \cos 3\pi = -1$$

and in the same way

(a) $\int_0^6 5 \cdot \delta(t - 3)\, dt = \ldots\ldots\ldots$

(b) $\int_2^5 e^{-2t} \cdot \delta(t - 4)\, dt = \ldots\ldots\ldots$

(c) $\int_0^\infty (3t^2 - 4t + 5) \cdot \delta(t - 2)\, dt = \ldots\ldots\ldots$

33

(a) $\int_0^6 5 \cdot \delta(t - 3)\, dt = 5 \times 1 = 5$

(b) $\int_2^5 e^{-2t} \cdot \delta(t - 4)\, dt = f(4) = \left[e^{-2t}\right]_{t=4} = e^{-8}$

(c) $\int_0^\infty (3t^2 - 4t + 5) \cdot \delta(t - 2)\, dt = 12 - 8 + 5 = 9$

Nothing could be easier. It all rests on the fact that, provided $p < a < q$

$$\int_p^q f(t) \cdot \delta(t - a)\, dt = \ldots\ldots\ldots$$

34

$$f(a)$$

Now let us consider the Laplace transform of $\delta(t - a)$.

On then to the next frame

---

35 **Laplace transform of $\delta(t-a)$**

We have already shown that

$$\int_p^q f(t) \cdot \delta(t - a)\, dt = f(a) \qquad p < a < q$$

Therefore, if $p = 0$ and $q = \infty$

$$\int_0^\infty f(t) \cdot \delta(t - a)\, dt = f(a)$$

Hence, if $f(t) = e^{-st}$, this becomes

$$\int_0^\infty e^{-st} \cdot \delta(t - a)\, dt = L\{\delta(t - a)\}$$

$$= \ldots\ldots\ldots$$

$$e^{-as}$$

i.e. the value of $f(t)$, i.e. e^{-st}, at $t = a$.

$L\{\delta(t - a)\} = e^{-as}$

It follows from this that the Laplace transform of the impulse function at the origin is . . . .

37

$$1$$

Because, for $a = 0$, $L\{\delta(t - a)\} = L\{\delta(t)\} = e^0 = 1$
$$\therefore \ L\{\delta(t)\} = 1$$

Finally, let us deal with the more general case of $L\{f(t) \cdot \delta(t - a)\}$. We have $L\{f(t) \cdot \delta(t - a)\} = \int_0^{\infty} e^{-st} \cdot f(t) \cdot \delta(t - a)\, dt$. Now the integrand $e^{-st} \cdot f(t) \cdot \delta(t - a) = 0$ for all values of t except at $t = a$ at which point $e^{-st} = e^{-as}$, and $f(t) = f(a)$.

$$\therefore \ L\{f(t) \cdot \delta(t - a)\} = f(a) \cdot e^{-as} \int_0^{\infty} \delta(t - a)\, dt$$
$$= f(a) \cdot e^{-as}(1)$$
$$\therefore \ L\{f(t) \cdot \delta(t - a)\} = f(a)e^{-as}$$

Another important result to note. Then let us deal with some examples

38

We have $L\{f(t) \cdot \delta(t - a)\} = f(a) \cdot e^{-as}$

Therefore

(a) $L\{6 \cdot \delta(t - 4)\}$ $a = 4$, $\therefore \ L\{6 \cdot \delta(t - 4)\} = 6e^{-4s}$
(b) $L\{t^3 \cdot \delta(t - 2)\}$ $a = 2$, $\therefore \ L\{t^3 \cdot \delta(t - 2)\} = 8e^{-2s}$

Similarly

(c) $L\{\sin 3t \cdot \delta(t - \pi/2)\} = $ . . . .

$$-e^{-\pi s/2}$$

Because

$$L\{\sin 3t \cdot \delta(t - \pi/2)\} = [\sin 3t]_{t=\pi/2} \cdot e^{-\pi s/2} = -e^{-\pi s/2}$$

and

(d) $L\{\cosh 2t \cdot \delta(t)\} = $ . . . .

40

$$\boxed{1}$$

Because

$$L\{\cosh 2t \cdot \delta(t)\} = [\cosh 2t]_{t=0} \cdot e^0 = \cosh 0 \cdot (1) = 1$$

So our main conclusions so far are as follows.

(1) $\displaystyle\int_p^q \delta(t-a)\,dt = \ldots\ldots\ldots$ provided $\ldots\ldots\ldots$

(2) $\displaystyle\int_p^q f(t) \cdot \delta(t-a)\,dt = \ldots\ldots\ldots$ provided $\ldots\ldots\ldots$

(3) $L\{\delta(t-a)\} = \ldots\ldots\ldots$

(4) $L\{\delta(t)\} = \ldots\ldots\ldots$

(5) $L\{f(t) \cdot \delta(t-a)\} = \ldots\ldots\ldots$

41

> (1) $\displaystyle\int_p^q \delta(t-a)\,dt = 1$ provided $p < a < q$
>
> (2) $\displaystyle\int_p^q f(t) \cdot \delta(t-a)\,dt = f(a)$ provided $p < a < q$
>
> (3) $L\{\delta(t-a)\} = e^{-as}$
>
> (4) $L\{\delta(t)\} = 1$
>
> (5) $L\{f(t) \cdot \delta(t-a)\} = f(a) \cdot e^{-as}$

Just check that you have noted this important list – the basis of all work on the Dirac delta function.

Now for one further example on this section

Example

Impulses of 1, 4, 7 units occur at $t = 1$, $t = 3$ and $t = 4$ respectively, in the directions shown.

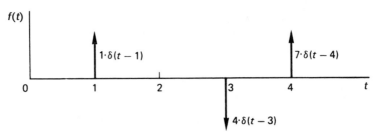

Write down an expression for $f(t)$ and determine its Laplace transform.

We have $f(t) = 1 \cdot \delta(t-1) - 4 \cdot \delta(t-3) + 7 \cdot \delta(t-4)$.

Then $L\{f(t)\} = \ldots\ldots\ldots$

$$\boxed{L\{f(t)\} = e^{-s} - 4e^{-3s} + 7e^{-4s}}$$

and that is all there is to that.

The derivative of the unit step function

One further consideration is interesting.

Consider some function $f(t)$ that is zero outside some finite interval $[a, b]$ of the real line. That is, $f(t) = 0$ for $t < a$ and $t > b$, then

$$\int_{-\infty}^{\infty} [u(t)f(t)]' \, dt = [u(t)f(t)]_{-\infty}^{\infty} = 0$$

where $u(t)$ is the unit step function and $f(t)$ is zero at the limits. Now

$$\int_{-\infty}^{\infty} [u(t)f(t)]' \, dt = \int_{-\infty}^{\infty} u'(t)f(t) \, dt + \int_{-\infty}^{\infty} u(t)f'(t) \, dt$$

and so

$$\int_{-\infty}^{\infty} u'(t)f(t) \, dt = -\int_{-\infty}^{\infty} u(t)f'(t) \, dt$$

This means that

$$
\begin{aligned}
\int_{-\infty}^{\infty} u'(t)f(t) \, dt &= -\int_{-\infty}^{\infty} u(t)f'(t) \, dt \\
&= -\int_{0}^{\infty} f'(t) \, dt && \text{Because the unit step} \\
&&& \text{is zero for negative } t \\
&= -\Big[f(t)\Big]_{0}^{\infty} \\
&= -f(\infty) + f(0) \\
&= f(0) && \text{Because } f(\infty) = 0 \text{ by} \\
&&& \text{definition} \\
&= \int_{-\infty}^{\infty} \delta(t)f(t) \, dt && \text{By the definition of} \\
&&& \text{the Dirac delta}
\end{aligned}
$$

and so $u'(t) = \delta(t)$ – *the unit impulse is equal to the derivative of the unit step function.*

Differential equations involving the unit impulse

43

Example 1

A system has the equation of motion

$$\ddot{x} + 6\dot{x} + 8x = g(t)$$

where $g(t)$ is an impulse of 4 units applied at $t = 5$. At $t = 0$, $x = 0$ and $\dot{x} = 3$. Determine an expression for the displacement x in terms of t.

The impulse of 4 units is applied at $t = 5$. $\therefore$ $g(t) = 4 \cdot \delta(t - 5)$.

$\therefore$ $\ddot{x} + 6\dot{x} + 8x = 4 \cdot \delta(t - 5)$ At $t = 0$, $x = 0$, $\dot{x} = 3$.

Taking Laplace transforms this differential equation becomes

. . . .

44

$$\left(s^2\bar{x} - sx_0 - x_1\right) + 6(s\bar{x} - x_0) + 8\bar{x} = 4e^{-5s}$$

Now $x_0 = 0$; $x_1 = 3$

$\therefore$ $s^2\bar{x} - 3 + 6s\bar{x} + 8\bar{x} = 4e^{-5s}$

$\therefore$ $(s^2 + 6s + 8)\bar{x} = 3 + 4e^{-5s}$

$\therefore$ $\bar{x} = (3 + 4e^{-5s})\dfrac{1}{(s+2)(s+4)}$

Writing $\dfrac{1}{(s+2)(s+4)}$ in partial fractions, we get

$$\bar{x} = \ldots \ldots \ldots \ldots$$

45

$$\bar{x} = (3 + 4e^{-5s})\left\{\frac{1}{2} \cdot \frac{1}{s+2} - \frac{1}{2} \cdot \frac{1}{s+4}\right\}$$

$$\therefore \quad \bar{x} = \frac{3}{2}\left\{\frac{1}{s+2} - \frac{1}{s+4}\right\} + 2\left\{\frac{e^{-5s}}{s+2} - \frac{e^{-5s}}{s+4}\right\}$$

Taking inverse transforms

$$x = \frac{3}{2}\left\{e^{-2t} - e^{-4t}\right\} + 2\left\{e^{-2(t-5)} \cdot u(t-5) - e^{-4(t-5)} \cdot u(t-5)\right\}$$

$$= \frac{3}{2}\left\{e^{-2t} - e^{-4t}\right\} + 2\left\{e^{-2t} \cdot e^{10} \cdot u(t-5) - e^{-4t} \cdot e^{20} \cdot u(t-5)\right\}$$

which simplifies to $x = \ldots \ldots \ldots \ldots$

$$x = e^{-2t} \left\{ \frac{3}{2} + 2e^{10} \cdot u(t-5) \right\} - e^{-4t} \left\{ \frac{3}{2} + 2e^{20} \cdot u(t-5) \right\}$$

Example 2

Solve the equation $\ddot{x} + 4\dot{x} + 13x = 2 \cdot \delta(t)$ where, at $t = 0$, $x = 2$ and $\dot{x} = 0$.

$$\ddot{x} + 4\dot{x} + 13x = 2 \cdot \delta(t) \qquad x_0 = 2; \; x_1 = 0$$

Expressing in Laplace transforms, we have

. . . .

$$\left(s^2 \bar{x} - sx_0 - x_1 \right) + 4(s\bar{x} - x_0) + 13\bar{x} = 2 \cdot (1)$$

Inserting the initial conditions and simplifying,

$$\bar{x} = \ldots \ldots \ldots \ldots$$

$$\bar{x} = (2s + 10)\frac{1}{s^2 + 4s + 13}$$

Rearranging the denominator by completing the square, this can be written

$$\bar{x} = (2s + 10)\frac{1}{(s+2)^2 + 9}$$

$$\therefore \; x = \ldots \ldots \ldots \ldots$$

$$x = 2e^{-2t}\{\cos 3t + \sin 3t\}$$

Because

$$\bar{x} = \frac{2(s+2)}{(s+2)^2 + 9} + \frac{6}{(s+2)^2 + 9}$$

$$\therefore \; x = 2e^{-2t}\cos 3t + 2e^{-2t}\sin 3t$$

$$\therefore \; x = 2e^{-2t}\{\cos 3t + \sin 3t\}$$

Now for one further example for you to work through on your own.

So move on

50

Example 3

The equation of motion of a system is

$\ddot{x} + 5\dot{x} + 4x = g(t)$ where $g(t) = 3 \cdot \delta(t - 2)$.

At $t = 0$, $x = 2$ and $\dot{x} = -2$. Determine an expression for the displacement x in terms of t.

We have $\ddot{x} + 5\dot{x} + 4x = 3 \cdot \delta(t - 2)$ with $x_0 = 2$ and $x_1 = -2$.

As before, you can express this in Laplace transforms, substitute the initial conditions, simplify to obtain an expression for x and finally take inverse transforms to determine the required expression for x.

Work right through it carefully. It is good review and there are no snags.

$$x = \ldots\ldots\ldots\ldots$$

51

$$\boxed{x = e^{-t}\{2 + e^2 \cdot u(t - 2)\} - e^8 \cdot e^{-4t} \cdot u(t - 2)}$$

Here is the working for you to check.

$$\ddot{x} + 5\dot{x} + 4x = 3 \cdot \delta(t - 2) \text{ with } x_0 = 2 \text{ and } x_1 = -2$$

$$(s^2\bar{x} - sx_0 - x_1) + 5(s\bar{x} - x_0) + 4\bar{x} = 3e^{-2s}$$

$$s^2\bar{x} - 2s + 2 + 5s\bar{x} - 10 + 4\bar{x} = 3e^{-2s}$$

$$(s^2 + 5s + 4)\bar{x} - 2s - 8 = 3e^{-2s}$$

$$\therefore \ (s + 1)(s + 4)\bar{x} = 2s + 8 + 3e^{-2s}$$

$$\therefore \ \bar{x} = \frac{2(s + 4)}{(s + 1)(s + 4)} + e^{-2s} \cdot \frac{3}{(s + 1)(s + 4)}$$

$$= \frac{2}{s + 1} + e^{-2s}\left\{\frac{1}{s + 1} - \frac{1}{s + 4}\right\}$$

$$\therefore \ \bar{x} = \frac{2}{s + 1} + \frac{e^{-2s}}{s + 1} - \frac{e^{-2s}}{s + 4}$$

$$\therefore \ x = 2e^{-t} + u(t - 2) \cdot e^{-(t-2)} - u(t - 2) \cdot e^{-4(t-2)}$$

$$= 2e^{-t} + u(t - 2) \cdot e^2 \cdot e^{-t} - u(t - 2) \cdot e^8 \cdot e^{-4t}$$

$$x = e^{-t}\{2 + e^2 \cdot u(t - 2)\} - e^8 \cdot e^{-4t} \cdot u(t - 2)$$

Harmonic oscillators

If the position of a system at time t is described by the expression $f(t)$ where $\boxed{52}$
$f(t)$ satisfies the differential equation

$af''(t) + bf(t) = 0$, $f(0) = \alpha$ and $f'(0) = \beta$

(and where a and b have the same sign)

then, taking Laplace transforms of both sides gives

$L\{af''(t) + bf(t)\} = L\{0\}$

That is

$a[s^2F(s) - s\alpha - \beta] + b[F(s)] = 0$

Collecting like terms gives

$(as^2 + b)F(s) = s\alpha + \beta$

giving

$F(s) = \dfrac{s\alpha + \beta}{as^2 + b}$

Therefore $F(s) = \dfrac{s(\alpha/a)}{s^2 + (b/a)} + \dfrac{\beta/a}{s^2 + (b/a)}$ and so

$f(t) = \dfrac{\alpha}{a}\cos\sqrt{\dfrac{b}{a}}t + \dfrac{\beta}{a}\sin\sqrt{\dfrac{b}{a}}t$

The system executes *simple harmonic, oscillatory motion with frequency*
$\sqrt{\dfrac{b}{a}}$ radians per unit of time and with period $\dfrac{2\pi}{\sqrt{b/a}} = 2\pi\sqrt{\dfrac{a}{b}}$. It is called
an **harmonic oscillator**. Let's try some examples.

Example 1

Find the solution to the harmonic oscillator

$f''(t) + 16f(t) = 0$ where $f(0) = 1$ and $f'(0) = 0$

Taking Laplace transforms gives

$$F(s) = \ldots\ldots\ldots$$

$\boxed{53}$

$$\boxed{F(s) = \dfrac{s}{s^2 + 16}}$$

Because

Taking Laplace transforms $L\{f''(t) + 16f(t)\} = L\{0\}$.

That is $s^2F(s) - s + 16F(s) = 0$ and so

$F(s) = \dfrac{s}{s^2 + 16}$

This means that

$$f(t) = \ldots\ldots\ldots$$

54

$$\boxed{f(t) = \cos 4t}$$

Because

$$F(s) = \frac{s}{s^2 + 16} = \frac{s}{s^2 + 4^2} \text{ so } f(t) = \cos 4t \text{ from the Table of Laplace transforms}$$

in Appendix 8-1 on page 254.

The motion of this system is then periodic with frequency 4 radians per unit of time and with period $2\pi/4 = \pi/2$ units of time.

Example 2

The frequency and period of the harmonic oscillator whose position $f(t)$ satisfies the differential equation

$$5f''(t) + 10f(t) = 0 \text{ where } f(0) = 0 \text{ and } f'(0) = 4$$

is given as

$$\text{frequency} \ldots\ldots\ldots \text{ radians per unit of time}$$
$$\text{and period} \ldots\ldots\ldots \text{ units of time}$$

55

$$\boxed{\text{frequency } \sqrt{2} \text{ and period } \sqrt{2}\pi}$$

Because

Taking Laplace transforms gives

$$L\{5f''(t) + 10f(t)\} = L\{0\} \text{ that is } 5s^2F(s) - 4 + 10F(s) = 0 \text{ so that}$$

$$F(s) = \frac{4}{5s^2 + 10} = \frac{4/5}{s^2 + 2}$$

and from the Table of Laplace transforms in Appendix 8-1 on page 254

$$f(t) = \frac{2\sqrt{2}}{5} \sin \sqrt{2}t$$

This is periodic with frequency $\sqrt{2}$ radians per unit of time and period $2\pi/\sqrt{2} = \sqrt{2}\pi$ units of time.

Notice that the amplitude of the motion is $\dfrac{2\sqrt{2}}{5}$.

56 **Damped motion**

Consider the equation

$$5f''(t) + 5f'(t) + 10f(t) = 0 \text{ where } f(0) = 0 \text{ and } f'(0) = 4$$

This is the same as the last equation in Frame 54 with an extra term added, namely $5f'(t)$. This term describes a particular effect on the system as you will see from the solution.

Solving the differential equation gives

$$f(t) = \ldots\ldots\ldots$$

57

$$f(t) = \frac{8}{5\sqrt{7}}e^{-t/2}\sin\left(\sqrt{7}t/2\right)$$

Because

Taking Laplace transforms gives

$L\{5f''(t) + 5f'(t) + 10f(t)\} = L\{0\}$ that is

$5\left(s^2 F(s) - 4\right) + 5sF(s) + 10F(s) = 0$

so that

$$F(s) = \frac{20}{5s^2 + 5s + 10} = \frac{4}{s^2 + s + 2} = \frac{4}{(s + 1/2)^2 + \left(\sqrt{7}/2\right)^2}$$

and from the Table of Laplace transforms in Appendix 8-1 on page 254

$$f(t) = \frac{8}{\sqrt{7}}e^{-t/2}\sin\left(\sqrt{7}t/2\right)$$

This is periodic with frequency 1 radian per unit of time and period 2π units of time but with an amplitude that is decreasing with time. The graph of this function is as follows

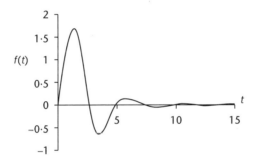

The effect of the $5f'(t)$ in the differential equation is to introduce **damping** into the oscillatory motion so causing the oscillations to decay. Let's try another example.

Example 3

Consider the equation

$5f''(t) + f'(t) + 10f(t) = 0$ where $f(0) = 0$ and $f'(0) = 4$

This equation is again similar to the previous equation but with a smaller damping term of $f'(t)$ instead of $5f'(t)$. Then here

$$f(t) = \ldots\ldots\ldots$$

58

$$f(t) = \frac{4}{\sqrt{1 \cdot 99}} e^{-0.1t} \sin \sqrt{1 \cdot 99}t$$

Because

Taking Laplace transforms gives

$$L\{5f''(t) + f'(t) + 10f(t)\} = L\{0\} \text{ that is}$$
$$5\left(s^2 F(s) - 4\right) + sF(s) + 10F(s) = 0$$

so that

$$F(s) = \frac{20}{5s^2 + 1s + 10} = \frac{4}{s^2 + 0 \cdot 2s + 2} = \frac{4}{(s + 0 \cdot 1)^2 + 1 \cdot 99}$$

and from the Table of Laplace transforms in Appendix 8-1 on page 254

$$f(t) = \frac{4}{\sqrt{1 \cdot 99}} e^{-0.1t} \sin \sqrt{1 \cdot 99}t$$

This is periodic with frequency $\sqrt{1 \cdot 99}$ radians per unit of time and period $2\pi/\sqrt{1 \cdot 99}$ units of time and with an amplitude that is decreasing with time. The graph of this function is as follows

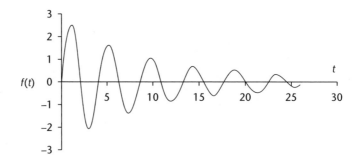

Again, the effect of the $f'(t)$ in the differential equation is to introduce damping into the oscillatory motion so causing it to decay. Also because the coefficient of $f'(t)$ is smaller in this example, the damping is less severe.

Forced harmonic motion with damping

59

The equation

$$f''(t) + f'(t) + f(t) = e^t \text{ where } f(0) = 0 \text{ and } f'(0) = 0$$

we know would represent damped harmonic motion were it not for the exponential on the right-hand side. To see the effect of the exponential we solve the equation.

Taking Laplace transforms we see that

$$F(s) = \dots\dots\dots$$

$$F(s) = \frac{1}{(s-1)(s^2+s+1)}$$

Because

$$L\{f''(t) + f'(t) + f(t)\} = L\{e^t\} \text{ that is } (s^2 + s + 1)F(s) = \frac{1}{s-1} \text{ so}$$

$$F(s) = \frac{1}{(s-1)(s^2+s+1)}$$

Separating into partial fractions gives

$$F(s) = \dots\dots\dots$$

$$F(s) = \frac{1}{3(s-1)} - \frac{s+2}{3(s^2+s+1)}$$

Because

$$\frac{1}{(s-1)(s^2+s+1)} = \frac{A}{(s-1)} + \frac{Bs+C}{(s^2+s+1)}$$
$$= \frac{A(s^2+s+1) + (Bs+C)(s-1)}{(s-1)(s^2+s+1)}$$

Equating numerators and then comparing coefficients of powers of s gives

$$1 = A(s^2+s+1) + (Bs+C)(s-1)$$

$[s^2]$: $0 = A + B$ (1)   So $(2)+(3)$: $1 = 2A - B$

$[s]$: $0 = A - B + C$ (2) $2 \times (1)$: $0 = 2A + 2B$

$[CT]$: $1 = A - C$ (3)   Therefore: $-1 = 3B$

so $B = -1/3 = -A$ and $C = -2/3$

Thus $F(s) = \dfrac{1}{(s-1)(s^2+s+1)} = \dfrac{1}{3(s-1)} - \dfrac{s+2}{3(s^2+s+1)}$

Consequently

$$f(t) = \dots\dots\dots$$

62

$$f(t) = \frac{e^t}{3} - \frac{1}{3}e^{-t/2}\left(\cos\frac{\sqrt{3}}{2}t + \sqrt{3}\sin\frac{\sqrt{3}}{2}t\right)$$

Because

$$F(s) = \frac{1}{3(s-1)} - \frac{s+2}{3(s^2+s+1)}$$

$$= \frac{1}{3(s-1)} - \frac{s+\frac{1}{2}}{3\left(\left(s+\frac{1}{2}\right)^2+\frac{3}{4}\right)} - \frac{\frac{3}{2}}{3\left(\left(s+\frac{1}{2}\right)^2+\frac{3}{4}\right)}$$

So

$$f(t) = \frac{e^t}{3} - \frac{1}{3}e^{-t/2}\left(\cos\frac{\sqrt{3}}{2}t + \sqrt{3}\sin\frac{\sqrt{3}}{2}t\right)$$

from the Table of Laplace transforms in Appendix 8-1 on page 254.

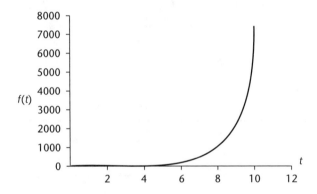

Notice that the term $\frac{1}{3}e^{-t/2}\left(\cos\frac{\sqrt{3}}{2}t + \sqrt{3}\sin\frac{\sqrt{3}}{2}t\right)$ represents damped harmonic motion and is called the **transient** term whereas the term $\frac{e^t}{3}$ represents a **steady-state** term, so called because as the transient term decays the steady-state term remains the dominant part of the solution. The steady-state solution is a direct consequence of the term on the right-hand side of the differential equation.

Try another one for yourself. The transient and steady-state terms of the system described by the differential equation

$$f''(t) + 2f'(t) + 5f(t) = e^{2t} \text{ where } f(0) = 0 \text{ and } f'(0) = 1$$

are   Transient term . . . .   Steady-state term . . . .

$$-\frac{1}{13}e^{-t}\cos 2t + \frac{5}{13}e^{-t}\sin 2t, \ \frac{1}{13}e^{2t}$$

Because

Taking Laplace transforms, $L\{f''(t) + 2f'(t) + 5f(t)\} = L\{e^{2t}\}$. That is

$$[s^2 F(s) - 1] + 2sF(s) + 5F(s) = \frac{1}{s-2}, \text{ that is}$$

$$(s^2 + 2s + 5)F(s) = 1 + \frac{1}{s-2} = \frac{s-1}{s-2}$$

So that $F(s) = \dfrac{s-1}{(s-2)(s^2+2s+5)} = \dfrac{A}{s-2} + \dfrac{Bs+C}{s^2+2s+5}$. Hence

$s - 1 = A(s^2 + 2s + 5) + (Bs + C)(s - 2)$. Equating powers of s gives

$[s^2]$: $0 = A + B$

$[s]$: $1 = 2A - 2B + C$

$[CT]$: $-1 = 5A - 2C$

Solving these three equations gives $A = 1/13$, $B = -1/13$ and $C = 9/13$ so that

$$F(s) = \frac{1}{13(s-2)} - \frac{s-9}{13(s^2+2s+5)}$$

$$= \frac{1}{13(s-2)} - \frac{s-9}{13\big((s+1)^2+2^2\big)}. \text{ That is}$$

$$F(s) = \frac{1}{13(s-2)} - \frac{s+1}{13\big((s+1)^2+2^2\big)} + \frac{10}{13\big((s+1)^2+2^2\big)}$$

Therefore

$$f(t) = \frac{1}{13}e^{2t} - \frac{1}{13}e^{-t}\cos 2t + \frac{5}{13}e^{-t}\sin 2t$$

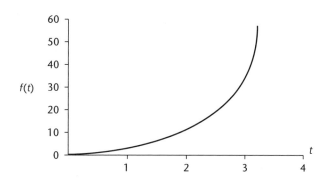

Next frame

64   Resonance

These differential equations with a function on the right-hand side are called **inhomogeneous differential equations**. They represent systems whose behaviour $f(t)$ is dictated by the structure of the left-hand side and the **forcing function** on the right-hand side. If an undamped and unforced system which exhibits periodic behaviour has a periodic forcing function applied that has the same period then **resonance** will occur and the system will undergo periodic behaviour with an increasing amplitude. An example will illustrate this.

The differential equation

$$f''(t) + f(t) = 0 \text{ where } f(0) = 0 \text{ and } f'(0) = 1$$

represents an undamped, unforced system with behaviour

$$f(t) = \ldots\ldots\ldots\ldots$$

65

$$\boxed{f(t) = \sin t}$$

Because

Taking the Laplace transform of both sides of the equation gives

$$L\{f''(t) + f(t)\} = L\{0\} \text{ that is } s^2F(s) - 1 + F(s) = 0 \text{ so that}$$

$$F(s) = \frac{1}{s^2 + 1} \text{ giving } f(t) = \sin t$$

If the forcing term $-2\sin t$ is applied to the right-hand side of the equation it has the same period as the natural frequency of the system being forced and so resonance will set in. The differential equation to solve is then

$$f''(t) + f(t) = -2\sin t \text{ where } f(0) = 0 \text{ and } f'(0) = 1$$

This has the solution $f(t) = \ldots\ldots\ldots\ldots$

66

$$f(t) = t \cos t$$

Because

Taking the Laplace transform of both sides of the equation gives

$$L\{f''(t) + f(t)\} = L\{-2\sin t\} \text{ that is } s^2 F(s) - 1 + F(s) = -\frac{2}{s^2 + 1}$$

so that $F(s) = \dfrac{1}{s^2 + 1} - \dfrac{2}{(s^2 + 1)^2}$ giving $F(s) = \dfrac{s^2 - 1}{(s^2 + 1)^2}$. Now, the

Laplace transform of $\cos t$ is $\dfrac{s}{s^2 + 1}$ and $\left(\dfrac{s}{s^2 + 1}\right)' = -\dfrac{s^2 - 1}{(s^2 + 1)^2}$.

Therefore $f(t) = t \cos t$

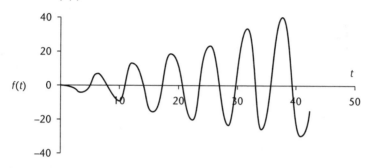

The system undergoes periodic behaviour with an increasing amplitude.

You have now reached the end of this Program and this brings you to the **Review summary** and the **Can You?** checklist. Following that is the **Test exercise**. Work through this *at your own pace*. A set of **Further problems** provides additional valuable practice.

 Review summary

67

1 *Periodic functions*

$$f(t) = f(t + nT) \qquad n = 1, 2, 3, \ldots \qquad \text{Period} = T.$$

2 *Laplace transform of a periodic function with period T*

$$L\{f(t)\} = \frac{1}{1 - e^{-Ts}} \int_0^T e^{-st} \cdot f(t)\, dt.$$

3 *Inverse transforms involving periodic functions*

e.g. $L^{-1}\left\{\dfrac{1 + 2e^{-3s} - 3e^{-2s}}{s(1 - e^{-3s})}\right\}$

Expand $(1 - e^{-3s})^{-1}$ as a binomial series, like

$$(1 - x)^{-1} = 1 + x + x^2 + x^3 + \ldots$$

Multiply out and take inverse transforms of each term in turn.

▶

4 *Dirac delta function* or unit impulse function

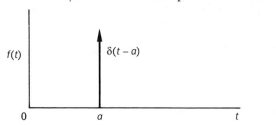

$$\delta(t - a) = 0 \qquad t \neq a$$
$$= \infty \qquad t = a.$$

5 *Delta function at the origin*

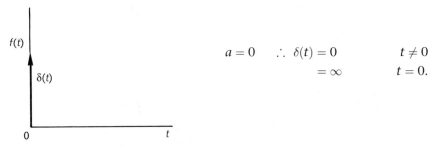

$$a = 0 \quad \therefore \ \delta(t) = 0 \qquad t \neq 0$$
$$= \infty \qquad t = 0.$$

6 *Area of pulse = 1*

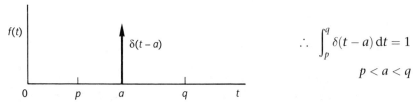

$$\therefore \ \int_{p}^{q} \delta(t - a)\,\mathrm{d}t = 1$$
$$p < a < q$$

7 *Integration of the impulse function*
$$\int_{p}^{q} f(t) \cdot \delta(t - a)\,\mathrm{d}t = f(a) \qquad p < a < q$$

8 *Laplace transform of* $\delta(t - a)$
$$L\{\delta(t - a)\} = e^{-as}$$
$$L\{\delta(t)\} = 1 \text{ because } a = 0$$
$$L\{f(t) \cdot \delta(t - a)\} = f(a) \cdot e^{-as}.$$

9 *Harmonic oscillators*
The equation of $af''(t) + bf(t) = 0$, $f(0) = \alpha$ and $f'(0) = \beta$, where a and b are of the same sign, represents a system undergoing simple harmonic motion and is referred to as an harmonic oscillator. The system oscillates with a frequency of $\sqrt{\dfrac{b}{a}}$ radians per unit of time and with period $\dfrac{2\pi}{\sqrt{b/a}} = 2\pi\sqrt{\dfrac{a}{b}}$ units of time. If a first derivative term is added to the left-hand side of the equation then, provided all three coefficients have the same sign, the system will undergo damped harmonic motion.

▶

10 *Forced harmonic motion*
Forced harmonic motion is achieved by the existence of a term on the right-hand side of the equation giving rise to transient and steady-state parts of the solution.

11 *Resonance*
Resonance is exhibited by a system undergoing periodic behaviour with a growing amplitude of vibration. Resonance occurs when a system, whose unforced behaviour is periodic, is forced with the same period.

☑ Can You?

Checklist 8 **68**

Check this list before and after you try the end of Program test.

On a scale of 1 to 5, how confident are you that you can: Frames

- Find the Laplace transforms of periodic functions? **1** to **14**
  Yes ☐ ☐ ☐ ☐ ☐ *No*

- Obtain the inverse Laplace transforms of transforms of periodic functions? **15** to **28**
  Yes ☐ ☐ ☐ ☐ ☐ *No*

- Describe and use the unit impulse to evaluate integrals? **29** to **34**
  Yes ☐ ☐ ☐ ☐ ☐ *No*

- Obtain the Laplace transform of the unit impulse? **35** to **42**
  Yes ☐ ☐ ☐ ☐ ☐ *No*

- Use the Laplace transform to solve differential equations involving the unit impulse? **43** to **51**
  Yes ☐ ☐ ☐ ☐ ☐ *No*

- Solve the equation and describe the behaviour of an harmonic oscillator? **52** to **66**
  Yes ☐ ☐ ☐ ☐ ☐ *No*

🚲 **Test exercise 8**

69   **1**   Determine the Laplace transform of the periodic function shown.

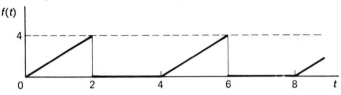

2   Evaluate

(a) $\displaystyle\int_0^4 e^{-3t} \cdot \delta(t-2)\, dt$

(b) $\displaystyle\int_0^\infty \sin 3t \cdot \delta(t-\pi)\, dt$

(c) $\displaystyle\int_1^3 (2t^2+3) \cdot \delta(t-2)\, dt.$

3   Determine (a) $L\{4 \cdot \delta(t-3)\}$,   (b) $L\{e^{-3t} \cdot \delta(t-2)\}$.

4   Sketch the graph of $f(t) = 3 \cdot \delta(t) + 4 \cdot \delta(t-2) - 3 \cdot \delta(t-4)$ and determine its Laplace transform.

5   Solve the equation $\ddot{x} + 6\dot{x} + 10x = 7 \cdot \delta(t)$ given that, at $t=0$, $x = -1$ and $\dot{x} = 0$.

6   The equation of motion of a system is
$$\ddot{x} + 3\dot{x} + 2x = 3 \cdot \delta(t-4).$$
At $t=0$, $x = 2$ and $\dot{x} = -4$. Determine an expression for the displacement x in terms of t.

7   Find the frequency, periodic time and solution for each of the following harmonic oscillators.
(a) $f''(t) + f(t) = 0$ given that $f(0) = 0$ and $f'(0) = 1$
(b) $6f''(t) + 2f'(t) + 9f(t) = 0$ given that $f(0) = 0$ and $f'(0) = 3$.

8   Find the transient and steady-state solutions of the forced harmonic oscillator
$f''(t) + 2f'(t) + 3f(t) = 4e^{5t}$ given that $f(0) = -2$ and $f'(0) = 6$.

Further problems 8

1 If $f(t) = a \sin t \quad 0 < t < \pi \\ = 0 \qquad\quad \pi < t < 2\pi \Big\} \; f(t + 2\pi) = f(t),$

70

   prove that $L\{f(t)\} = \dfrac{a}{(s^2 + 1)(1 - e^{-\pi s})}$.

2 If $f(t) = a \sin t \quad 0 < t < \pi \quad f(t + \pi) = f(t)$, determine $L\{f(t)\}$.

3 Find the Laplace transforms of the following periodic functions.

   (a) $f(t) = t \qquad 0 < t < T \qquad\quad f(t + T) = f(t)$

   (b) $f(t) = e^t \qquad 0 < t < 2\pi \qquad f(t + 2\pi) = f(t)$

   (c) $f(t) = t \qquad 0 < t < 1 \\ = 0 \qquad\quad 1 < t < 2 \Big\} \; f(t + 2) = f(t)$

   (d) $f(t) = t^2 \qquad 0 < t < 2 \\ = 4 \qquad\quad 2 < t < 3 \Big\} \; f(t + 3) = f(t)$

4 A mass M is attached to a spring of stiffness $\omega^2 M$ and is set in motion at $t = 0$ by an impulsive force P. The equation of motion is

   $$M\ddot{x} + M\omega^2 x = P \cdot \delta(t).$$

   Obtain an expression for x in terms of t.

5 An impulsive voltage E is applied at $t = 0$ to a series circuit containing inductance L and capacitance C. Initially, the current and charge are zero. The current i at time t is given by

   $$L\frac{di}{dt} + \frac{q}{C} = E \cdot \delta(t)$$

   where q is the instantaneous value of the charge on the capacitor. Since $i = \dfrac{dq}{dt}$, determine an expression for the current i in the circuit at time t.

6 A system has the equation of motion

   $$\ddot{x} + 5\dot{x} + 6x = F(t)$$

   where, at $t = 0$, $x = 0$ and $\dot{x} = 2$. If $F(t)$ is an impulse of 20 units applied at $t = 4$, determine an expression for x in terms of t.

7 Find the frequency, periodic time and solution for each of the following harmonic oscillators.

   (a) $12f''(t) + f(t) = 0$ given that $f(0) = -1$ and $f'(0) = 2$

   (b) $f''(t) + 12f(t) = 0$ given that $f(0) = 2$ and $f'(0) = -1$.

8 Solve for each of the following harmonic oscillators.

   (a) $4.6f''(t) + 2.2f(t) = 0$ given that $f(0) = 1.6$ and $f'(0) = -3.1$

   (b) $\sqrt{2}f''(t) + \sqrt{3}f(t) = 0$ given that $f(0) = 0$ and $f'(0) = \pi$.

9 Find the transient and steady-state solutions of the forced harmonic oscillator

   $$4f''(t) + 3f'(t) + 2f(t) = e^t$$

   given that $f(0) = 0$ and $f'(0) = 6$.

Appendix 8-1: Table of Laplace Transforms

$F(s)$	$f(t)$	
$\dfrac{a}{s}$	a	
$\dfrac{1}{s+a}$	e^{-at}	
$\dfrac{n!}{s^{n+1}}$	t^n	(n a positive integer)
$\dfrac{1}{s^n}$	$\dfrac{t^{n-1}}{(n-1)!}$	(n a positive integer)
$\dfrac{a}{s^2+a^2}$	$\sin at$	
$\dfrac{s}{s^2+a^2}$	$\cos at$	
$\dfrac{a}{s^2-a^2}$	$\sinh at$	
$\dfrac{s}{s^2-a^2}$	$\cosh at$	

Z transforms

Learning outcomes

When you have completed this Program you will be able to:

- Define the Z transform of a sequence and derive transforms of specified sequences
- Make reference to a table of standard Z transforms
- Recognise the Z transform as being a linear transform and so obtain the transform of linear combinations of standard sequences
- Apply the first and second shift theorems, the translation theorem, the initial and final value theorems and the derivative theorem
- Use partial fractions to derive the inverse transforms
- Solve linear, first-order, constant coefficient recurrence relations
- Demonstrate the relationship between the Laplace transform and the Z transform

Introduction

1

The Laplace transform deals with continuous functions and can be used to solve many differential equations that arise in science and engineering. There are occasions, however, when we have to deal with discrete functions – *sequences* – and their associated **difference equations**. For example, the central processing unit of your computer can only handle information in the form of pulses of electricity. This information transmission is called **digital** transmission. There are, however, times when information is fed into the computer in the form of a continuously varying signal called an **analog** signal. For instance, a mouse can be moved about the flat surface of your desk in a continuous manner but the central processing unit will only recognize position on the screen to the nearest pixel. The analog signal coming from the mouse needs to be converted into a digital signal for recognition by the computer's central processing unit. This conversion of a signal from analog to digital is achieved by a device called a **demodulator** that *samples* the analog signal at regular intervals of time and outputs the sampled values as the digital signal – as a sequence of values. The Z transform, which is allied to the Laplace transform, deals with such sequences and the recurrence relations – or difference equations – that arise.

Sequences

2

The sequence $\ldots, 3^{-2}, 3^{-1}, 3^0, 3, 3^2, 3^3, \ldots$ has a general term of the form 3^k and as a shorthand notation we use $\{3^k\}_{-\infty}^{\infty}$ to represent this sequence and to indicate that the powers range from $-\infty$ to ∞. The sum

$$\sum_{k=-\infty}^{\infty} \left(\frac{3}{z}\right)^k = \ldots + \left(\frac{3}{z}\right)^{-1} + \left(\frac{3}{z}\right)^0 + \left(\frac{3}{z}\right)^1 + \left(\frac{3}{z}\right)^2 + \ldots$$

is called the Z **transform** of the sequence, $Z\{3^k\}_{-\infty}^{\infty}$, and is denoted by $F(z)$, where the complex number z is chosen to ensure that the sum is finite. We say that

$$\{3^k\}_{-\infty}^{\infty} \text{ and } Z\{3^k\}_{-\infty}^{\infty} = F(z) = \sum_{k=-\infty}^{\infty} \left(\frac{3}{z}\right)^k \text{ form a } Z \text{ transform pair.}$$

▶

For our purposes we shall consider only *causal sequences* of the form $\{x_k\}_0^\infty$ where $x_k = 0$ for $k < 0$ which for brevity we shall denote by $\{x_k\}$ with corresponding Z transform

$$Z\{x_k\} = F(z) = \sum_{k=0}^{\infty} \frac{x_k}{z^k}.$$

Notice that this is the *definition* of the Z transform of the sequence $\{x_k\}$. For example, the *unit impulse* sequence $\{\delta_k\} = \{1, 0, 0, 0, \ldots\}$ has the Z transform

$$Z\{\delta_k\} = \ldots\ldots\ldots\ldots \text{ valid for } \ldots\ldots\ldots\ldots \text{ values of } z$$

<hr>

$Z\{\delta_k\} = 1$ valid for all values of z

3

Because

$$Z\{\delta_k\} = \sum_{k=0}^{\infty} \frac{\delta_k}{z_k}$$

$$= 1 + \frac{0}{z} + \frac{0}{z^2} + \ldots = 1$$

Try another.

The sequence $\{u_k\} = \{1, 1, 1, \ldots\} = \{1\}$ is called the *unit step* sequence and has the Z transform

$$\ldots\ldots\ldots\ldots \text{ provided } |z| \ldots\ldots\ldots\ldots$$

Next frame

<hr>

$\dfrac{z}{z-1}$ provided $|z| > 1$

4

Because

$$Z\{u_k\} = F(z)$$

$$= \sum_{k=0}^{\infty} \frac{u_k}{z^k} = \sum_{k=0}^{\infty} \frac{1}{z^k}$$

$$= 1 + \frac{1}{z} + \frac{1}{z^2} + \frac{1}{z^3} + \frac{1}{z^4} + \ldots$$

Comparing this to the series expansion of $\dfrac{1}{1-x} = 1 + x + x^2 + x^3 + \ldots$ which is valid for $|x| < 1$ then

$$F(z) = \frac{1}{1 - \dfrac{1}{z}} \text{ provided } \left|\frac{1}{z}\right| < 1$$

$$= \frac{z}{z-1} \text{ provided } |z| > 1$$

And another.

Given the causal sequence $\{x_k\} = \{1, a, a^2, a^3, a^4, \ldots\} = \{a^k\}$

the Z transform is $\ldots\ldots\ldots\ldots$

Next frame

5

$$\boxed{\frac{z}{z-a} \text{ provided } |z| > a}$$

Because

$$Z\{a^k\} = \sum_{k=0}^{\infty} \frac{a^k}{z^k}$$

$$= \sum_{k=0}^{\infty} \left(\frac{a}{z}\right)^k$$

$$= 1 + \frac{a}{z} + \left(\frac{a}{z}\right)^2 + \left(\frac{a}{z}\right)^3 + \dots$$

Comparing this to the series expansion of $\dfrac{1}{1-x} = 1 + x + x^2 + x^3 + \dots$ which is valid for $|x| < 1$ then

$$F(z) = 1 + \frac{a}{z} + \left(\frac{a}{z}\right)^2 + \left(\frac{a}{z}\right)^3 + \dots$$

$$= \frac{1}{1 - \dfrac{a}{z}} \text{ provided } \left|\frac{a}{z}\right| < 1.$$

That is, multiplying numerator and denominator by z

$$F(z) = \frac{z}{z-a} \text{ provided } |z| > |a|$$

Therefore $\{a^k\}$ and $F(z) = \dfrac{z}{z-a}$, $(|z| > |a|)$ form a Z transform pair.

Let's try another. The sequence $\{x_k\} = \{0, 1, 2, 3, 4, \dots\} = \{k\}$ has the Z transform

$$Z\{k\} = F(z) = \dots\dots\dots$$

Answer in the next frame

6

$$\boxed{F(z) = \frac{1}{z} + \frac{2}{z^2} + \frac{3}{z^3} + \frac{4}{z^4} + \dots}$$

Because

$$Z\{k\} = F(z)$$

$$= \sum_{k=0}^{\infty} \frac{x_k}{z^k}$$

$$= \sum_{k=0}^{\infty} \frac{k}{z^k}$$

$$= 0 + \frac{1}{z} + \frac{2}{z^2} + \frac{3}{z^3} + \frac{4}{z^4} + \dots$$

By comparing this sequence with the derivative of $(1-x)^{-1}$ and its series representation, this sequence can be written as a rational expression in z as $F(z) = \dots\dots\dots$

7

$$F(z) = \frac{z}{(z-1)^2}$$

Because

$$F(z) = 0 + \frac{1}{z} + \frac{2}{z^2} + \frac{3}{z^3} + \frac{4}{z^4} + \cdots$$

Comparing this with the series expansion

$$1 + 2x + 3x^2 + 4x^3 + \cdots = \frac{d}{dx}(1 + x + x^2 + x^3 + \cdots)$$

$$= \frac{d}{dx}(1-x)^{-1} = \frac{1}{(1-x)^2}$$

then we can see that by multiplying $F(z)$ by z

$$zF(z) = 1 + \frac{2}{z} + \frac{3}{z^2} + \frac{4}{z^3} + \cdots = \frac{1}{(1-1/z)^2}$$

so, dividing both sides by z gives

$$F(z) = \frac{1}{z(1-1/z)^2} = \frac{z}{(z-1)^2}$$

Next frame

Table of **Z** transforms

We list the results that we have obtained so far as well as some additional ones for future reference.

8

Sequence	Transform $F(z)$	Permitted values of z
$\{\delta_k\} = \{1, 0, 0, \ldots\}$	1	All values of z
$\{u_k\} = \{1, 1, 1, \ldots\}$	$\dfrac{z}{z-1}$	$\|z\| > 1$
$\{k\} = \{0, 1, 2, 3, \ldots\}$	$\dfrac{z}{(z-1)^2}$	$\|z\| > 1$
$\{k^2\} = \{0, 1, 4, 9, \ldots\}$	$\dfrac{z(z+1)}{(z-1)^3}$	$\|z\| > 1$
$\{k^3\} = \{0, 1, 8, 27, \ldots\}$	$\dfrac{z(z^2 + 4z + 1)}{(z-1)^4}$	$\|z\| > 1$
$\{a^k\} = \{1, a, a^2, a^3, \ldots\}$	$\dfrac{z}{(z-a)}$	$\|z\| > \|a\|$
$\{ka^k\} = \{0, a, 2a^2, 3a^3, \ldots\}$	$\dfrac{az}{(z-a)^2}$	$\|z\| > \|a\|$

Next frame

Properties of *Z* transforms

9

1 *Linearity*

The *Z* transform is a linear transform. That is, if a and b are constants then

$$Z(a\{x_k\} + b\{y_k\}) = aZ\{x_k\} + bZ\{y_k\}$$

For example, the *Z* transform of the sequence $\{k\}$ is $Z\{k\} = \ldots\ldots\ldots\ldots$ and the *Z* transform of the sequence $\{e^{-2k}\}$ is $Z\{e^{-2k}\} = \ldots\ldots\ldots\ldots$

10

$$Z\{k\} = \frac{z}{(z-1)^2} \text{ and } Z\{e^{-2k}\} = \frac{z}{z - e^{-2}}$$

Because

$$Z\{k\} = \frac{z}{(z-1)^2} \text{ from the table and, also from the table,}$$

$$Z\{a^k\} = \frac{z}{z-a} \text{ so when } a = e^{-2},$$

$$Z\{e^{-2k}\} = \frac{z}{z - e^{-2}}$$

Consequently, the *Z* transform of $3\{k\} - 5\{e^{-2k}\}$ is $\ldots\ldots\ldots\ldots$

11

$$\frac{-5z^3 + 13z^2 - z(3e^{-2} + 5)}{(z-1)^2(z - e^{-2})}$$

Because

$$Z(3\{k\} - 5\{e^{-2k}\}) = 3Z\{k\} - 5Z\{e^{-2k}\}$$

$$= \frac{3z}{(z-1)^2} - \frac{5z}{(z - e^{-2})}$$

$$= \frac{3z(z - e^{-2}) - 5z(z-1)^2}{(z-1)^2(z - e^{-2})}$$

$$\frac{3z^2 - 3ze^{-2} - 5z^3 + 10z^2 - 5z}{(z-1)^2(z - e^{-2})}$$

$$= \frac{-5z^3 + 13z^2 - z(3e^{-2} + 5)}{(z-1)^2(z - e^{-2})}$$

2 *First shift theorem* (shifting to the left)

If $Z\{x_k\} = F(z)$ then

$$Z\{x_{k+m}\} = z^m F(z) - \left[z^m x_0 + z^{m-1} x_1 + \ldots + z x_{m-1}\right]$$

is the Z transform of the sequence that has been shifted by m places to the left. For example

$$Z\{x_{k+1}\} = zF(z) - zx_0$$
$$Z\{x_{k+2}\} = z^2 F(z) - z^2 x_0 - zx_1$$

These will be used later when solving difference equations. Note the similarity between these results and the Laplace transforms for the first and second derivatives for continuous functions.

For example, given that $Z\{4^k\} = \dfrac{z}{z-4}$ then

$$Z\{4^{k+3}\} = \ldots\ldots\ldots\ldots$$

12

$$\boxed{\dfrac{64z}{z-4}}$$

Because

$$Z\{x_{k+m}\} = z^m F(z) - \left[z^m x_0 + z^{m-1} x_1 + \ldots + z x_{m-1}\right]$$

so

$$Z\{4^{k+3}\} = z^3 Z\{4^k\} - \left[z^3 4^0 + z^2 4^1 + z 4^2\right] \text{ where } Z\{4^k\} = \dfrac{z}{z-4}$$

$$= z^3 \dfrac{z}{z-4} - \left[z^3 + 4z^2 + 16z\right]$$

$$= \dfrac{z^4}{z-4} - \left[z^3 + 4z^2 + 16z\right]$$

$$= \dfrac{z^4 - (z^3 + 4z^2 + 16z)(z-4)}{z-4}$$

$$= \dfrac{z^4 - (z^4 - 64z)}{z-4}$$

$$= \dfrac{64z}{z-4}$$

In this way we have derived the Z transform of the sequence $\{64, 256, 1024, \ldots\}$ by shifting the sequence $\{1, 4, 16, 64, 256, \ldots\}$ three places to the left and losing the first three terms.

Try another. Given that $Z\{k\} = \dfrac{z}{(z-1)^2}$ then

$$Z\{(k+1)\} = \ldots\ldots\ldots\ldots$$

13

$$\boxed{\dfrac{z^2}{(z-1)^2}}$$

Because

$$Z\{x_{k+m}\} = z^m F(z) - \left[z^m x_0 + z^{m-1} x_1 + \ldots + z x_{m-1}\right]$$

so

$$Z\{k+1\} = z\frac{z}{(z-1)^2} - [z \times 0]$$

$$= \frac{z^2}{(z-1)^2}$$

3 *Second shift theorem* (shifting to the right)

If $Z\{x_k\} = F(z)$ then

$$Z\{x_{k-m}\} = z^{-m} F(z)$$

the Z transform of the sequence that has been shifted by m places to the right.

For example, given that $Z\{x_k\} = \dfrac{z}{z-1}$ then

$$Z\{x_{k-3}\} = \ldots\ldots\ldots\ldots$$

14

$$\boxed{\dfrac{1}{z^2(z-1)}}$$

Because

$$Z\{x_{k-m}\} = z^{-m} F(z)$$

so

$$Z\{x_{k-3}\} = z^{-3}\frac{z}{z-1}$$

$$= \frac{1}{z^2(z-1)}$$

In this way we have derived the Z transform of the sequence $\{0, 0, 0, 1, 1, 1, \ldots\}$ by shifting the sequence $\{1, 1, 1, 1, \ldots\}$ three places to the right and defining the first three terms as zeros.

Try this one. The sequence $\{x_k\}$ with Z transform

$$Z\{x_k\} = \frac{1}{(z-a)}, \text{ where } a \text{ is a constant, is } \{\ldots\ldots\ldots\ldots\}$$

15

$$\boxed{\{a^{k-1}\}}$$

Because

From the table of transforms the nearest transform to the one in question is $\dfrac{z}{(z-a)}$ which is the Z transform of $\{a^k\}$. Now

$$\frac{1}{(z-a)} = \frac{1}{z} \times \frac{z}{(z-a)}$$
$$= z^{-1}F(z) \quad \text{where } F(z) = Z\{a^k\}$$

and so

$$\frac{1}{(z-a)} = Z\{a^{k-1}\}$$

which is the Z transform of $\{a^k\}$, shifted one place to the right.

4 *Translation*

If the sequence $\{x_k\}$ has the Z transform $Z\{x_k\} = F(z)$ then the sequence $\{a^k x_k\}$ has the Z transform $Z\{a^k x_k\} = F(a^{-1}z)$.

For example, $Z\{k\} = \dfrac{z}{(z-1)^2}$ so that $Z\{2^k k\} = \dots\dots\dots$

16

$$\boxed{\dfrac{2z}{(z-2)^2}}$$

Because

Since $Z\{k\} = \dfrac{z}{(z-1)^2} = F(z)$ then by the translation property

$$Z\{2^k k\} = F(2^{-1}z)$$
$$= \frac{2^{-1}z}{(2^{-1}z - 1)^2}$$
$$= \frac{2z}{(z-2)^2}$$

▶

5 *Final value theorem*

For the sequence $\{x_k\}$ with Z transform $F(z)$

$$\underset{k\to\infty}{Lim}\, x_k = \underset{z\to 1}{Lim}\left\{\left(\frac{z-1}{z}\right)F(z)\right\} \text{ provided that } \underset{k\to\infty}{Lim}\, x_k \text{ exists.}$$

For example, the sequence $\left\{\left(\frac{1}{2}\right)^k\right\}$ has the Z transform

$$F(z) = \frac{z}{z-\frac{1}{2}} = \frac{2z}{2z-1}.$$

Now

$$\underset{z\to 1}{Lim}\left\{\left(\frac{z-1}{z}\right)F(z)\right\} = \underset{z\to 1}{Lim}\left\{\frac{2(z-1)}{2z-1}\right\} = 0$$

and

$$\underset{k\to\infty}{Lim}\left\{\left(\frac{1}{2}\right)^k\right\} = 0 \text{ which confirms the final value theorem.}$$

Using the final value theorem the final value of the sequence with the Z transform

$$F(z) = \frac{10z^2 + 2z}{(z-1)(5z-1)^2} \text{ is } \ldots\ldots\ldots\ldots$$

17

$$\boxed{0.75}$$

Because

$$\underset{z\to 1}{Lim}\left\{\left(\frac{z-1}{z}\right)F(z)\right\} = \underset{z\to 1}{Lim}\left\{\left(\frac{z-1}{z}\right)\frac{10z^2+2z}{(z-1)(5z-1)^2}\right\}$$

$$= \underset{z\to 1}{Lim}\left\{\frac{10z+2}{(5z-1)^2}\right\}$$

$$= \frac{12}{16}$$

$$= 0.75$$

6 *The initial value theorem*

For the sequence $\{x_k\}$ with Z transform $F(z)$

$$x_0 = \underset{z\to\infty}{Lim}\left\{F(z)\right\}$$

For example, the sequence $\{a^k\}$ has the Z transform $F(z) = \dfrac{z}{z-a}$ and

$\underset{z\to\infty}{Lim}\, F(z) = \underset{z\to\infty}{Lim}\dfrac{z}{z-a} = \underset{z\to\infty}{Lim}\dfrac{1}{1} = 1$ by L'Hôpital's rule. Furthermore $x_0 = a^0 = 1$

so demonstrating the validity of the theorem.

▶

7 *The derivative of the transform*

If $Z\{x_k\} = F(z)$ then $-zF'(z) = Z\{kx_k\}$

This is easily proved.

$$F(z) = \sum_{k=0}^{\infty} x_k z^{-k} \text{ and so } F'(z) = \sum_{k=0}^{\infty} x_k(-k)z^{-k-1} = -\frac{1}{z}\sum_{k=0}^{\infty} x_k k z^{-k}$$

$$= -\frac{1}{z}Z\{kx_k\}$$

and so $-zF'(z) = Z\{kx_k\}$

For example, the sequence $\{a^k\}$ has the Z transform $F(z) = \dfrac{z}{z-a}$ and so the sequence $\{ka^k\}$ has Z transform

$$Z\{kx_k\} = -zF'(z) = \ldots\ldots\ldots\ldots$$

18

$$Z\{kx_k\} = \frac{az}{(z-a)^2}$$

Because

$$-zF'(z) = -z\left(\frac{z}{z-a}\right)' = -z\left(\frac{z-a-z}{(z-a)^2}\right) = \frac{az}{(z-a)^2}$$

Notice that this is in agreement with the Table of transforms in Frame 8.

Next frame

Inverse transforms

19

If the sequence $\{x_k\}$ has Z transform $Z\{x_k\} = F(z)$, the inverse transform is defined as

$$Z^{-1}F(z) = \{x_k\}$$

There are many times when, given the Z transform of a sequence, it is not possible to immediately read off the sequence from the Table of transforms. Instead some manipulation may be required and, as with Laplace transforms, very often this involves using partial fractions.

Example

The sequence $\{x_k\}$ has Z transform $F(z) = \dfrac{z}{z^2 - 5z + 6}$. To find the inverse transform, and hence the sequence, we recognise that the denominator can be factored and separated into partial fractions as

$$F(z) = \ldots\ldots\ldots\ldots$$

20

$$F(z) = \frac{3}{z-3} - \frac{2}{z-2}$$

Because

$$F(z) = \frac{z}{z^2 - 5z + 6}$$

$$= \frac{z}{(z-2)(z-3)}$$

$$= \frac{A}{z-2} + \frac{B}{z-3}$$

$$= \frac{A(z-3) + B(z-2)}{(z-2)(z-3)}$$

Equating numerators gives $z = A(z-3) + B(z-2)$, giving $A + B = 1$ and $-3A - 2B = 0$. From these two equations we find that $A = -2$ and $B = 3$. So

$$F(z) = \frac{3}{z-3} - \frac{2}{z-2}$$

The nearest Z transform in the table to either of these two partial fractions is $Z\{a^k\} = \frac{z}{z-a}$. Therefore if we write

$$F(z) = \frac{3}{z-3} - \frac{2}{z-2}$$

$$= \frac{3}{z} \times \frac{z}{z-3} - \frac{2}{z} \times \frac{z}{z-2}$$

so

$$Z^{-1}F(z) = \ldots\ldots\ldots\ldots$$

21

$$Z^{-1}F(z) = \{3^k - 2^k\}$$

Because

$$F(z) = \frac{3}{z} \times \frac{z}{z-3} - \frac{2}{z} \times \frac{z}{z-2}$$

$$= 3 \times z^{-1} Z\{3^k\} - 2 \times z^{-1} Z\{2^k\}$$

and so

$$Z^{-1}F(z) = 3 \times \{3^{k-1}\} - 2 \times \{2^{k-1}\} \text{ by the second shift theorem}$$

$$= \{3^k\} - \{2^k\}$$

$$= \{3^k - 2^k\} \text{ giving } x_k = 3^k - 2^k$$

There is a simpler way of doing this without employing the second shift theorem. Recognizing that z appears in the numerator of $F(z)$, we consider instead the partial fraction breakdown of $\dfrac{F(z)}{z}$

$$\frac{F(z)}{z} = \ldots\ldots\ldots\ldots$$

$$\boxed{\frac{1}{z-3} - \frac{1}{z-2}}$$

Because

$$\frac{F(z)}{z} = \frac{1}{z} \times \frac{z}{z^2 - 5z + 6}$$

$$= \frac{1}{z^2 - 5z + 6}$$

$$= \frac{1}{(z-2)(z-3)}$$

$$= \frac{A}{z-2} + \frac{B}{z-3}$$

$$= \frac{A(z-3) + B(z-2)}{(z-2)(z-3)}$$

Equating numerators gives $1 = A(z-3) + B(z-2)$, giving

[z]: $A + B = 0$

[CT]: $-3A - 2B = 1$ with solution $A = -1$ and $B = 1$. So that

$$\frac{F(z)}{z} = \frac{1}{z-3} - \frac{1}{z-2} \text{ that is}$$

$$F(z) = \frac{z}{z-3} - \frac{z}{z-2}$$

$$= Z\{3^k\} - Z\{2^k\} \text{ and so}$$

$$Z^{-1}F(z) = \{3^k\} - \{2^k\}$$

$$= \{3^k - 2^k\}$$

Thus the use of the second shift theorem is avoided.

So try one yourself. The sequence $\{x_k\}$ has Z transform

$$F(z) = \frac{5z}{(z^2 - 4z + 4)(z + 2)}$$

therefore $\{x_k\} = \ldots\ldots\ldots\ldots$

23

$$\{x_k\} = \left\{ \frac{5k}{4} - \frac{5}{16} \times \left(2^k + (-2)^k \right) \right\}$$

Because

$$\frac{F(z)}{z} = \frac{1}{z} \times \frac{5z}{(z^2 - 4z + 4)(z + 2)}$$

$$= \frac{5}{(z - 2)^2(z + 2)}$$

$$= \frac{A}{(z - 2)^2} + \frac{B}{z - 2} + \frac{C}{z + 2}$$

$$= \frac{A(z + 2) + B(z - 2)(z + 2) + C(z - 2)^2}{(z - 2)^2(z + 2)}$$

Equating numerators gives $5 = A(z + 2) + B(z^2 - 4) + C(z^2 - 4z + 4)$, giving

[z^2]: $\qquad\qquad B + C = 0$

[z]: $\qquad\qquad A - 4C = 0$

[CT]: $\qquad 2A - 4B + 4C = 5$

with solution $A = 5/4$, $B = -5/16$ and $C = 5/16$, so

$$\frac{F(z)}{z} = \frac{5/4}{(z - 2)^2} - \frac{5/16}{z - 2} + \frac{5/16}{z + 2} \text{ giving}$$

$$F(z) = \frac{5}{8} \times \frac{2z}{(z - 2)^2} - \frac{5}{16} \times \frac{z}{z - 2} + \frac{5}{16} \times \frac{z}{z + 2} \text{ and so}$$

$$Z^{-1}F(z) = \frac{5}{8} \times \{k2^k\} - \frac{5}{16} \times \{2^k\} + \frac{5}{16} \times \left\{ (-2)^k \right\}$$

$$= \left\{ \frac{5}{16} \left[(2k - 1)2^k + (-2)^k \right] \right\}$$

Next frame

Recurrence relations

24

Sometimes adjacent terms of a sequence are related to each other. For example the terms of the sequence

$$\{x_k\} = \{2^k\}$$

are such that $x_{k+1} = 2^{k+1} = 2 \times 2^k = 2x_k$. That is

$$x_{k+1} = 2x_k$$

This equation holds true for all adjacent terms of the sequence – it *recurs* for all values of k. The equation is called a **linear, first-order, constant coefficient recurrence relation**. The order of the equation is given by the maximum shift between related terms – here it is 1. Clearly, the recurrence relation

$$x_{k+2} - x_{k+1} - x_k = 1 \text{ is of order} \ldots\ldots\ldots\ldots$$

$$\boxed{2}$$

25

Because

The maximum shift between terms in the relation is 2 – that is from k to $k+2$.

Initial terms

A recurrence relation can be used to generate the terms of a sequence provided initial terms are given – equal in number to the order of the equation. For example, given the sequence $\{x_k\}$ where $x_{k+1} = 3x_k$ with the initial term $x_0 = 2$ generates the sequence of terms

$$\{x_k\} = \{2, \ldots, \ldots, \ldots, \ldots\}$$

26

$$\boxed{\{x_k\} = \{2, 6, 18, 54, \ldots\}}$$

Because

Since $x_{k+1} = 3x_k$ where $x_0 = 2$ then

$$x_1 = 3x_0 = 3 \times 2 = 6$$
$$x_2 = 3x_1 = 3 \times 6 = 18$$
$$x_3 = 3x_2 = 3 \times 18 = 54$$

Similarly, if another sequence has terms that satisfy the second-order recurrence relation

$$x_{k+2} - 3x_{k+1} + 2x_k = 1 \text{ where } x_0 = 0 \text{ and } x_1 = 1$$

then the first five terms of the sequence are

$$\{x_k\} = \{0, 1, \ldots, \ldots, \ldots, \ldots\}$$

27

$$\boxed{\{x_k\} = \{0, 1, 4, 11, 26, \ldots\}}$$

Because

Since $x_{k+2} - 3x_{k+1} + 2x_k = 1$ where $x_0 = 0$ and $x_1 = 1$ then

$$x_2 - 3x_1 + 2x_0 = 1 \text{ that is } x_2 - 3 \times 1 + 2 \times 0 = 1 \quad \text{and so } x_2 = 4$$
$$x_3 - 3x_2 + 2x_1 = 1 \text{ that is } x_3 - 3 \times 4 + 2 \times 1 = 1 \quad \text{and so } x_3 = 11$$
$$x_4 - 3x_3 + 2x_2 = 1 \text{ that is } x_4 - 3 \times 11 + 2 \times 4 = 1 \text{ and so } x_4 = 26$$

Try another yourself.

The sequence $\{x_k\}$ has terms that satisfy the second-order recurrence relation

$$x_{k+2} - x_k = 1 \text{ where } x_0 = 0 \text{ and } x_1 = -1$$

The first six terms of this sequence are

$$\{x_k\} = \{0, -1, \ldots, \ldots, \ldots, \ldots, \ldots\}$$

28

$$\{x_k\} = \{0, \ -1, 1, 0, 2, 1, \ \ldots\}$$

Because

Since $x_{k+2} - x_k = 1$ where $x_0 = 0$ and $x_1 = -1$ then

$x_2 - x_0 = 1$ that is $x_2 - 0 = 1$ and so $x_2 = 1$

$x_3 - x_1 = 1$ that is $x_3 + 1 = 1$ and so $x_3 = 0$

$x_4 - x_2 = 1$ that is $x_4 - 1 = 1$ and so $x_4 = 2$

$x_5 - x_3 = 1$ that is $x_5 - 0 = 1$ and so $x_5 = 1$

Therefore $\{x_k\} = \{0, \ -1, 1, 0, 2, 1, \ \ldots\}$

Next frame

Solving the recurrence relation

29

If a sequence $\{x_k\}$ satisfies a recurrence relation with given initial conditions then the general term of the sequence can be found by using the Z transform where $Z\{x_k\} = F(z)$. This is referred to as *solving the recurrence relation*. For example, solve the recurrence relation

$$x_{k+2} - 3x_{k+1} + 2x_k = 1 \text{ where } x_0 = 0 \text{ and } x_1 = 1$$

Because this recurrence relation is true for all values of k it can itself be used to form a sequence $\{y_k\}$, namely

$$\{y_k\} = \{x_{k+2} - 3x_{k+1} + 2x_k\} = \{1\}$$

Now, taking the Z transform of both sides of this equation gives

$$Z\{y_k\} = Z\{x_{k+2} - 3x_{k+1} + 2x_k\} = Z\{1\} \text{ that is}$$
$$Z\{x_{k+2}\} - 3Z\{x_{k+1}\} + 2\{x_k\} = Z\{1\}$$

Using the first shift theorem and $Z\{x_k\} = F(z)$ this then becomes

$$\left(z^2 F(z) - z^2 x_0 - z x_1\right) - 3(z F(z) - z x_0) + 2F(z) = \frac{z}{z-1}$$

Collecting like terms and substituting for the initial terms $x_0 = 0$ and $x_1 = 1$ gives

$$\left(z^2 - 3z + 2\right)F(z) - z = \frac{z}{z-1} \text{ so } \left(z^2 - 3z + 2\right)F(z) = z + \frac{z}{z-1} = \frac{z^2}{z-1}$$

That is $F(z) = \dfrac{z^2}{(z-1)(z^2 - 3z + 2)} = \dfrac{z^2}{(z-1)^2(z-2)}$

and so $\dfrac{F(z)}{z} = \dfrac{z}{(z-1)^2(z-2)}$

This has the partial fraction breakdown

$$\frac{F(z)}{z} = \frac{\ldots\ldots}{(z-1)^2} \cdots \frac{\ldots\ldots}{z-1} \cdots \frac{\ldots\ldots}{z-2}$$

$$\frac{F(z)}{z} = -\frac{1}{(z-1)^2} - \frac{2}{z-1} + \frac{2}{z-2}$$

Because

$$\frac{F(z)}{z} = \frac{z}{(z-1)^2(z-2)}$$

$$= \frac{A}{(z-1)^2} + \frac{B}{z-1} + \frac{C}{z-2}$$

$$= \frac{A(z-2) + B(z-1)(z-2) + C(z-1)^2}{(z-1)^2(z-2)}$$

and so

$$z = A(z-2) + B(z-1)(z-2) + C(z-1)^2 \text{ giving}$$

$[z^2]:$ $\qquad\qquad B + C = 0$

$[z^1]:$ $\qquad A - 3B - 2C = 1$

$[CT]:$ $\qquad -2A + 2B + C = 0$

with solution $A = -1$, $B = -2$ and $C = 2$

Therefore

$$\frac{F(z)}{z} = -\frac{1}{(z-1)^2} - \frac{2}{z-1} + \frac{2}{z-2}$$

Taking the inverse Z transform of $F(z)$ yields the sequence

$$Z^{-1}F(z) = \dots\dots\dots$$

31

$$Z^{-1}F(z) = \{-k - 2 + 2^{k+1}\}$$

Because

$$\frac{F(z)}{z} = -\frac{1}{(z-1)^2} - \frac{2}{z-1} + \frac{2}{z-2} \quad \text{and so}$$

$$F(z) = -\frac{z}{(z-1)^2} - \frac{2z}{z-1} + \frac{2z}{z-2}$$

Therefore

$$Z^{-1}F(z) = -Z^{-1}\left(\frac{z}{(z-1)^2}\right) - 2Z^{-1}\left(\frac{z}{z-1}\right) + 2Z^{-1}\left(\frac{z}{z-2}\right)$$

$$= \{-k - 2x_k + 2(2^k)\}$$

$$= \{-k - 2 + 2^{k+1}\} \text{ since } x_k = 1$$

Indeed, $\{x_k\} = \{-k - 2 + 2^{k+1}\}$ is the solution to the recurrence relation as can be seen by substituting back

$$x_{k+2} - 3x_{k+1} + 2x_k$$

$$= \left(-[k+2] - 2 + 2^{[k+2]+1}\right) - 3\left(-[k+1] - 2 + 2^{[k+1]+1}\right)$$

$$\quad + 2\left(-k - 2 + 2^{k+1}\right)$$

$$= \left(-k - 4 + 8 \times 2^k\right) - 3\left(-k - 3 + 4 \times 2^k\right) + 2\left(-k - 2 + 2 \times 2^k\right)$$

$$= -k - 4 + 8 \times 2^k + 3k + 9 - 12 \times 2^k - 2k - 4 + 4 \times 2^k$$

$$= 1$$

Try one yourself.

The solution of the second-order recurrence relation

$$x_{k+2} - x_k = 1 \text{ where } x_0 = 0 \text{ and } x_1 = -1 \text{ is } x_k = \ldots\ldots\ldots$$

32

$$x_k = \begin{cases} k/2 & k \text{ even} \\ (k-3)/2 & k \text{ odd} \end{cases}$$

Because

Taking the Z transform of the recurrence relation gives

$Z\{x_{k+2} - x_k\} = Z\{1\}$. That is, $Z\{x_{k+2}\} - Z\{x_k\} = Z\{1\}$ so that

$$\left(z^2F(z) - z^2x_0 - zx_1\right) - F(z) = \frac{z}{z-1}.$$

Substituting for $x_0 = 0$ and $x_1 = -1$ gives

$$F(z) = \ldots\ldots\ldots$$

33

$$F(z) = \frac{-z+2}{(z+1)(z-1)^2}$$

Because

$$\left(z^2 F(z) - z^2 x_0 - z x_1\right) - F(z) = \frac{z}{z-1} \text{ where } x_0 = 0 \text{ and } x_1 = -1 \text{ giving}$$

$$\left(z^2 - 1\right)F(z) + z = \frac{z}{z-1} \text{ so}$$

$$F(z) = \frac{z}{(z^2-1)(z-1)} - \frac{z}{(z^2-1)} \text{ so}$$

$$\frac{F(z)}{z} = \frac{1}{(z+1)(z-1)^2} - \frac{1}{(z+1)(z-1)}$$

$$= \frac{1-(z-1)}{(z+1)(z-1)^2}$$

$$= \frac{-z+2}{(z+1)(z-1)^2}$$

Separating into partial fractions gives

$$\frac{F(z)}{z} = \ldots\ldots\ldots$$

34

$$\frac{F(z)}{z} = \frac{3}{4}\frac{z}{z+1} - \frac{3}{4}\frac{z}{z-1} + \frac{1}{2}\frac{z}{(z-1)^2}$$

Because

$$\frac{F(z)}{z} = \frac{-z+2}{(z+1)(z-1)^2}$$

$$= \frac{A}{z+1} + \frac{B}{z-1} + \frac{C}{(z-1)^2}$$

$$= \frac{A(z-1)^2 + B(z+1)(z-1) + C(z+1)}{(z+1)(z-1)^2}$$

Equating numerators and comparing coefficients of powers of z gives

$[z^2]$: $\quad A + B = 0$

$[z]$: $\quad -2A + C = -1$

$[CT]$: $\quad A - B + C = 2$ with solution $A = 3/4$, $B = -3/4$ and $C = 1/2$

so that $F(z) = \dfrac{3}{4}\dfrac{z}{z+1} - \dfrac{3}{4}\dfrac{z}{z-1} + \dfrac{1}{2}\dfrac{z}{(z-1)^2}$

By inverting the transform we find that

$$x_k = \ldots\ldots\ldots$$

35

$$x_k = \begin{cases} k/2 & k \text{ even} \\ (k-3)/2 & k \text{ odd} \end{cases}$$

Because

$$F(z) = \frac{3}{4}\frac{z}{z+1} - \frac{3}{4}\frac{z}{z-1} + \frac{1}{2}\frac{z}{(z-1)^2}$$

and

$$Z^{-1}\left\{\frac{z}{z+1}\right\} = \left\{(-1)^k\right\} \text{ so } Z^{-1}\left\{(3/4)\frac{z}{z+1}\right\} = (3/4)\left\{(-1)^k\right\}$$

$$Z^{-1}\left\{\frac{z}{z-1}\right\} = \left\{1^k\right\} \text{ so } Z^{-1}\left\{(-3/4)\frac{z}{z-1}\right\} = (-3/4)\left\{1^k\right\}$$

$$Z^{-1}\left\{\frac{z}{(z-1)^2}\right\} = \{k\} \text{ so } Z^{-1}\left\{(1/2)\frac{z}{(z-1)^2}\right\} = (1/2)\{k\}$$

Therefore $\{x_k\} = \left\{(3/4)(-1)^k - (3/4) + (k/2)\right\}$

so that $x_k = \begin{cases} k/2 & k \text{ even} \\ (k-3)/2 & k \text{ odd} \end{cases}$

Next frame

Sampling

36

If a continuous function $f(t)$ of time t progresses from $t = 0$ onwards and is measured at every time interval T then what will result is the sequence of values

$$\{f(kT)\} = \{f(0), f(T), f(2T), f(3T), \ldots\}$$

A new, piecewise continuous function $f^*(t)$ can then be created from the sequence of sampled values such that

$$f^*(t) = \begin{cases} f(kT) & \text{if } t = kT \\ 0 & \text{otherwise} \end{cases}$$

The graph of this new function consists of a series of spikes at the regular intervals $t = kT$

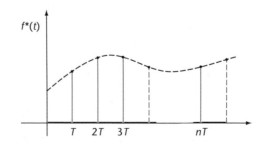

▶

This function can alternatively be described in terms of the delta function $\delta(t)$ as

$$f^*(t) = f(0)\delta(t) + f(T)\delta(t - T) + f(2T)\delta(t - 2T) + f(3T)\delta(t - 3T) + \dots$$

$$= \sum_{k=0}^{\infty} f(kT)\delta(t - kT)$$

The Laplace transform of $f^*(t)$ is then given as

$$F^*(s) = L\{f^*(t)\}$$

$$= \int_0^{\infty} \{f(0)\delta(t) + f(T)\delta(t - T) + f(2T)\delta(t - 2T) + \dots\}e^{-st}\,dt$$

$$= f(0) + f(T)e^{-sT} + f(2T)e^{-2sT} + f(3T)e^{-3sT} + \dots$$

$$= \sum_{k=0}^{\infty} f(kT)e^{-ksT}$$

Define a new variable $z = e^{sT}$ and we see that

$$L\{f^*(t)\} = \sum_{k=0}^{\infty} f(kT)z^{-k} = \sum_{k=0}^{\infty} \frac{f(kT)}{z^k}$$

which is the Z transform of the sequence $\{f(kT)\}$.

Example 1

The function $f(t) = e^{-at}$ is sampled every interval of T.

The Z transform of the sampled function is then ....

37

$$F(z) = \frac{z}{z - e^{-aT}}$$

Because

Defining $f^*(t) = \sum_{k=0}^{\infty} f(kT)\delta(t - kT) = \sum_{k=0}^{\infty} e^{-akT}\delta(t - kT)$ then the Laplace transform of $f^*(t)$ is given as

$$F^*(s) = \sum_{k=0}^{\infty} e^{-kaT}e^{-ksT}$$

This means that the Z transform of $\{f(kT)\}$ is

$$F(z) = \sum_{k=0}^{\infty} \frac{e^{-kaT}}{z^k} = \frac{1}{1 - \frac{e^{-aT}}{z}} = \frac{z}{z - e^{-aT}}$$

Notice that this agrees with the Z transform of the sequence $\{b^k\}$ (which is $\frac{z}{z - b}$) when b is replaced by e^{-aT}.
Try another.

Example 2

The function $f(t) = t$ is sampled every interval of T.

The Z transform of the sampled function is then ....

38

$$\boxed{F(z) = \frac{Tz}{(z-1)^2}}$$

Because

The Z transform of $\{f(kT)\}$ is $F(z) = \sum_{k=0}^{\infty} \frac{f(kT)}{z^k}$. Here $f(kT) = kT$ and so

$$
\begin{aligned}
F(z) &= \sum_{k=0}^{\infty} \frac{kT}{z^k} \\
&= T\left(\frac{1}{z} + \frac{2}{z^2} + \frac{3}{z^3} + \ldots\right) \\
&= \frac{T}{z}\left(1 + 2z^{-1} + 3z^{-2} + 4z^{-3} + \ldots\right) \\
&= -Tz\frac{d}{dz}\left(1 + z^{-1} + z^{-2} + z^{-3} + \ldots\right\} \\
&= -Tz\frac{d}{dz}\left(1 - \frac{1}{z}\right)^{-1} = \frac{T}{z}\left(1 - \frac{1}{z}\right)^{-2} = \frac{Tz}{(z-1)^2}
\end{aligned}
$$

Example 3

The function $f(t) = \cos t$ is sampled every interval of T.

The Z transform of the sampled function is then ....

39

Because

$$f(t) = \cos t = \frac{e^{iT} + e^{-iT}}{2} \quad \text{and the } Z \text{ transform of } \{e^{-kaT}\} \text{ is}$$

$$F(z) = \frac{z}{z - e^{-aT}}.$$

Therefore the Z transform of $\dfrac{e^{iT} + e^{-iT}}{2}$ is

$$
\begin{aligned}
\frac{1}{2}\left(\frac{z}{z - e^{-iT}} + \frac{z}{z - e^{iT}}\right) &= \frac{1}{2}\left(\frac{z(z - e^{iT}) + z(z - e^{-iT})}{(z - e^{-iT})(z - e^{iT})}\right) \\
&= \frac{1}{2}\left(\frac{2z^2 - z(e^{iT} + e^{-iT})}{z^2 - [e^{iT} + e^{-iT}]z + 1}\right) \\
&= \frac{z(z - \cos T)}{z^2 - 2z\cos T + 1}
\end{aligned}
$$

▶

And that is the end of the Program on *Z* transforms. All that remain are the **Review summary** and the **Can You?** checklist. Read through these closely and make sure that you understand all the workings of this Program. Then try the **Test exercise**; there is no need to hurry, take your time and work through the questions carefully. The **Further problems** then provide a valuable collection of additional exercises for you to try.

Review summary

40

1 *Sequences*
The sequence $\ldots, x_{-2}, x_{-1}, x_0, x_1, x_2, \ldots$ is represented by the notation $\{x_k\}_{-\infty}^{\infty}$. The sequence $\{x_k\}_0^{\infty}$ is called a causal sequence and is denoted simply by $\{x_k\}$.

2 *Z transform*
The *Z* transform of the causal sequence $\{x_k\}$ is

$$Z\{x_k\} = \sum_{k=0}^{\infty}\left(\frac{x_k}{z^k}\right) = F(z) \text{ where the value of } z \text{ is chosen to ensure that the sum converges.}$$

$\{x_k\}$ and $Z\{x_k\}$ form a *Z* transform pair.

3 *Table of Z transforms*

Sequence	Transform $F(z)$	Permitted values of z
$\{\delta_k\} = \{1, 0, 0, \ldots\}$	1	All values of z
$\{x_k\} = \{1, 1, 1, \ldots\}$	$\dfrac{z}{z-1}$	$\|z\| > 1$
$\{k\} = \{0, 1, 2, 3, \ldots\}$	$\dfrac{z}{(z-1)^2}$	$\|z\| > 1$
$\{k^2\} = \{0, 1, 4, 9, \ldots\}$	$\dfrac{z(z+1)}{(z-1)^3}$	$\|z\| > 1$
$\{k^3\} = \{0, 1, 8, 27, \ldots\}$	$\dfrac{z(z^2 + 4z + 1)}{(z-1)^4}$	$\|z\| > 1$
$\{a^k\} = \{1, a, a^2, a^3, \ldots\}$	$\dfrac{z}{(z-a)}$	$\|z\| > \|a\|$
$\{ka^k\} = \{0, a, 2a^2, 3a^3, \ldots\}$	$\dfrac{az}{(z-a)^2}$	$\|z\| > \|a\|$

4 *Linearity*
The *Z* transform is a linear transform. That is, if a and b are constants then

$$Z(a\{x_k\} + b\{y_k\}) = aZ\{x_k\} + bZ\{y_k\}.$$

▶

5 *First shift theorem* (shifting to the left)
If $Z\{x_k\} = F(z)$ then

$$Z\{x_{k+m}\} = z^m F(z) - \left[z^m x_0 + z^{m-1} x_1 + \ldots + z x_{m-1}\right]$$

the Z transform of the sequence that has been shifted by m places to the left.

6 *Second shift theorem* (shifting to the right)
If $Z\{x_k\} = F(z)$ then

$$Z\{x_{k-m}\} = z^{-m} F(z)$$

the Z transform of the sequence that has been shifted by m places to the right.

7 *Translation*
If the sequence $\{x_k\}$ has the Z transform $Z\{x_k\} = F(z)$ then the sequence $\{a^k x_k\}$ has the Z transform $Z\{a^k x_k\} = F(a^{-1} z)$.

8 *Final value theorem*
For the sequence $\{x_k\}$ with Z transform $F(z)$

$$\underset{k \to \infty}{Lim}\, x_k = \underset{z \to 1}{Lim} \left\{ \left(\frac{z - 1}{z} \right) F(z) \right\} \text{ provided that } \underset{k \to \infty}{Lim}\, x_k \text{ exists.}$$

9 *The initial value theorem*
For the sequence $\{x_k\}$ with Z transform $F(z)$

$$x_0 = \underset{z \to \infty}{Lim}\, \{F(z)\}.$$

10 *The derivative of the transform*
If $Z\{x_k\} = F(z)$ then $-z F'(z) = Z\{k x_k\}$.

11 *Inverse transformations*
If the sequence $\{x_k\}$ has Z transform $Z\{x_k\} = F(z)$, the inverse transform is defined as

$$Z^{-1} F(z) = \{x_k\}.$$

12 *Recurrence relations*
A recurrence relation expresses the relationship that adjacent terms of a series hold to each other. The order of the equation is given by the maximum shift between related terms.

Initial terms
A recurrence relation can be used to generate the terms of a sequence provided initial terms are given – equal in number to the order of the equation.

Solving the recurrence relation
If a sequence $\{x_k\}$ satisfies a recurrence relation with given initial conditions then the general term of the sequence can be found by using the Z transform where $Z\{x_k\} = F(z)$. This is referred to as *solving the recurrence relation*.

▶

13 *Sampling*

If a continuous function $f(t)$ is sampled at equal intervals, the resulting sequence has a Z transform that is related to the Laplace transform of the piecewise function created from the sequence of sample values.

$$L\{f^*(t)\} = \sum_{k=0}^{\infty} f(kT)z^{-k} = \sum_{k=0}^{\infty} \frac{f(kT)}{z^k} = Z\{f(kT)\}$$

where

$$\{f(kT)\} = \{f(0), f(T), f(2T), f(3T),\ldots\},$$

$$f^*(t) = \begin{cases} f(kT) & \text{if } t = k \\ 0 & \text{otherwise} \end{cases}$$

and

$$z = e^{sT}.$$

☑ Can You?

Checklist 9

41

Check this list before and after you try the end of Program test.

On a scale of 1 to 5 how confident are you that you can: Frames

- Define the Z transform of a sequence and derive transforms of specified sequences?   1 to 7

  Yes ☐ ☐ ☐ ☐ ☐ *No*

- Make reference to a table of standard Z transforms?   8

  Yes ☐ ☐ ☐ ☐ ☐ *No*

- Recognize the Z transform as being a linear transform and so obtain the transform of linear combinations of standard sequences?   9 to 11

  Yes ☐ ☐ ☐ ☐ ☐ *No*

- Apply the first and second shift theorems, the translation theorem, the initial and final value theorems and the derivative theorem?   11 to 18

  Yes ☐ ☐ ☐ ☐ ☐ *No*

- Use partial fractions to derive the inverse transforms?   19 to 23

  Yes ☐ ☐ ☐ ☐ ☐ *No*

▶

- Solve linear, constant coefficient recurrence relations? `24` to `35`
  Yes ☐ ☐ ☐ ☐ ☐ *No*

- Demonstrate the relationship between the Laplace transform and the Z transform? `36` to `39`
  Yes ☐ ☐ ☐ ☐ ☐ *No*

🚲 Test exercise 9

42

1 Find the Z transform of the causal sequence $\{x_k\}$ where $x_k = (-1)^k$.

2 Find the Z transform of the causal sequence $\{x_k\}$ where $x_k = 4k - 2a^k$.

3 Find the Z transform of the causal sequences:
   (a) $\{k - 3\}$
   (b) $\{5^{k+2}\}$

4 Find the inverse Z transformation of
   $$F(z) = \frac{z^2(z-3)}{(z^2 - 2z + 1)(z - 2)}.$$

5 Solve the recurrence relation
   $$x_{k+2} - 4x_{k+1} + 4x_k = 3 \text{ where } x_0 = 1 \text{ and } x_1 = 0.$$

6 The function $f(t) = \sin t$ is sampled at equal intervals of $t = T$. Find the Z transform of the resulting sequence of values.

🚲 Further problems 9

43

1 Find the Z transform of the causal sequence $\{x_k\}$ where $x_k = (-a)^k$ where $a > 0$.

2 Solve each of the following recurrence relations.
   (a) $x_{k+2} + 5x_{k+1} + 6x_k = 1$ where $x_0 = 0$ and $x_1 = 1$
   (b) $3x_{k+2} - 7x_{k+1} + 2x_k = k$ where $x_0 = 1$ and $x_1 = 0$
   (c) $x_{k+2} - 9x_k = 2k$ where $x_0 = 1$ and $x_1 = 1$.

3 Given that $y_{k+1} = v_k$ and $v_{k+1} = w_k$ where $w_k = x_k - y_k$, show that $y_{k+2} + y_k = x_k$ and solve for y_k when $\{x_k\} = \{\delta_k\}$, the unit impulse sequence where $y_0 = 0$, $y_1 = 1$.

▶

4 If

$$p_{k+1} = q_k$$

$$q_{k+1} = r_k$$

$r_k = x_k - \alpha q_k - \beta p_k$ where α and β are constants, show that

$$p_{k+2} + \alpha p_{k+1} + \beta p_k = x_k$$

Solve this recurrence relation when $p_0 = 1$, $p_1 = 0$ for

(a) $\alpha = 4$, $\beta = 4$ and $\{x_k\} = \{\delta_k\}$, the unit impulse sequence

(b) $\alpha = 4$, $\beta = 4$ and $\{x_k\} = \{u_k\}$ the unit step sequence.

5 Find the Z transform of each of the following sequences.

(a) $\{1, 0, 1, 0, 1, 0, \ldots\}$

(b) $\{0, 1, 0, 1, 0, 1, \ldots\}$

(c) $\{1, 0, 1, 1, 0, 0, 0, 1\}$

(d) $\{1, 1, 1, 0, 0, 0, 1, 1\}$

(e) $\{0, 0, 0, 1, 1, 1, 0, 0, 0, 1, 1\}$

(f) $\{1, 1, 0, 0, 0, 1, 1\}$

Note that the last four of these are finite sequences.

6 Find the inverse transform of

(a) $F(z) = \dfrac{z}{(z+1)(z+2)(z+3)}$

(b) $F(z) = \dfrac{z^2}{(z+1)(z+2)(z+3)}$

(c) $F(z) = \dfrac{z(3z+1)}{(z-2)(z-3)}$

(d) $F(z) = \dfrac{z^2}{2 - 3z + z^2}.$

7 Given

$$F(z) = \frac{3z^2}{z^2 - z + 1}$$

show that

$$Z^{-1}F(z) = \{3, 3, -3, -3, \ldots\}.$$

Hint: Use long division on $F(z)$.

8 Given

$$F(z) = \left(1 + \frac{2}{z}\right)^{-3}$$

show that

$$Z^{-1}F(z) = \{1, -6, 24, -48, \ldots\}.$$

Hint: Use the binomial theorem on $F(z)$.

▶

9   Find the final value of the sequence $\{x_k\}$ with Z transform

$$F(z) = \frac{4z^2 - z}{2z^2 - 3z + 1}.$$

10   What is the initial value of the sequence whose Z transform is given by

$$F(z) = \frac{2z^2 - z + 1}{5 - 3z - 7z^2}?$$

11   Given the sequence of n terms $\{x_k\}$ for $0 \le k \le n - 1$ with Z transform $F_n(z)$, show that the Z transform of the sequence formed by continually repeating the terms $\{x_k\}$ is given as

$$F(z) = \frac{F_n(z)}{1 - z^{-n}}.$$

12   Using the result of Question 11, show that the Z transform of the sequence obtained by continually repeating the three term sequence $\{1, 0, -1\}$ is

$$F(z) = \frac{z^2}{z^2 + 1}.$$

13   Find the Z transforms of the sequence of values obtained when $f(t)$ is sampled at regular intervals of $t = T$ where

   (a) $f(t) = \sinh t$
   (b) $f(t) = \cosh at$
   (c) $f(t) = e^{-at} \cosh bt$.

Matrix algebra

Frames
1 to 112

Learning outcomes

When you have completed this Program you will be able to:

- Determine whether a matrix is singular or non-singular
- Determine the rank of a matrix
- Determine the consistency of a set of linear equations and hence demonstrate the uniqueness of their solution
- Obtain the solution of a set of simultaneous linear equations by using matrix inversion, by row transformation, by Gaussian elimination and by triangular decomposition
- Obtain the eigenvalues and corresponding eigenvectors of a square matrix
- Demonstrate the validity of the Cayley–Hamilton theorem
- Solve systems of first-order ordinary differential equations using eigenvalue and eigenvector methods
- Construct the modal matrix from the eigenvectors of a matrix and the spectral matrix from the eigenvalues
- Solve systems of second-order ordinary differential equations using diagonalization
- Use matrices to represent transformations between coordinate systems

Singular and non-singular matrices

1

Every square matrix **A** has associated with it a number called the determinant of **A** and denoted by $|\mathbf{A}|$. If $|\mathbf{A}| \neq 0$ then **A** is called a *non-singular* matrix. Otherwise if $|\mathbf{A}| = 0$, then **A** is called a *singular* matrix.

Example 1

Is $\mathbf{A} = \begin{pmatrix} 1 & 2 & 8 \\ 4 & 7 & 6 \\ 9 & 5 & 3 \end{pmatrix}$ singular or non-singular?

$$|\mathbf{A}| = \begin{vmatrix} 1 & 2 & 8 \\ 4 & 7 & 6 \\ 9 & 5 & 3 \end{vmatrix}$$

$$= 1\begin{vmatrix} 7 & 6 \\ 5 & 3 \end{vmatrix} - 2\begin{vmatrix} 4 & 6 \\ 9 & 3 \end{vmatrix} + 8\begin{vmatrix} 4 & 7 \\ 9 & 5 \end{vmatrix}$$

$$= (21 - 30) - 2(12 - 54) + 8(20 - 63)$$

$$= -9 + 84 - 344$$

$$= -269$$

Because $|\mathbf{A}| \neq 0$ then **A** is non-singular.

Example 2

Is $\mathbf{A} = \begin{pmatrix} 3 & 9 & 2 \\ 1 & 5 & 6 \\ 2 & 7 & 4 \end{pmatrix}$ singular or non-singular?

A is . . . .

2

$$\boxed{\text{singular}}$$

Because

$$|\mathbf{A}| = \begin{vmatrix} 3 & 9 & 2 \\ 1 & 5 & 6 \\ 2 & 7 & 4 \end{vmatrix}$$

$$= 3(20 - 42) - 9(4 - 12) + 2(7 - 10)$$

$$= -66 + 72 - 6$$

$$= 0$$

Because $|\mathbf{A}| = 0$ then $|\mathbf{A}|$ is singular.

▶

Exercise

Determine whether each of the following is singular or non-singular.

1 $|\mathbf{A}| = \begin{pmatrix} 4 & 5 \\ 2 & 3 \end{pmatrix}$
  2 $|\mathbf{B}| = \begin{pmatrix} 3 & -4 \\ -6 & 8 \end{pmatrix}$

3 $|\mathbf{C}| = \begin{pmatrix} 4 & 1 & -2 \\ 1 & 7 & 3 \\ 5 & 8 & 1 \end{pmatrix}$
  4 $|\mathbf{D}| = \begin{pmatrix} 3 & 2 & 4 \\ 5 & 1 & 6 \\ 2 & 0 & 3 \end{pmatrix}$

3

1 non-singular	**2** singular
3 singular	**4** non-singular

Because

Straightforward evaluation of the relevant determinants gives

1 $|\mathbf{A}| = 2$ **2** $|\mathbf{B}| = 0$

3 $|\mathbf{C}| = 0$ **4** $|\mathbf{D}| = -5$

Closely related to the notion of the singularity or otherwise of a square matrix is the notion of **rank** of a general $n \times m$ matrix.

Rank of a matrix

The rank of an $n \times m$ matrix **A** is the order of the largest square, non-singular sub-matrix. That is, the largest square sub-matrix whose determinant is non-zero. If $n = m$, so making **A** itself square, then this sub-matrix could be the matrix **A** itself.

Example

To find the rank of the matrix $\mathbf{A} = \begin{pmatrix} 3 & 4 & 5 \\ 1 & 2 & 3 \\ 4 & 5 & 6 \end{pmatrix}$ we note that

$$|\mathbf{A}| = \begin{vmatrix} 3 & 4 & 5 \\ 1 & 2 & 3 \\ 4 & 5 & 6 \end{vmatrix} = \ldots\ldots\ldots\ldots$$

4

$$\boxed{0}$$

Because

$$|\mathbf{A}| = \begin{vmatrix} 3 & 4 & 5 \\ 1 & 2 & 3 \\ 4 & 5 & 6 \end{vmatrix}$$

$$= 3(12 - 15) - 4(6 - 12) + 5(5 - 8)$$

$$= -9 + 24 - 15 = 0$$

Therefore we can say that the rank of **A** is . . . .

5

$$\boxed{\text{not } 3}$$

Because

$|\mathbf{A}| = 0$ and therefore **A** is singular.

Now try a sub-matrix of order 2.

$$\begin{vmatrix} 3 & 4 \\ 1 & 2 \end{vmatrix} = 6 - 4 = -2 \neq 0. \text{ Therefore the rank of } \mathbf{A} \text{ is } \ldots \ldots \ldots$$

6

$$\boxed{2}$$

Because

The largest square, non-singular sub-matrix of **A** has order 2 therefore **A** has rank 2.

This method of finding the rank of a matrix can be a very hit and miss affair and a better, more systematic method is to use **elementary operations** and the notion of an **equivalent matrix**.

Next frame

Elementary operations and equivalent matrices

7

Each of the following row operations on matrix **A** produces a *row equivalent matrix* **B**, where the order and rank of **B** is the same as that of **A**. We write **A** ∼ **B**.

1 Interchanging two rows
2 Multiplying each element of a row by the same non-zero scalar quantity
3 Adding or subtracting corresponding elements from those of another row

are operations called *elementary row operations*. There is a corresponding set of three *elementary column operations* that can be used to form *column equivalent matrices*.

▶

Example 1

Given $\mathbf{A} = \begin{pmatrix} 3 & 4 & 5 \\ 1 & 2 & 3 \\ 4 & 5 & 6 \end{pmatrix}$ then

$\begin{pmatrix} 3 & 4 & 5 \\ 1 & 2 & 3 \\ 4 & 5 & 6 \end{pmatrix} \sim \begin{pmatrix} 0 & -2 & -4 \\ 1 & 2 & 3 \\ 4 & 5 & 6 \end{pmatrix}$ by subtracting 3 times each element of row 2 from row 1

$\sim \begin{pmatrix} 0 & -2 & -4 \\ 1 & 2 & 3 \\ 0 & -3 & -6 \end{pmatrix}$ by subtracting 4 times each element of row 2 from row 3

$\sim \begin{pmatrix} 0 & -3 & -6 \\ 1 & 2 & 3 \\ 0 & -3 & -6 \end{pmatrix}$ by multiplying each element of row 1 by 3/2

$\sim \begin{pmatrix} 0 & -3 & -6 \\ 1 & 2 & 3 \\ 0 & 0 & 0 \end{pmatrix}$ by subtracting corresponding elements of row 1 from row 3

$= \mathbf{B}$

The row of zeros in matrix $\mathbf{B}$ means that its determinant is zero and so its rank is not 3. The largest sub-matrix with non-zero determinant has order 2 and so the rank of $\mathbf{B}$ is 2. Because matrix $\mathbf{B}$ is row equivalent to matrix $\mathbf{A}$ we can say that the rank of $\mathbf{A}$ is also 2.

Example 2

Determine the rank of $\mathbf{A} = \begin{pmatrix} 1 & 2 & 8 \\ 4 & 7 & 6 \\ 9 & 5 & 3 \end{pmatrix}$

By taking 4 times the elements of row 1 from row 2 we obtain the equivalent matrix . . . .

8

$$\begin{pmatrix} 1 & 2 & 8 \\ 0 & -1 & -26 \\ 9 & 5 & 3 \end{pmatrix}$$

By taking 9 times the elements of row 1 from row 3 we obtain the equivalent matrix . . . .

9

$$\begin{pmatrix} 1 & 2 & 8 \\ 0 & -1 & -26 \\ 0 & -13 & -69 \end{pmatrix}$$

By multiplying the elements of row 2 by -13 we obtain the equivalent matrix . . . .

10

$$\begin{pmatrix} 1 & 2 & 8 \\ 0 & 13 & 338 \\ 0 & -13 & -69 \end{pmatrix}$$

By adding corresponding elements of row 2 to row 3 we obtain the equivalent matrix . . . .

11

$$\begin{pmatrix} 1 & 2 & 8 \\ 0 & 13 & 269 \\ 0 & 0 & -69 \end{pmatrix}$$

Because all the elements below the main diagonal of this matrix are zero we call the matrix an *upper triangular matrix*. By inspection we can see that the determinant of this triangular matrix is non-zero, being the product of its three diagonal elements $1 \times 13 \times (-69) = -897$. Therefore its rank is 3 and so the rank of matrix **A** is also 3.

Try another one for yourself.

Example 3

The rank of $\mathbf{A} = \begin{pmatrix} 3 & 9 & 2 \\ 1 & 5 & 6 \\ 2 & 7 & 4 \end{pmatrix}$ is . . . .

12

$$\boxed{2}$$

Because

$$\mathbf{A} = \begin{pmatrix} 3 & 9 & 2 \\ 1 & 5 & 6 \\ 2 & 7 & 4 \end{pmatrix} \sim \begin{pmatrix} 0 & -6 & -16 \\ 1 & 5 & 6 \\ 2 & 7 & 4 \end{pmatrix} \quad \text{Subtracting 3 times row 2 from row 1}$$

$$\sim \begin{pmatrix} 0 & -6 & -16 \\ 1 & 5 & 6 \\ 0 & -3 & -8 \end{pmatrix} \quad \text{Subtracting 2 times row 2 from row 3}$$

$$\sim \begin{pmatrix} 0 & 3 & 8 \\ 1 & 5 & 6 \\ 0 & -3 & -8 \end{pmatrix} \quad \text{Multiplying row 1 by } -1/2$$

$$\sim \begin{pmatrix} 0 & 3 & 8 \\ 1 & 5 & 6 \\ 0 & 0 & 0 \end{pmatrix} \quad \text{Adding row 1 to row 3}$$

$$\sim \begin{pmatrix} 1 & 5 & 6 \\ 0 & 3 & 8 \\ 0 & 0 & 0 \end{pmatrix} \quad \text{Interchanging rows 1 and 2}$$

▶

and $\begin{vmatrix} 1 & 5 & 6 \\ 0 & 3 & 8 \\ 0 & 0 & 0 \end{vmatrix} = 0.$ So the rank of this matrix is not 3. The largest

square sub-matrix of this matrix with non-zero determinant is, by inspection, of order 2 and so the rank of this matrix, and hence the rank of the equivalent matrix **A** is 2.

Finally try a non-square matrix.

Example 4

The rank of $\mathbf{A} = \begin{pmatrix} 2 & 2 & 3 & 1 \\ 0 & 8 & 2 & 4 \\ 1 & 7 & 3 & 2 \end{pmatrix}$ is ....

$\boxed{3}$ $\boxed{13}$

Because

$$\mathbf{A} = \begin{pmatrix} 2 & 2 & 3 & 1 \\ 0 & 8 & 2 & 4 \\ 1 & 7 & 3 & 2 \end{pmatrix} \sim \begin{pmatrix} 0 & -12 & -3 & -3 \\ 0 & 8 & 2 & 4 \\ 1 & 7 & 3 & 2 \end{pmatrix}$$ Subtracting 2 times row 3 from row 1

$$\sim \begin{pmatrix} 0 & -8 & -2 & -2 \\ 0 & 8 & 2 & 4 \\ 1 & 7 & 3 & 2 \end{pmatrix}$$ Multiplying row 1 by 2/3

$$\sim \begin{pmatrix} 0 & 0 & 0 & 2 \\ 0 & 8 & 2 & 4 \\ 1 & 7 & 3 & 2 \end{pmatrix}$$ Adding row 2 to row 1

It is possible to find a 3×3 sub-matrix of this matrix that has non-zero determinant, namely

$$\begin{pmatrix} 0 & 0 & 2 \\ 8 & 2 & 4 \\ 7 & 3 & 2 \end{pmatrix} \text{ where } \begin{vmatrix} 0 & 0 & 2 \\ 8 & 2 & 4 \\ 7 & 3 & 2 \end{vmatrix} = 2(24 - 14) = 20.$$

Consequently, this matrix and hence matrix **A** has rank 3.

Consistency of a set of equations

14

In solving sets of simultaneous equations, we can express the equations in matrix form. For example

$$a_{11}x_1 + a_{12}x_2 + a_{13}x_3 = b_1$$
$$a_{21}x_1 + a_{22}x_2 + a_{23}x_3 = b_2$$
$$a_{31}x_1 + a_{32}x_2 + a_{33}x_3 = b_3$$

can be written in the form

$$\begin{pmatrix} a_{11} & a_{12} & a_{13} \\ a_{21} & a_{22} & a_{23} \\ a_{31} & a_{32} & a_{33} \end{pmatrix} \begin{pmatrix} x_1 \\ x_2 \\ x_3 \end{pmatrix} = \begin{pmatrix} b_1 \\ b_2 \\ b_3 \end{pmatrix}$$

i.e. $\mathbf{Ax} = \mathbf{b}$

The set of three equations is said to be *consistent* if solutions for x_1, x_2, x_3 exist and *inconsistent* if no such solutions can be found.

In practice, we can solve the equations by operating on the *augmented coefficient matrix*, i.e. we write the constant terms as a fourth column of the coefficient matrix to form $\mathbf{A}_b$.

$$\mathbf{A}_b = \begin{pmatrix} a_{11} & a_{12} & a_{13} & b_1 \\ a_{21} & a_{22} & a_{23} & b_2 \\ a_{31} & a_{32} & a_{33} & b_3 \end{pmatrix}$$

which, of course, is a (3×4) matrix.

The general test for consistency is then:

A set of n simultaneous equations in n unknowns is consistent if the rank of the coefficient matrix $\mathbf{A}$ is equal to the rank of the augmented matrix $\mathbf{A}_b$.

If the rank of $\mathbf{A}$ is less than the rank of $\mathbf{A}_b$, then the equations are inconsistent and have no solution.

Make a note of this test. It can save time in working

15

Example

If $\begin{pmatrix} 1 & 3 \\ 2 & 6 \end{pmatrix} \begin{pmatrix} x_1 \\ x_2 \end{pmatrix} = \begin{pmatrix} 4 \\ 5 \end{pmatrix}$ then

$\mathbf{A} = \begin{pmatrix} 1 & 3 \\ 2 & 6 \end{pmatrix}$ and $\mathbf{A}_b = \begin{pmatrix} 1 & 3 & 4 \\ 2 & 6 & 5 \end{pmatrix}$

Rank of $\mathbf{A}$: $\begin{vmatrix} 1 & 3 \\ 2 & 6 \end{vmatrix} = 6 - 6 = 0$ $\therefore$ rank of $\mathbf{A} = 1$

Rank of $\mathbf{A}_b$: $\begin{vmatrix} 1 & 3 \\ 2 & 6 \end{vmatrix} = 0$ as before

but $\begin{vmatrix} 3 & 4 \\ 6 & 5 \end{vmatrix} = 15 - 24 = -9$ $\therefore$ rank of $\mathbf{A}_b = 2$

In this case, rank of $\mathbf{A} <$ rank of $\mathbf{A}_b$

$\therefore$ ....

16

$$\boxed{\text{no solution exists}}$$

Remember that, for consistency,

$$\text{rank of } \mathbf{A} = \ldots\ldots\ldots\ldots$$

17

$$\boxed{\text{rank of } \mathbf{A}_b}$$

Uniqueness of solutions

1 With a set of n equations in n unknowns, the equations are consistent if the coefficient matrix $\mathbf{A}$ and the augmented matrix $\mathbf{A}_b$ are each of rank n. There is then a *unique* solution for the n equations.

Note that if the rank of $\mathbf{A} = n$ then $\mathbf{A}$ is a non-singular sub-matrix of $\mathbf{A}_b$ and so the rank of $\mathbf{A}_b = n$ also. Therefore there is no need to test for the rank of $\mathbf{A}_b$ in this case.

2 If the rank of $\mathbf{A}$ and that of $\mathbf{A}_b$ is m, where $m < n$, then the matrix $\mathbf{A}$ is singular, i.e. $|\mathbf{A}| = \mathbf{0}$, and there will be an *infinite number* of solutions for the equations.

3 As we have already seen, if the rank of $\mathbf{A}$ < the rank of $\mathbf{A}_b$, then *no solution* exists.

Copy these up in your record book; they are important

18

Writing the results in a slightly different way:

With a set of n equations in n unknowns, checking the rank of the coefficient matrix $\mathbf{A}$ and that of the augmented matrix $\mathbf{A}_b$ enables us to see whether

(a) a unique solution exists

$$\text{rank } \mathbf{A} = \text{rank } \mathbf{A}_b = n$$

(b) an infinite number of solutions exist

$$\text{rank } \mathbf{A} = \text{rank } \mathbf{A}_b = m < n$$

(c) no solution exists

$$\text{rank } \mathbf{A} < \text{rank } \mathbf{A}_b$$

Example

$$\begin{pmatrix} -4 & 5 \\ -8 & 10 \end{pmatrix} \begin{pmatrix} x_1 \\ x_2 \end{pmatrix} = \begin{pmatrix} -3 \\ -6 \end{pmatrix}$$

Finding the rank of $\mathbf{A}$ and of $\mathbf{A}_b$ leads us to the conclusion that

$$\ldots\ldots\ldots\ldots$$

19

there is an infinite
number of solutions

Because

$$\mathbf{A} = \begin{pmatrix} -4 & 5 \\ -8 & 10 \end{pmatrix} \text{ and } \mathbf{A}_b = \begin{pmatrix} -4 & 5 & -3 \\ -8 & 10 & -6 \end{pmatrix}$$

Rank of $\mathbf{A}$: $\begin{vmatrix} -4 & 5 \\ -8 & 10 \end{vmatrix} = -40 + 40 = 0$ ∴ Rank of $\mathbf{A} = 1$

Rank of $\mathbf{A}_b$: $\begin{vmatrix} -4 & 5 \\ -8 & 10 \end{vmatrix} = 0;$ $\begin{vmatrix} 5 & -3 \\ 10 & -6 \end{vmatrix} = 0;$ $\begin{vmatrix} -4 & -3 \\ -8 & -6 \end{vmatrix} = 0$

∴ Rank of $\mathbf{A}_b = 1$

∴ Rank of $\mathbf{A}$ = rank of $\mathbf{A}_b = 1$

But there are two equations in two unknowns, i.e. $n = 2$

∴ Rank of $\mathbf{A}$ = rank of $\mathbf{A}_b = 1 < n$

∴ Infinite number of solutions.

The solutions can be written as x_1 arbitrary and $x_2 = \dfrac{4x_1 - 3}{5}$.

You will recall that, for a unique solution of n equations in n unknowns

. . . .

20

rank $\mathbf{A}$ = rank $\mathbf{A}_b = n$

Now for some examples for you to try. In each of the following cases, apply the rank tests to determine the nature of the solutions. Do not solve the sets of equations.

Example 1

$$\begin{pmatrix} 1 & 2 & -1 \\ 3 & 4 & 2 \\ 1 & 4 & 3 \end{pmatrix} \begin{pmatrix} x_1 \\ x_2 \\ x_3 \end{pmatrix} = \begin{pmatrix} 1 \\ -2 \\ 3 \end{pmatrix}$$

$$\mathbf{A} = \begin{pmatrix} 1 & 2 & -1 \\ 3 & 4 & 2 \\ 1 & 4 & 3 \end{pmatrix} \text{ and } \mathbf{A}_b = \begin{pmatrix} 1 & 2 & -1 & 1 \\ 3 & 4 & 2 & -2 \\ 1 & 4 & 3 & 3 \end{pmatrix}$$

Finish it off and we find that . . . .

$$\boxed{\text{a unique solution exists}}$$

Because

$n = 3$;   rank of $\mathbf{A} = 3$; rank of $\mathbf{A}_b = 3$.

$\therefore$ rank of $\mathbf{A} = $ rank of $\mathbf{A}_b = 3 = n$ $\therefore$ Solution unique

And this one.

Example 2

$$\begin{pmatrix} 2 & -1 & 7 \\ 4 & 2 & 2 \\ 3 & 1 & 3 \end{pmatrix} \begin{pmatrix} x_1 \\ x_2 \\ x_3 \end{pmatrix} = \begin{pmatrix} 2 \\ 5 \\ 1 \end{pmatrix}$$

This time we find that . . . .

$$\boxed{\text{no solution is possible}}$$

Because

$$\mathbf{A} = \begin{pmatrix} 2 & -1 & 7 \\ 4 & 2 & 2 \\ 3 & 1 & 3 \end{pmatrix}; \qquad \mathbf{A}_b = \begin{pmatrix} 2 & -1 & 7 & 2 \\ 4 & 2 & 2 & 5 \\ 3 & 1 & 3 & 1 \end{pmatrix}$$

$n = 3$;   rank of $\mathbf{A} = 2$;   rank of $\mathbf{A}_b = 3$

$\therefore$ rank of $\mathbf{A} < $ rank of $\mathbf{A}_b$

$\therefore$ No solution exists

and finally

Example 3

$$\begin{pmatrix} 1 & 2 & -3 \\ 1 & 3 & 4 \\ 2 & 5 & 1 \end{pmatrix} \begin{pmatrix} x_1 \\ x_2 \\ x_3 \end{pmatrix} = \begin{pmatrix} 1 \\ 2 \\ 3 \end{pmatrix}$$

In this case, we find that . . . .

23

$$\boxed{\text{infinite number of solutions possible}}$$

Because

$$\mathbf{A} = \begin{pmatrix} 1 & 2 & -3 \\ 1 & 3 & 4 \\ 2 & 5 & 1 \end{pmatrix} \quad \text{and} \quad \mathbf{A}_b = \begin{pmatrix} 1 & 2 & -3 & 1 \\ 1 & 3 & 4 & 2 \\ 2 & 5 & 1 & 3 \end{pmatrix}$$

Rank of **A**:

$$\mathbf{A} = \begin{pmatrix} 1 & 2 & -3 \\ 1 & 3 & 4 \\ 2 & 5 & 1 \end{pmatrix} \sim \begin{pmatrix} 1 & 2 & -3 \\ 0 & 1 & 7 \\ 2 & 5 & 1 \end{pmatrix} \quad \text{Subtracting row 1 from row 2}$$

$$\sim \begin{pmatrix} 1 & 2 & -3 \\ 0 & 1 & 7 \\ 0 & 1 & 7 \end{pmatrix} \quad \begin{array}{l} \text{Subtracting 2 times row 1} \\ \text{from row 2} \end{array}$$

$$\sim \begin{pmatrix} 1 & 2 & -3 \\ 0 & 1 & 7 \\ 0 & 0 & 0 \end{pmatrix} \quad \text{Subtracting row 2 from row 3}$$

and so rank of **A** is 2 by inspection.

Rank of **A**$_b$:

$$\mathbf{A}_b = \begin{pmatrix} 1 & 2 & -3 & 1 \\ 1 & 3 & 4 & 2 \\ 2 & 5 & 1 & 3 \end{pmatrix} \sim \begin{pmatrix} 1 & 2 & -3 & 1 \\ 0 & 1 & 7 & 1 \\ 2 & 5 & 1 & 3 \end{pmatrix} \quad \begin{array}{l} \text{Subtracting row 1} \\ \text{from row 2} \end{array}$$

$$\sim \begin{pmatrix} 1 & 2 & -3 & 1 \\ 0 & 1 & 7 & 1 \\ 0 & 1 & 7 & 1 \end{pmatrix} \quad \begin{array}{l} \text{Subtracting 2 times} \\ \text{row 1 from row 2} \end{array}$$

$$\sim \begin{pmatrix} 1 & 2 & -3 & 1 \\ 0 & 1 & 7 & 1 \\ 0 & 0 & 0 & 0 \end{pmatrix} \quad \begin{array}{l} \text{Subtracting row 2} \\ \text{from row 3} \end{array}$$

and so rank of **A**$_b$ is 2 by inspection.

Therefore rank of **A** = rank of **A**$_b$ = $2 < n$ (that is 3), therefore there is an infinite number of solutions.

Now let us move on to a new section of the work

Solution of sets of equations

1 *Inverse method*

24

Let us work through an example by way of explanation.

Example 1

To solve
$$3x_1 + 2x_2 - x_3 = 4$$
$$2x_1 - x_2 + 2x_3 = 10$$
$$x_1 - 3x_2 - 4x_3 = 5.$$

We first write this in matrix form, which is ....

25

$$\begin{pmatrix} 3 & 2 & -1 \\ 2 & -1 & 2 \\ 1 & -3 & -4 \end{pmatrix} \begin{pmatrix} x_1 \\ x_2 \\ x_3 \end{pmatrix} = \begin{pmatrix} 4 \\ 10 \\ 5 \end{pmatrix}$$

Then if $\mathbf{A} = \begin{pmatrix} 3 & 2 & -1 \\ 2 & -1 & 2 \\ 1 & -3 & -4 \end{pmatrix}$ then $\begin{pmatrix} x_1 \\ x_2 \\ x_3 \end{pmatrix} = \mathbf{A}^{-1} \begin{pmatrix} 4 \\ 10 \\ 5 \end{pmatrix}$

where $\mathbf{A}^{-1}$ is the *inverse* of $\mathbf{A}$.

To find $\mathbf{A}^{-1}$

(a) Form the determinant of $\mathbf{A}$ and evaluate it.
$$|\mathbf{A}| = \begin{vmatrix} 3 & 2 & -1 \\ 2 & -1 & 2 \\ 1 & -3 & -4 \end{vmatrix} = 3(4+6) - 2(-8-2) - 1(-6+1) = 55$$

(b) Form a new matrix $\mathbf{C}$ consisting of the cofactors of the elements in $\mathbf{A}$.
 The cofactor of any one element is its minor together with its 'place sign'

i.e. $\mathbf{C} = \begin{pmatrix} A_{11} & A_{12} & A_{13} \\ A_{21} & A_{22} & A_{23} \\ A_{31} & A_{32} & A_{33} \end{pmatrix}$

where A_{11} is the cofactor of a_{11} in $\mathbf{A}$.

$A_{11} = \begin{vmatrix} -1 & 2 \\ -3 & -4 \end{vmatrix} = 10; \quad A_{12} = -\begin{vmatrix} 2 & 2 \\ 1 & -4 \end{vmatrix} = 10;$

$A_{13} = \begin{vmatrix} 2 & -1 \\ 1 & -3 \end{vmatrix} = -5$

$A_{21} = -\begin{vmatrix} 2 & -1 \\ -3 & -4 \end{vmatrix} = 11; \quad A_{22} = \begin{vmatrix} 3 & -1 \\ 1 & -4 \end{vmatrix} = -11;$

$A_{23} = -\begin{vmatrix} 3 & 2 \\ 1 & -3 \end{vmatrix} = 11$

$A_{31} = \dots\dots; \quad A_{32} = \dots\dots; \quad A_{33} = \dots\dots$

26

$$A_{31} = \begin{vmatrix} 2 & -1 \\ -1 & 2 \end{vmatrix} = 3; \quad A_{32} = -\begin{vmatrix} 3 & -1 \\ 2 & 2 \end{vmatrix} = -8; \quad A_{33} = \begin{vmatrix} 3 & 2 \\ 2 & -1 \end{vmatrix} = -7$$

So $\quad \mathbf{C} = \begin{pmatrix} 10 & 10 & -5 \\ 11 & -11 & 11 \\ 3 & -8 & -7 \end{pmatrix}$

We now write the transpose of $\mathbf{C}$, i.e. $\mathbf{C}^{\text{T}}$ in which we write rows as columns and columns as rows.

$$\mathbf{C}^{\text{T}} = \ldots\ldots\ldots$$

27

$$\mathbf{C}^{\text{T}} = \begin{pmatrix} 10 & 11 & 3 \\ 10 & -11 & -8 \\ -5 & 11 & -7 \end{pmatrix}$$

This is called the *adjoint* (adj) of the original matrix $\mathbf{A}$

i.e. $\text{adj } \mathbf{A} = \mathbf{C}^{\text{T}}$

Then the inverse of $\mathbf{A}$, i.e. $\mathbf{A}^{-1}$ is given by

$$\mathbf{A}^{-1} = \frac{1}{|\mathbf{A}|} \times \mathbf{C}^{\text{T}} = \frac{1}{55} \begin{pmatrix} 10 & 11 & 3 \\ 10 & -11 & -8 \\ -5 & 11 & -7 \end{pmatrix}$$

As a check that all the calculations have been done correctly and without error, the product of matrix $\mathbf{A}$ with its adjoint should be equal to the unit matrix multiplied by the determinant of $\mathbf{A}$. That is

$$\mathbf{A} \times \text{adj } \mathbf{A} = \det \mathbf{A} \times \mathbf{I}$$

For this case

$$\mathbf{A} \times \text{adj } \mathbf{A} = \begin{pmatrix} 3 & 2 & -1 \\ 2 & -1 & 2 \\ 1 & -3 & -4 \end{pmatrix} \begin{pmatrix} 10 & 11 & 3 \\ 10 & -11 & -8 \\ -5 & 11 & -7 \end{pmatrix}$$

$$= \begin{pmatrix} 55 & 0 & 0 \\ 0 & 55 & 0 \\ 0 & 0 & 55 \end{pmatrix}$$

$$= \det \mathbf{A} \times \mathbf{I}$$

Thus all is well. We can now continue to find the solution.

So $\begin{pmatrix} x_1 \\ x_2 \\ x_3 \end{pmatrix} = \mathbf{A}^{-1} \begin{pmatrix} 4 \\ 10 \\ 5 \end{pmatrix}$ becomes

$$\begin{pmatrix} x_1 \\ x_2 \\ x_3 \end{pmatrix} = \frac{1}{55} \begin{pmatrix} 10 & 11 & 3 \\ 10 & -11 & -8 \\ -5 & 11 & -7 \end{pmatrix} \begin{pmatrix} 4 \\ 10 \\ 5 \end{pmatrix} = \ldots\ldots\ldots$$

$$\boxed{x_1 = 3; \quad x_2 = -2; \quad x_3 = 1}$$

Because

$$\begin{pmatrix} x_1 \\ x_2 \\ x_3 \end{pmatrix} = \frac{1}{55} \begin{pmatrix} 10 & 11 & 3 \\ 10 & -11 & -8 \\ -5 & 11 & -7 \end{pmatrix} \begin{pmatrix} 4 \\ 10 \\ 5 \end{pmatrix}$$

$$= \frac{1}{55} \begin{pmatrix} 40 & +110 & +15 \\ 40 & -110 & -40 \\ -20 & +110 & -35 \end{pmatrix} = \begin{pmatrix} 3 \\ -2 \\ 1 \end{pmatrix}$$

$$\therefore \ x_1 = 3; \quad x_2 = -2; \quad x_3 = 1$$

The method is the same every time.

To solve $\mathbf{Ax} = \mathbf{b}$ $\quad \mathbf{x} = \mathbf{A}^{-1}\mathbf{b}$

To find $\mathbf{A}^{-1}$

(1) Evaluate $|\mathbf{A}|$

If $|\mathbf{A}| \neq 0$ then proceed to (2)

If $|\mathbf{A}| = 0$ then there is no inverse and hence no unique solution. Later we shall discover how to determine whether there is an infinity of solutions or none.

(2) Form $\mathbf{C}$, the matrix of cofactors of $\mathbf{A}$

(3) Write $\mathbf{C}^{\mathrm{T}}$, the transpose of $\mathbf{C}$

(4) Then $\mathbf{A}^{-1} = \dfrac{1}{|\mathbf{A}|} \times \mathbf{C}^{\mathrm{T}}$.

Now apply the method to Example 2.

Example 2

$$4x_1 + 5x_2 + x_3 = 2$$
$$x_1 - 2x_2 - 3x_3 = 7$$
$$3x_1 - x_2 - 2x_3 = 1.$$

$x_1 = \ldots\ldots\ldots$; $\quad x_2 = \ldots\ldots\ldots$; $\quad x_3 = \ldots\ldots\ldots$

29

$$\boxed{x_1 = -2; \quad x_2 = 3; \quad x_3 = -5}$$

Here is the complete working.

$$\mathbf{A} = \begin{pmatrix} 4 & 5 & 1 \\ 1 & -2 & -3 \\ 3 & -1 & -2 \end{pmatrix} \quad \therefore \quad |\mathbf{A}| = \begin{vmatrix} 4 & 5 & 1 \\ 1 & -2 & -3 \\ 3 & -1 & -2 \end{vmatrix} = -26$$

$$\mathbf{C} = \begin{pmatrix} A_{11} & A_{12} & A_{13} \\ A_{21} & A_{22} & A_{23} \\ A_{31} & A_{32} & A_{33} \end{pmatrix}$$

$$A_{11} = \begin{vmatrix} -2 & -3 \\ -1 & -2 \end{vmatrix} = 1 \qquad A_{12} = -\begin{vmatrix} 1 & -3 \\ 3 & -2 \end{vmatrix} = -7 \quad A_{13} = \begin{vmatrix} 1 & -2 \\ 3 & -1 \end{vmatrix} = 5$$

$$A_{21} = -\begin{vmatrix} 5 & 1 \\ -1 & -2 \end{vmatrix} = 9 \qquad A_{22} = \begin{vmatrix} 4 & 1 \\ 3 & -2 \end{vmatrix} = -11 \quad A_{23} = -\begin{vmatrix} 4 & 5 \\ 3 & -1 \end{vmatrix} = 19$$

$$A_{31} = \begin{vmatrix} 5 & 1 \\ -2 & -3 \end{vmatrix} = -13 \quad A_{32} = -\begin{vmatrix} 4 & 1 \\ 1 & -3 \end{vmatrix} = 13 \quad A_{33} = \begin{vmatrix} 4 & 5 \\ 1 & -2 \end{vmatrix} = -13$$

$$\therefore \quad \mathbf{C} = \begin{pmatrix} 1 & -7 & 5 \\ 9 & -11 & 19 \\ -13 & 13 & -13 \end{pmatrix} \quad \therefore \quad \mathbf{C}^T = \begin{pmatrix} 1 & 9 & -13 \\ -7 & -11 & 13 \\ 5 & 19 & -13 \end{pmatrix}$$

$$\mathbf{A}^{-1} = \frac{1}{|\mathbf{A}|} \times \mathbf{C}^T = -\frac{1}{26}\begin{pmatrix} 1 & 9 & -13 \\ -7 & -11 & 13 \\ 5 & 19 & -13 \end{pmatrix}$$

$$\begin{pmatrix} x_1 \\ x_2 \\ x_3 \end{pmatrix} = \mathbf{A}^{-1}\begin{pmatrix} 2 \\ 7 \\ 1 \end{pmatrix} = -\frac{1}{26}\begin{pmatrix} 1 & 9 & -13 \\ -7 & -11 & 13 \\ 5 & 19 & -13 \end{pmatrix}\begin{pmatrix} 2 \\ 7 \\ 1 \end{pmatrix}$$

$$= -\frac{1}{26}\begin{pmatrix} 2 & +63 & -13 \\ -14 & -77 & +13 \\ 10 & +133 & -13 \end{pmatrix}$$

$$= -\frac{1}{26}\begin{pmatrix} 52 \\ -78 \\ 130 \end{pmatrix} = -\begin{pmatrix} 2 \\ -3 \\ 5 \end{pmatrix}$$

$$\therefore \quad x_1 = -2; \quad x_2 = 3; \quad x_3 = -5$$

With a set of four equations with four unknowns, the method becomes somewhat tedious as there are then sixteen cofactors to be evaluated and each one is a third-order determinant! There are, however, other methods that can be applied – so let us see method 2.

2 *Row transformation method*

30

Elementary row transformations that can be applied are as follows

(a) Interchange any two rows.
(b) Multiply (or divide) every element in a row by a non-zero scalar (constant) *k*.
(c) Add to (or subtract from) all the elements of any row *k* times the corresponding elements of any other row.

Equivalent matrices
Two matrices, **A** and **B**, are said to be equivalent if **B** can be obtained from **A** by a sequence of elementary transformations.

Solutions of equations
The method is best described by working through a typical example.

Example 1

Solve
$$\begin{aligned} 2x_1 + x_2 + x_3 &= 5 \\ x_1 + 3x_2 + 2x_3 &= 1 \\ 3x_1 - 2x_2 - 4x_3 &= -4. \end{aligned}$$

This can be written
$$\begin{pmatrix} 2 & 1 & 1 \\ 1 & 3 & 2 \\ 3 & -2 & -4 \end{pmatrix} \begin{pmatrix} x_1 \\ x_2 \\ x_3 \end{pmatrix} = \begin{pmatrix} 5 \\ 1 \\ -4 \end{pmatrix}$$

and for convenience we introduce the unit matrix
$$\begin{pmatrix} 2 & 1 & 1 \\ 1 & 3 & 2 \\ 3 & -2 & -4 \end{pmatrix} \begin{pmatrix} x_1 \\ x_2 \\ x_3 \end{pmatrix} = \begin{pmatrix} 1 & 0 & 0 \\ 0 & 1 & 0 \\ 0 & 0 & 1 \end{pmatrix} \begin{pmatrix} 5 \\ 1 \\ -4 \end{pmatrix}$$

where $\begin{pmatrix} 1 & 0 & 0 \\ 0 & 1 & 0 \\ 0 & 0 & 1 \end{pmatrix}$ may be regarded as the coefficient of $\begin{pmatrix} 5 \\ 1 \\ -4 \end{pmatrix}$

We then form the combined coefficient matrix
$$\begin{pmatrix} 2 & 1 & 1 & 1 & 0 & 0 \\ 1 & 3 & 2 & 0 & 1 & 0 \\ 3 & -2 & -4 & 0 & 0 & 1 \end{pmatrix}$$

and work on this matrix from now on.

On then to the next frame

31

The rest of the working is mainly concerned with applying row transformations to convert the left-hand half of the matrix to a unit matrix and the right-hand side to the inverse, eventually obtaining

$$\begin{pmatrix} 1 & 0 & 0 & a & b & c \\ 0 & 1 & 0 & d & e & f \\ 0 & 0 & 1 & g & h & i \end{pmatrix}$$

with $a, b, c, \ldots g, h, i$ being evaluated in the process.

The following notation will be helpful to denote the transformation used:

$(1) \sim (2)$ denotes 'interchange rows 1 and 2'

$(3) - 2(1)$ denotes 'subtract twice row 1 from row 3', etc.

So off we go.

$$(1) \sim (2) \begin{pmatrix} 1 & 3 & 2 & 0 & 1 & 0 \\ 2 & 1 & 1 & 1 & 0 & 0 \\ 3 & -2 & -4 & 0 & 0 & 1 \end{pmatrix}$$

$$\begin{matrix} (2) - 2(1) \\ (3) - 3(1) \end{matrix} \begin{pmatrix} 1 & 3 & 2 & 0 & 1 & 0 \\ 0 & -5 & -3 & 1 & -2 & 0 \\ 0 & -11 & -10 & 0 & -3 & 1 \end{pmatrix}$$

$$(3) - 2(2) \begin{pmatrix} 1 & 3 & 2 & 0 & 1 & 0 \\ 0 & -5 & -3 & 1 & -2 & 0 \\ 0 & -1 & -4 & -2 & 1 & 1 \end{pmatrix}$$

$$-(2) \sim -(3) \begin{pmatrix} 1 & 3 & 2 & 0 & 1 & 0 \\ 0 & 1 & 4 & 2 & -1 & -1 \\ 0 & 5 & 3 & -1 & 2 & 0 \end{pmatrix}$$

$$(3) - 5(2) \begin{pmatrix} 1 & 3 & 2 & 0 & 1 & 0 \\ 0 & 1 & 4 & 2 & -1 & -1 \\ 0 & 0 & -17 & -11 & 7 & 5 \end{pmatrix}$$

$$\begin{matrix} (1) - 3(2) \\ (3) \div (-17) \end{matrix} \begin{pmatrix} 1 & 0 & -10 & -6 & 4 & 3 \\ 0 & 1 & 4 & 2 & -1 & -1 \\ 0 & 0 & 1 & 11/17 & -7/17 & -5/17 \end{pmatrix}$$

$$\begin{matrix} (1) + 10(3) \\ (2) - 4(3) \end{matrix} \begin{pmatrix} 1 & 0 & 0 & 8/17 & -2/17 & 1/17 \\ 0 & 1 & 0 & -10/17 & 11/17 & 3/17 \\ 0 & 0 & 1 & 11/17 & -7/17 & -5/17 \end{pmatrix}$$

We now have

$$\begin{pmatrix} 1 & 0 & 0 \\ 0 & 1 & 0 \\ 0 & 0 & 1 \end{pmatrix} \begin{pmatrix} x_1 \\ x_2 \\ x_3 \end{pmatrix} = \frac{1}{17} \begin{pmatrix} 8 & -2 & 1 \\ -10 & 11 & 3 \\ 11 & -7 & -5 \end{pmatrix} \begin{pmatrix} 5 \\ 1 \\ -4 \end{pmatrix}$$

$$\therefore x_1 = \ldots\ldots\ldots\ldots; \quad x_2 = \ldots\ldots\ldots\ldots; \quad x_3 = \ldots\ldots\ldots\ldots$$

$$\boxed{x_1 = 2; \qquad x_2 = -3; \qquad x_3 = 4}$$

$$\begin{pmatrix} x_1 \\ x_2 \\ x_3 \end{pmatrix} = \frac{1}{17} \begin{pmatrix} 40 & -2 & -4 \\ -50 & +11 & -12 \\ 55 & -7 & +20 \end{pmatrix} = \frac{1}{17} \begin{pmatrix} 34 \\ -51 \\ 68 \end{pmatrix} = \begin{pmatrix} 2 \\ -3 \\ 4 \end{pmatrix}$$

$$x_1 = 2; \qquad x_2 = -3; \qquad x_3 = 4$$

Of course, there is no set pattern of how to carry out the row transformations. It depends on one's ingenuity and every case is different. Here is a further example.

Example 2

$$2x_1 - x_2 - 3x_3 = 1$$
$$x_1 + 2x_2 + x_3 = 3$$
$$2x_1 - 2x_2 - 5x_3 = 2.$$

First write the set of equations in matrix form – with the unit matrix included. This gives ....

$$\boxed{\begin{pmatrix} 2 & -1 & -3 \\ 1 & 2 & 1 \\ 2 & -2 & -5 \end{pmatrix} \begin{pmatrix} x_1 \\ x_2 \\ x_3 \end{pmatrix} = \begin{pmatrix} 1 & 0 & 0 \\ 0 & 1 & 0 \\ 0 & 0 & 1 \end{pmatrix} \begin{pmatrix} 1 \\ 3 \\ 2 \end{pmatrix}}$$

The combined coefficient matrix is now ....

$$\begin{pmatrix} 2 & -1 & -3 & 1 & 0 & 0 \\ 1 & 2 & 1 & 0 & 1 & 0 \\ 2 & -2 & -5 & 0 & 0 & 1 \end{pmatrix}$$

If we start off by interchanging the top two rows, we obtain a 1 at the beginning of the top row which is a help.

$$(1) \sim (2) \quad \begin{pmatrix} 1 & 2 & 1 & 0 & 1 & 0 \\ 2 & -1 & -3 & 1 & 0 & 0 \\ 2 & -2 & -5 & 0 & 0 & 1 \end{pmatrix}$$

Now, if we subtract $2 \times$ row 1 from row 2

and $\qquad\qquad 2 \times$ row 1 from row 3, we get

....

35

$$\begin{pmatrix} 1 & 2 & 1 & 0 & 1 & 0 \\ 0 & -5 & -5 & 1 & -2 & 0 \\ 0 & -6 & -7 & 0 & -2 & 1 \end{pmatrix}$$

Continuing with the same line of reasoning, we then have

$$(2) - (3) \quad \begin{pmatrix} 1 & 2 & 1 & 0 & 1 & 0 \\ 0 & 1 & 2 & 1 & 0 & -1 \\ 0 & -6 & -7 & 0 & -2 & 1 \end{pmatrix}$$

$$(3) + 6(2) \quad \begin{pmatrix} 1 & 2 & 1 & 0 & 1 & 0 \\ 0 & 1 & 2 & 1 & 0 & -1 \\ 0 & 0 & 5 & 6 & -2 & -5 \end{pmatrix}$$

$$\begin{matrix}(1) - 2(2)\\ (3) \div 5\end{matrix} \quad \begin{pmatrix} 1 & 0 & -3 & -2 & 1 & 2 \\ 0 & 1 & 2 & 1 & 0 & -1 \\ 0 & 0 & 1 & \frac{6}{5} & -\frac{2}{5} & -1 \end{pmatrix}$$ Notice the three diagonal
1s appearing at the
left-hand end

What do you suggest we should do now?

. . . .

36

Add three times row 3 to row 1
and subtract twice row 3 from row 2

Right. That gives

$$\begin{matrix}(1) + 3(3)\\ (2) - 3(3)\end{matrix} \quad \begin{pmatrix} 1 & 0 & 0 & \frac{8}{5} & -\frac{1}{5} & -1 \\ 0 & 1 & 0 & -\frac{7}{5} & \frac{4}{5} & 1 \\ 0 & 0 & 1 & \frac{6}{5} & -\frac{2}{5} & -1 \end{pmatrix}$$

$$\therefore \begin{pmatrix} 1 & 0 & 0 \\ 0 & 1 & 0 \\ 0 & 0 & 1 \end{pmatrix}\begin{pmatrix} x_1 \\ x_2 \\ x_3 \end{pmatrix} = \frac{1}{5}\begin{pmatrix} 8 & -1 & -5 \\ -7 & 4 & 5 \\ 6 & -2 & -5 \end{pmatrix}\begin{pmatrix} 1 \\ 3 \\ 2 \end{pmatrix}$$

Now you can finish it off.

$$x_1 = \ldots\ldots\ldots\ldots; \quad x_2 = \ldots\ldots\ldots\ldots; \quad x_3 = \ldots\ldots\ldots\ldots$$

37

$$x_1 = -1; \quad x_2 = 3; \quad x_3 = -2$$

Because

$$\begin{pmatrix} x_1 \\ x_2 \\ x_3 \end{pmatrix} = \frac{1}{5}\begin{pmatrix} 8 - 3 - 10 \\ -7 + 12 + 10 \\ 6 - 6 - 10 \end{pmatrix} = \frac{1}{5}\begin{pmatrix} -5 \\ 15 \\ -10 \end{pmatrix} = \begin{pmatrix} -1 \\ 3 \\ -2 \end{pmatrix}$$

Let us now look at a somewhat similar method with rather fewer steps involved.

So move on

3 *Gaussian elimination method* 38

Once again we will demonstrate the method by a typical example.

Example 1

$$2x_1 - 3x_2 + 2x_3 = 9$$
$$3x_1 + 2x_2 - x_3 = 4$$
$$x_1 - 4x_2 + 2x_3 = 6.$$

We start off as usual

$$\begin{pmatrix} 2 & -3 & 2 \\ 3 & 2 & -1 \\ 1 & -4 & 2 \end{pmatrix} \begin{pmatrix} x_1 \\ x_2 \\ x_3 \end{pmatrix} = \begin{pmatrix} 9 \\ 4 \\ 6 \end{pmatrix}$$

We then form the *augmented coefficient matrix* by including the constants as an extra column on the right-hand side of the matrix

$$\begin{pmatrix} 2 & -3 & 2 & \vdots & 9 \\ 3 & 2 & -1 & \vdots & 4 \\ 1 & -4 & 2 & \vdots & 6 \end{pmatrix}$$

Now we operate on the rows to convert the first three columns into an upper triangular matrix

$$(1) \sim (3) \quad \begin{pmatrix} 1 & -4 & 2 & 6 \\ 3 & 2 & -1 & 4 \\ 2 & -3 & 2 & 9 \end{pmatrix} \qquad (2) \sim (3) \quad \begin{pmatrix} 1 & -4 & 2 & 6 \\ 2 & -3 & 2 & 9 \\ 3 & 2 & -1 & 4 \end{pmatrix}$$

$$\begin{matrix} (2) - 2(1) \\ (3) - 3(1) \end{matrix} \begin{pmatrix} 1 & -4 & 2 & 6 \\ 0 & 5 & -2 & -3 \\ 0 & 14 & -7 & -14 \end{pmatrix} \qquad \begin{matrix} (2) \div 5 \\ (3) \div 7 \end{matrix} \begin{pmatrix} 1 & -4 & 2 & 6 \\ 0 & 1 & -\frac{2}{5} & -\frac{3}{5} \\ 0 & 2 & -1 & -2 \end{pmatrix}$$

$$(3) - 2(2) \quad \begin{pmatrix} 1 & -4 & 2 & 6 \\ 0 & 1 & -\frac{2}{5} & -\frac{3}{5} \\ 0 & 0 & -\frac{1}{5} & -\frac{4}{5} \end{pmatrix} \qquad (3) \times (-5) \quad \begin{pmatrix} 1 & -4 & 2 & 6 \\ 0 & 1 & -\frac{2}{5} & -\frac{3}{5} \\ 0 & 0 & 1 & 4 \end{pmatrix}$$

The first three columns now form an upper triangular matrix which has been our purpose. If we now detach the fourth column back to its original position on the right-hand side of the matrix equation, we have

. . . .

39

$$\begin{pmatrix} 1 & -4 & 2 \\ 0 & 1 & -\frac{2}{5} \\ 0 & 0 & 1 \end{pmatrix} \begin{pmatrix} x_1 \\ x_2 \\ x_3 \end{pmatrix} = \begin{pmatrix} 6 \\ -\frac{3}{5} \\ 4 \end{pmatrix}$$

Expanding from the bottom row, working upwards

$$x_3 = 4 \qquad\qquad\qquad \therefore\ x_3 = 4$$
$$x_2 - \tfrac{2}{5}x_3 = -\tfrac{3}{5} \quad \therefore\ x_2 = -\tfrac{3}{5} + \tfrac{8}{5} = 1 \quad \therefore\ x_2 = 1$$
$$x_1 - 4x_2 + 2x_3 = 6 \quad \therefore\ x_1 - 4 + 8 = 6 \qquad \therefore\ x_1 = 2$$
$$\therefore\ x_1 = 2; \quad x_2 = 1; \quad x_3 = 4$$

It is a very useful method and entails fewer tedious steps, and can be used to solve efficiently higher-order sets of equations and non-square systems. It can also solve a sequence of problems with the same coefficient matrix $\mathbf{A}$ by using the augmented matrix $(\mathbf{A}\mathbf{b}_1\mathbf{b}_2 \ldots \mathbf{b}_n)$.

Example 2

$$x_1 + 3x_2 - 2x_3 + x_4 = -1$$
$$2x_1 - 2x_2 + x_3 - 2x_4 = 1$$
$$x_1 + x_2 - 3x_3 + x_4 = 6$$
$$3x_1 - x_2 + 2x_3 - x_4 = 3.$$

First we write this in matrix form and compile the augmented matrix which is

. . . .

40

$$\begin{pmatrix} 1 & 3 & -2 & 1 & \vdots & -1 \\ 2 & -2 & 1 & -2 & \vdots & 1 \\ 1 & 1 & -3 & 1 & \vdots & 6 \\ 3 & -1 & 2 & -1 & \vdots & 3 \end{pmatrix}$$

Next we operate on rows to convert the left-hand side to an upper triangular matrix. There is no set way of doing this. Use any trickery to save yourself unnecessary work.

So now you can go ahead and complete the transformations and obtain

$$x_1 = \ldots\ldots\ldots\ldots; \quad x_2 = \ldots\ldots\ldots\ldots$$
$$x_3 = \ldots\ldots\ldots\ldots; \quad x_4 = \ldots\ldots\ldots\ldots$$

$$x_1 = 2; \quad x_2 = -3; \quad x_3 = -1; \quad x_4 = 4$$

Here is one way. You may well have taken quite a different route.

$$\begin{pmatrix} 1 & 3 & -2 & 1 & \vdots & -1 \\ 2 & -2 & 1 & -2 & \vdots & 1 \\ 1 & 1 & -3 & 1 & \vdots & 6 \\ 3 & -1 & 2 & -1 & \vdots & 3 \end{pmatrix}$$

$$\begin{matrix} (2) - 2(1) \\ (3) - (1) \\ (4) - [(1) + (2)] \end{matrix} \begin{pmatrix} 1 & 3 & -2 & 1 & \vdots & -1 \\ 0 & -8 & 5 & -4 & \vdots & 3 \\ 0 & -2 & -1 & 0 & \vdots & 7 \\ 0 & -2 & 3 & 0 & \vdots & 3 \end{pmatrix}$$

$$\begin{matrix} (2) - 4(4) \\ (3) - (4) \end{matrix} \begin{pmatrix} 1 & 3 & -2 & 1 & \vdots & -1 \\ 0 & 0 & -7 & -4 & \vdots & -9 \\ 0 & 0 & -4 & 0 & \vdots & 4 \\ 0 & -2 & 3 & 0 & \vdots & 3 \end{pmatrix}$$

$$\begin{matrix} (2) \sim (4) \\ (3) \div 4 \end{matrix} \begin{pmatrix} 1 & 3 & -2 & 1 & \vdots & -1 \\ 0 & -2 & 3 & 0 & \vdots & 3 \\ 0 & 0 & -1 & 0 & \vdots & 1 \\ 0 & 0 & -7 & -4 & \vdots & -9 \end{pmatrix}$$

$$\begin{matrix} (4) - 7(3) \end{matrix} \begin{pmatrix} 1 & 3 & -2 & 1 & \vdots & -1 \\ 0 & -2 & 3 & 0 & \vdots & 3 \\ 0 & 0 & -1 & 0 & \vdots & 1 \\ 0 & 0 & 0 & -4 & \vdots & -16 \end{pmatrix}$$

Returning the right-hand column to its original position

$$\begin{pmatrix} 1 & 3 & -2 & 1 \\ 0 & -2 & 3 & 0 \\ 0 & 0 & -1 & 0 \\ 0 & 0 & 0 & -4 \end{pmatrix} \begin{pmatrix} x_1 \\ x_2 \\ x_3 \\ x_4 \end{pmatrix} = \begin{pmatrix} -1 \\ 3 \\ 1 \\ -16 \end{pmatrix}$$

Expanding from the bottom row, we have

$$-4x_4 = -16 \qquad\qquad\qquad\qquad\qquad\qquad \therefore\ x_4 = 4$$

$$-x_3 = 1 \qquad\qquad\qquad\qquad\qquad\qquad\qquad \therefore\ x_3 = -1$$

$$-2x_2 + 3x_3 = 3 \quad \therefore\ -2x_2 = 6 \qquad\qquad \therefore\ x_2 = -3$$

$$x_1 + 3x_2 - 2x_3 + x_4 = -1 \quad \therefore\ x_1 - 9 + 2 + 4 = -1 \quad \therefore\ x_1 = 2$$

$$\therefore\ x_1 = 2; \quad x_2 = -3; \quad x_3 = -1; \quad x_4 = 4$$

42

We still have a further method for solving sets of simultaneous equations.

4 Triangular decomposition method

A square matrix $\mathbf{A}$ can usually be written as a product of a lower-triangular matrix $\mathbf{L}$ and an upper-triangular matrix $\mathbf{U}$, where $\mathbf{A} = \mathbf{LU}$.

For example, if $\mathbf{A} = \begin{pmatrix} 1 & 2 & 3 \\ 3 & 5 & 8 \\ 4 & 9 & 10 \end{pmatrix}$, $\mathbf{A}$ can be expressed as

$$\mathbf{A} = \mathbf{LU} = \begin{pmatrix} l_{11} & 0 & 0 \\ l_{21} & l_{22} & 0 \\ l_{31} & l_{32} & l_{33} \end{pmatrix} \begin{pmatrix} u_{11} & u_{12} & u_{13} \\ 0 & u_{22} & u_{23} \\ 0 & 0 & u_{33} \end{pmatrix}$$

$$(\mathbf{L}) \qquad\qquad (\mathbf{U})$$

$$= \begin{pmatrix} l_{11}u_{11} & l_{11}u_{12} & l_{11}u_{13} \\ l_{21}u_{11} & l_{21}u_{12} + l_{22}u_{22} & l_{21}u_{13} + l_{22}u_{23} \\ l_{31}u_{11} & l_{31}u_{12} + l_{32}u_{22} & l_{31}u_{13} + l_{32}u_{23} + l_{33}u_{33} \end{pmatrix}$$

Note that, in $\mathbf{L}$ and $\mathbf{U}$, elements occur in the major diagonal in each case. These are related in the product and whatever values we choose to put for u_{11}, u_{22}, $u_{33}\dots$ then the corresponding values of l_{11}, l_{22}, $l_{33}\dots$ will be determined – and vice versa.

For convenience, we put $u_{11} = u_{22} = u_{33}\dots = 1$

Then $\mathbf{A} = \mathbf{LU} = \begin{pmatrix} l_{11} & l_{11}u_{12} & l_{11}u_{13} \\ l_{21} & l_{21}u_{12} + l_{22} & l_{21}u_{13} + l_{22}u_{23} \\ l_{31} & l_{31}u_{12} + l_{32} & l_{31}u_{13} + l_{32}u_{23} + l_{33} \end{pmatrix}$

In our example, $\mathbf{A} = \begin{pmatrix} 1 & 2 & 3 \\ 3 & 5 & 8 \\ 4 & 9 & 10 \end{pmatrix}$

$\therefore l_{11} = 1;$ $\quad l_{11}u_{12} = 2 \therefore u_{12} = 2;$ $\quad l_{11}u_{13} = 3 \therefore u_{13} = 3$

$\quad l_{21} = 3;$ $\quad$ Similarly $l_{22} = \dots\dots\dots;$ $\quad u_{23} = \dots\dots\dots$

$\quad l_{31} = 4;$ $\quad\quad l_{32} = \dots\dots\dots;$ $\quad l_{33} = \dots\dots\dots$

43

$$\boxed{l_{22} = -1; \quad u_{23} = 1; \quad l_{32} = 1; \quad l_{33} = -3}$$

Because

$l_{21}u_{12} + l_{22}u_{22} = 5$ that is $3 \times 2 + l_{22} \times 1 = 5$ and so $l_{22} = -1$
$l_{21}u_{13} + l_{22}u_{23} = 8$ that is $3 \times 3 + (-1) \times u_{23} = 8$ and so $u_{23} = 1$
$l_{31}u_{12} + l_{32}u_{22} = 9$ that is $4 \times 2 + l_{32} \times 1 = 9$ and so $l_{32} = 1$
$l_{31}u_{13} + l_{32}u_{23} + l_{33}u_{33} = 10$ that is $4 \times 3 + 1 \times 1 + l_{33} \times 1 = 10$
and so $l_{33} = -3$

Now we substitute all these values back into the upper and lower triangular matrices and obtain

$$\mathbf{A} = \mathbf{LU} = \dots\dots\dots$$

$$\mathbf{A} = \mathbf{LU} = \begin{pmatrix} 1 & 0 & 0 \\ 3 & -1 & 0 \\ 4 & 1 & -3 \end{pmatrix} \begin{pmatrix} 1 & 2 & 3 \\ 0 & 1 & 1 \\ 0 & 0 & 1 \end{pmatrix}$$

We have thus expressed the given matrix **A** as the product of lower and upper triangular matrices. Let us now see how we use them.

Example 1

$$x_1 + 2x_2 + 3x_3 = 16$$
$$3x_1 + 5x_2 + 8x_3 = 43$$
$$4x_1 + 9x_2 + 10x_3 = 57.$$

i.e. $\begin{pmatrix} 1 & 2 & 3 \\ 3 & 5 & 8 \\ 4 & 9 & 10 \end{pmatrix} \begin{pmatrix} x_1 \\ x_2 \\ x_3 \end{pmatrix} = \begin{pmatrix} 16 \\ 43 \\ 57 \end{pmatrix}$ i.e. $\mathbf{Ax} = \mathbf{b}.$

We have seen above that **A** can be written as **LU** where

$$\mathbf{A} = \mathbf{LU} = \begin{pmatrix} 1 & 0 & 0 \\ 3 & -1 & 0 \\ 4 & 1 & -3 \end{pmatrix} \begin{pmatrix} 1 & 2 & 3 \\ 0 & 1 & 1 \\ 0 & 0 & 1 \end{pmatrix}$$

To solve $\mathbf{Ax} = \mathbf{b}$, we have $\mathbf{LUx} = \mathbf{b}$ i.e. $\mathbf{L(Ux)} = \mathbf{b}$

Putting $\mathbf{Ux} = \mathbf{y}$, we solve $\mathbf{Ly} = \mathbf{b}$ to obtain **y**

and then $\mathbf{Ux} = \mathbf{y}$ to obtain **x**.

(a) Solving $\mathbf{Ly} = \mathbf{b}$ $\begin{pmatrix} 1 & 0 & 0 \\ 3 & -1 & 0 \\ 4 & 1 & -3 \end{pmatrix} \begin{pmatrix} y_1 \\ y_2 \\ y_3 \end{pmatrix} = \begin{pmatrix} 16 \\ 43 \\ 57 \end{pmatrix}$

Expanding from the top $y_1 = 16$; $3y_1 - y_2 = 43$ $\therefore y_2 = 5$; and
$4y_1 + y_2 - 3y_3 = 57$ $\therefore 64 + 5 - 3y_3 = 57$ $\therefore y_3 = 4$

$$\therefore \begin{pmatrix} y_1 \\ y_2 \\ y_3 \end{pmatrix} = \begin{pmatrix} 16 \\ 5 \\ 4 \end{pmatrix}$$

(b) Solving $\mathbf{Ux} = \mathbf{y}$ $\begin{pmatrix} 1 & 2 & 3 \\ 0 & 1 & 1 \\ 0 & 0 & 1 \end{pmatrix} \begin{pmatrix} x_1 \\ x_2 \\ x_3 \end{pmatrix} = \begin{pmatrix} 16 \\ 5 \\ 4 \end{pmatrix}$

Expanding from the bottom, we then have

$$x_1 = \ldots\ldots\ldots\ldots; \qquad x_2 = \ldots\ldots\ldots\ldots; \qquad x_3 = \ldots\ldots\ldots\ldots$$

45

$$\boxed{x_1 = 2; \quad x_2 = 1; \quad x_3 = 4}$$

Note:
1 If $l_{ii} = 0$, then either decomposition is not possible, or, if $\mathbf{A}$ is singular, i.e. $|\mathbf{A}| = 0$, there is an infinite number of possible decompositions.
2 Instead of putting $u_{11} = u_{22} = u_{33} \ldots = 1$, we could have used the alternative substitution $l_{11} = l_{22} = l_{33} \ldots = 1$ and obtained values of u_{11}, u_{22}, $u_{33} \ldots$ etc. The working is as before.
3 One advantage of employing **LU** decomposition over Gaussian elimination is in the solution of a sequence of problems in which the same coefficient matrix occurs.

Now for another example.

46   **Example 2**

$$x_1 + 3x_2 + 2x_3 = 19$$
$$2x_1 + x_2 + x_3 = 13$$
$$4x_1 + 2x_2 + 3x_3 = 31.$$

$$\therefore \begin{pmatrix} 1 & 3 & 2 \\ 2 & 1 & 1 \\ 4 & 2 & 3 \end{pmatrix} \begin{pmatrix} x_1 \\ x_2 \\ x_3 \end{pmatrix} = \begin{pmatrix} 19 \\ 13 \\ 31 \end{pmatrix} \quad \text{i.e. } \mathbf{Ax = b}$$

$$\mathbf{A = LU} = \begin{pmatrix} l_{11} & 0 & 0 \\ l_{21} & l_{22} & 0 \\ l_{31} & l_{32} & l_{33} \end{pmatrix} \begin{pmatrix} 1 & u_{12} & u_{13} \\ 0 & 1 & u_{23} \\ 0 & 0 & 1 \end{pmatrix}$$

$$= \begin{pmatrix} l_{11} & l_{11}u_{12} & l_{11}u_{13} \\ l_{21} & l_{21}u_{12} + l_{22} & l_{21}u_{13} + l_{22}u_{23} \\ l_{31} & l_{31}u_{12} + l_{32} & l_{31}u_{13} + l_{32}u_{23} + l_{33} \end{pmatrix}$$

$$= \begin{pmatrix} 1 & 3 & 2 \\ 2 & 1 & 1 \\ 4 & 2 & 3 \end{pmatrix}$$

Now we have to find the values of the various elements. The usual order of doing this is shown by the diagram.

▶

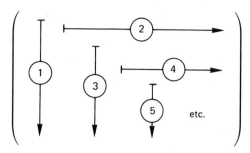

That is, first we can write down values for l_{11}, l_{21}, l_{31} from the left-hand column; then follow this by finding u_{12}, u_{13} from the top row; and proceed for the others.

So, completing the two triangular matrices, we have

$$\mathbf{A} = \mathbf{LU} = \ldots\ldots\ldots$$

47

$$\mathbf{A} = \mathbf{LU} = \begin{pmatrix} 1 & 0 & 0 \\ 2 & -5 & 0 \\ 4 & -10 & 1 \end{pmatrix} \begin{pmatrix} 1 & 3 & 2 \\ 0 & 1 & \frac{3}{5} \\ 0 & 0 & 1 \end{pmatrix}$$

As we stated before: $\mathbf{Ax} = \mathbf{b}$; $\mathbf{L(Ux)} = \mathbf{b}$. Put $\mathbf{Ux} = \mathbf{y}$

then   (a) solve $\mathbf{Ly} = \mathbf{b}$ to obtain $\mathbf{y}$

and   (b) solve $\mathbf{Ux} = \mathbf{y}$ to obtain $\mathbf{x}$.

Solving $\mathbf{Ly} = \mathbf{b}$ gives $\begin{pmatrix} y_1 \\ y_2 \\ y_3 \end{pmatrix} = \begin{pmatrix} \ldots \\ \ldots \\ \ldots \end{pmatrix}$

48

$$\begin{pmatrix} y_1 \\ y_2 \\ y_3 \end{pmatrix} = \begin{pmatrix} 19 \\ 5 \\ 5 \end{pmatrix}$$

Because

$$\begin{pmatrix} 1 & 0 & 0 \\ 2 & -5 & 0 \\ 4 & -10 & 1 \end{pmatrix} \begin{pmatrix} y_1 \\ y_2 \\ y_3 \end{pmatrix} = \begin{pmatrix} 19 \\ 13 \\ 31 \end{pmatrix}$$

Expanding from the top gives

$y_1 = 19$; $y_2 = 5$; $y_3 = 5$.

(b) Now solve $\mathbf{Ux} = \mathbf{y}$ from which $x_1 = \ldots\ldots\ldots$; $x_2 = \ldots\ldots\ldots$;
   $x_3 = \ldots\ldots\ldots$

49

$$x_1 = 3; \quad x_2 = 2; \quad x_3 = 5$$

Because we have

$$\mathbf{Ux} = y$$

i.e. $\begin{pmatrix} 1 & 3 & 2 \\ 0 & 1 & \frac{3}{5} \\ 0 & 0 & 1 \end{pmatrix} \begin{pmatrix} x_1 \\ x_2 \\ x_3 \end{pmatrix} = \begin{pmatrix} 19 \\ 5 \\ 5 \end{pmatrix}$

Expanding from the bottom $x_3 = 5$; $x_2 + \dfrac{3}{5}x_3 = 5$ $\therefore x_2 = 2$

and $x_1 + 3x_2 + 2x_3 = 19$ $\therefore x_1 + 6 + 10 = 19$ $\therefore x_1 = 3$

$$\therefore x_1 = 3; \quad x_2 = 2; \quad x_3 = 5$$

We can of course apply the same method to a set of four equations.

Example 3

$$x_1 + 2x_2 - x_3 + 3x_4 = 9$$
$$2x_1 - x_2 + 3x_3 + 2x_4 = 23$$
$$3x_1 + 3x_2 + x_3 + x_4 = 5$$
$$4x_1 + 5x_2 - 2x_3 + 2x_4 = -2.$$

i.e. $\begin{pmatrix} 1 & 2 & -1 & 3 \\ 2 & -1 & 3 & 2 \\ 3 & 3 & 1 & 1 \\ 4 & 5 & -2 & 2 \end{pmatrix} \begin{pmatrix} x_1 \\ x_2 \\ x_3 \\ x_4 \end{pmatrix} = \begin{pmatrix} 9 \\ 23 \\ 5 \\ -2 \end{pmatrix}$ i.e. $\mathbf{Ax} = \mathbf{b}$

$$\mathbf{A} = \mathbf{LU} = \begin{pmatrix} l_{11} & 0 & 0 & 0 \\ l_{21} & l_{22} & 0 & 0 \\ l_{31} & l_{32} & l_{33} & 0 \\ l_{41} & l_{42} & l_{43} & l_{44} \end{pmatrix} \begin{pmatrix} 1 & u_{12} & u_{13} & u_{14} \\ 0 & 1 & u_{23} & u_{24} \\ 0 & 0 & 1 & u_{34} \\ 0 & 0 & 0 & 1 \end{pmatrix} = \begin{pmatrix} 1 & 2 & -1 & 3 \\ 2 & -1 & 3 & 2 \\ 3 & 3 & 1 & 1 \\ 4 & 5 & -2 & 2 \end{pmatrix}$$

$$\mathbf{A} = \begin{pmatrix} l_{11} & l_{11}u_{12} & l_{11}u_{13} & l_{11}u_{14} \\ l_{21} & l_{21}u_{12} + l_{22} & l_{21}u_{13} + l_{22}u_{23} & l_{21}u_{14} + l_{22}u_{24} \\ l_{31} & l_{31}u_{12} + l_{32} & l_{31}u_{13} + l_{32}u_{23} + l_{33} & l_{31}u_{14} + l_{32}u_{24} + l_{33}u_{34} \\ l_{41} & l_{41}u_{12} + l_{42} & l_{41}u_{13} + l_{42}u_{23} + l_{43} & l_{41}u_{14} + l_{42}u_{24} + l_{43}u_{34} + l_{44} \end{pmatrix}$$

Now we have to find the values of the individual elements. It is easy enough if we follow the order indicated in the diagram earlier. So the two triangular matrices are

$$\mathbf{A} = \mathbf{LU} = (\dots\dots\dots)(\dots\dots\dots)$$

$$\mathbf{A} = \mathbf{LU} = \begin{pmatrix} 1 & 0 & 0 & 0 \\ 2 & -5 & 0 & 0 \\ 3 & -3 & 1 & 0 \\ 4 & -3 & -1 & -\frac{66}{5} \end{pmatrix} \begin{pmatrix} 1 & 2 & -1 & 3 \\ 0 & 1 & -1 & \frac{4}{5} \\ 0 & 0 & 1 & -\frac{28}{5} \\ 0 & 0 & 0 & 1 \end{pmatrix}$$

As usual $\mathbf{Ax} = \mathbf{b}$; $\mathbf{L(Ux)} = \mathbf{b}$. Put $\mathbf{Ux} = \mathbf{y}$ $\therefore$ $\mathbf{Ly} = \mathbf{b}$

(a) Solving $\mathbf{Ly} = \mathbf{b}$

$$\begin{pmatrix} 1 & 0 & 0 & 0 \\ 2 & -5 & 0 & 0 \\ 3 & -3 & 1 & 0 \\ 4 & -3 & -1 & -\frac{66}{5} \end{pmatrix} \begin{pmatrix} y_1 \\ y_2 \\ y_3 \\ y_4 \end{pmatrix} = \begin{pmatrix} 9 \\ 23 \\ 5 \\ -2 \end{pmatrix}$$

$$\therefore \begin{pmatrix} y_1 \\ y_2 \\ y_3 \\ y_4 \end{pmatrix} = \begin{pmatrix} \ldots \\ \ldots \\ \ldots \\ \ldots \end{pmatrix}$$

$$\begin{pmatrix} y_1 \\ y_2 \\ y_3 \\ y_4 \end{pmatrix} = \begin{pmatrix} 9 \\ -1 \\ -25 \\ 5 \end{pmatrix}$$

(b) Solving $\mathbf{Ux} = \mathbf{y}$

$$\begin{pmatrix} 1 & 2 & -1 & 3 \\ 0 & 1 & -1 & \frac{4}{5} \\ 0 & 0 & 1 & -\frac{28}{5} \\ 0 & 0 & 0 & 1 \end{pmatrix} \begin{pmatrix} x_1 \\ x_2 \\ x_3 \\ x_4 \end{pmatrix} = \begin{pmatrix} 9 \\ -1 \\ -25 \\ 5 \end{pmatrix}$$

which finally gives

$$x_1 = \ldots\ldots\ldots\ldots; \quad x_2 = \ldots\ldots\ldots\ldots$$
$$x_3 = \ldots\ldots\ldots\ldots; \quad x_4 = \ldots\ldots\ldots\ldots$$

52

$$\boxed{x_1 = 1; \quad x_2 = -2; \quad x_3 = 3; \quad x_4 = 5}$$

Comparison of methods

Inverse method

This is an elementary method but it is very inefficient when the number of equations to solve increases beyond three.

Row transformation method

An efficient method but each case is different and relies on ingenuity to see the way forward.

Gaussian elimination method

The most efficient method and should be used in most cases. It must be used when there is a singular or non-square system.

Triangular decomposition method

An alternative to Gaussian elimination in some cases.

Now let us proceed to something rather different,
so move on to the next frame for a new start

Eigenvalues and eigenvectors

53

Matrices commonly appear in technological problems, for example those involving coupled oscillations and vibrations, and give rise to equations of the form

$$\mathbf{Ax} = \lambda\mathbf{x}$$

where $\mathbf{A} = (a_{ij})$ is a square matrix, $\mathbf{x}$ is a column matrix (x_i) and λ is a scalar quantity, i.e. a number.

For non-trivial solutions, i.e. for $\mathbf{x} \neq \mathbf{0}$, the values of λ are called the *eigenvalues, characteristic values,* or *latent roots* of the matrix $\mathbf{A}$ and the corresponding solutions of the given equations $\mathbf{Ax} = \lambda\mathbf{x}$ are called the *eigenvectors,* or *characteristic vectors* of $\mathbf{A}$.

▶

The set of equations

$$\begin{pmatrix} a_{11} & a_{12} & \cdots & a_{1n} \\ a_{21} & a_{22} & \cdots & a_{2n} \\ \vdots & \vdots & & \vdots \\ a_{n1} & a_{n2} & \cdots & a_{nn} \end{pmatrix} \begin{pmatrix} x_1 \\ x_2 \\ \vdots \\ x_n \end{pmatrix} = \lambda \begin{pmatrix} x_1 \\ x_2 \\ \vdots \\ x_n \end{pmatrix}$$

then simplifies to

$$\begin{pmatrix} (a_{11} - \lambda) & a_{12} & \cdots & a_{1n} \\ a_{21} & (a_{22} - \lambda) & \cdots & a_{2n} \\ \vdots & \vdots & & \vdots \\ a_{n1} & a_{n2} & & (a_{nn} - \lambda) \end{pmatrix} \begin{pmatrix} x_1 \\ x_2 \\ \vdots \\ x_n \end{pmatrix} = \begin{pmatrix} 0 \\ 0 \\ \vdots \\ 0 \end{pmatrix}$$

That is, $\mathbf{Ax} = \lambda\mathbf{x}$ becomes $\mathbf{Ax} - \lambda\mathbf{x} = \mathbf{0}$

$$\text{i.e.} \quad (\mathbf{A} - \lambda\mathbf{I})\mathbf{x} = \mathbf{0}$$

the unit matrix $\mathbf{I}$ being introduced since we can subtract only a matrix from another matrix.

For this set of homogeneous linear equations (right-hand side constant terms all zero) to have non-trivial solutions

$|\mathbf{A} - \lambda\mathbf{I}|$ **must be zero**

This is called the *characteristic determinant* of $\mathbf{A}$ and $|\mathbf{A} - \lambda\mathbf{I}| = 0$ is the *characteristic equation*, the solution of which gives the values of λ , i.e. the eigenvalues of $\mathbf{A}$.

Example 1

54

Find the eigenvalues and corresponding eigenvectors of

$$\mathbf{Ax} = \lambda\mathbf{x} \text{ where } \mathbf{A} = \begin{pmatrix} 2 & 3 \\ 4 & 1 \end{pmatrix}.$$

The characteristic equation is $|\mathbf{A} - \lambda\mathbf{I}| = 0$

$$\text{i.e.} \quad \begin{vmatrix} 2 - \lambda & 3 \\ 4 & 1 - \lambda \end{vmatrix} = 0, \text{ which, when expanded, gives}$$

$$\lambda_1 = \ldots\ldots\ldots\ldots \quad \text{and} \quad \lambda_2 = \ldots\ldots\ldots\ldots$$

55

$$\lambda_1 = -2 \quad \text{and} \quad \lambda_2 = 5$$

Because

$$(2 - \lambda)(1 - \lambda) - 12 = 0 \quad \therefore \quad 2 - 3\lambda + \lambda^2 - 12 = 0$$

$$\lambda^2 - 3\lambda - 10 = 0 \quad (\lambda - 5)(\lambda + 2) = 0 \quad \therefore \quad \lambda = -2 \text{ or } 5$$

Now we substitute each value of λ in turn in the equation

$$(\mathbf{A} - \lambda\mathbf{I})\mathbf{x} = 0$$

With $\lambda = -2$

$$\left\{ \begin{pmatrix} 2 & 3 \\ 4 & 1 \end{pmatrix} - (-2)\begin{pmatrix} 1 & 0 \\ 0 & 1 \end{pmatrix} \right\} \begin{pmatrix} x_1 \\ x_2 \end{pmatrix} = \begin{pmatrix} 0 \\ 0 \end{pmatrix}$$

$$\left\{ \begin{pmatrix} 2 & 3 \\ 4 & 1 \end{pmatrix} + \begin{pmatrix} 2 & 0 \\ 0 & 2 \end{pmatrix} \right\} \begin{pmatrix} x_1 \\ x_2 \end{pmatrix} = \begin{pmatrix} 0 \\ 0 \end{pmatrix}$$

$$\begin{pmatrix} 4 & 3 \\ 4 & 3 \end{pmatrix} \begin{pmatrix} x_1 \\ x_2 \end{pmatrix} = \begin{pmatrix} 0 \\ 0 \end{pmatrix}$$

Multiplying out the left-hand side, we get

. . . .

56

$$4x_1 + 3x_2 = 0$$

from which we get $x_2 = -\frac{4}{3}x_1$ i.e. not specific values for x_1 and x_2, but a relationship between them. Whatever value we assign to x_1 we obtain a corresponding value of x_2.

$$\mathbf{x}_1 = \begin{pmatrix} x_1 \\ x_2 \end{pmatrix} = \begin{pmatrix} 3 \\ -4 \end{pmatrix} \text{ or } \begin{pmatrix} 6 \\ -8 \end{pmatrix} \text{ or } \begin{pmatrix} 9 \\ -12 \end{pmatrix}, \text{ etc.}$$

The most convenient way to do this is to choose $x_1 = 1$ and then scale $\mathbf{x}_1$ to obtain integer elements. So here we find for $x_1 = 1$ then $x_2 = -4/3$ so $\mathbf{x}_1$ is of the form

$$\begin{pmatrix} 1 \\ -\dfrac{4}{3} \end{pmatrix}$$

This is now scaled up by multiplying by 3 to give

$$\mathbf{x}_1 = \alpha \begin{pmatrix} 3 \\ -4 \end{pmatrix} \text{ where } \alpha \text{ is a constant multiplier.}$$

The simplest result, with $\alpha = 1$, is the one normally quoted.

$$\therefore \text{ for } \lambda_1 = -2, \quad \mathbf{x}_1 = \begin{pmatrix} 3 \\ -4 \end{pmatrix}$$

Similarly, for $\lambda_2 = 5$, the corresponding eigenvector is . . . .

$$\mathbf{x_2} = \begin{pmatrix} 1 \\ 1 \end{pmatrix}$$

Because, with $\lambda_2 = 5$, $(\mathbf{A} - \lambda\mathbf{I})\mathbf{x} = \mathbf{0}$ becomes

$$\left\{ \begin{pmatrix} 2 & 3 \\ 4 & 1 \end{pmatrix} - 5\begin{pmatrix} 1 & 0 \\ 0 & 1 \end{pmatrix} \right\}\begin{pmatrix} x_1 \\ x_2 \end{pmatrix} = \begin{pmatrix} 0 \\ 0 \end{pmatrix}$$

$$\left\{ \begin{pmatrix} 2 & 3 \\ 4 & 1 \end{pmatrix} - \begin{pmatrix} 5 & 0 \\ 0 & 5 \end{pmatrix} \right\}\begin{pmatrix} x_1 \\ x_2 \end{pmatrix} = \begin{pmatrix} 0 \\ 0 \end{pmatrix}$$

$$\begin{pmatrix} -3 & 3 \\ 4 & -4 \end{pmatrix}\begin{pmatrix} x_1 \\ x_2 \end{pmatrix} = \begin{pmatrix} 0 \\ 0 \end{pmatrix}$$

$\therefore \ -3x_1 + 3x_2 = 0$ i.e. $x_2 = x_1$

$\therefore$ with $\lambda_2 = 5$, the corresponding eigenvector is $\mathbf{x_2} = \beta\begin{pmatrix} 1 \\ 1 \end{pmatrix}$

Again, taking $\beta = 1$, for $\lambda_2 = 5$, $\mathbf{x_2} = \begin{pmatrix} 1 \\ 1 \end{pmatrix}$

So the required eigenvectors are

$\mathbf{x_1} = \begin{pmatrix} 3 \\ -4 \end{pmatrix}$ corresponding to $\lambda_1 = -2$

$\mathbf{x_2} = \begin{pmatrix} 1 \\ 1 \end{pmatrix}$ corresponding to $\lambda_2 = 5$.

Example 2

Determine the eigenvalues and corresponding eigenvectors of

$$\mathbf{Ax} = \lambda\mathbf{x} \text{ where } \mathbf{A} = \begin{pmatrix} 3 & 10 \\ 2 & 4 \end{pmatrix}.$$

The characteristic equation is $|\mathbf{A} - \lambda\mathbf{I}| = 0$, which in this case can be written as
. . . .

$$\begin{vmatrix} 3 - \lambda & 10 \\ 2 & 4 - \lambda \end{vmatrix} = 0$$

Expanding the determinant and solving the equation gives

$$\lambda_1 = \dots\dots\dots; \quad \lambda_2 = \dots\dots\dots$$

59

$$\boxed{\lambda_1 = -1; \quad \lambda_2 = 8}$$

Because the equation is $(3 - \lambda)(4 - \lambda) - 20 = 0$ $\quad \therefore \quad \lambda^2 - 7\lambda - 8 = 0$

$\therefore \quad (\lambda + 1)(\lambda - 8) = 0 \quad \therefore \quad \lambda = -1$ or 8

(a) With $\lambda_1 = -1$, we solve $(\mathbf{A} - \lambda\mathbf{I})\mathbf{x} = \mathbf{0}$ to obtain an eigenvector, which is

. . . .

60

$$\boxed{\mathbf{x_1} = \begin{pmatrix} 5 \\ -2 \end{pmatrix}}$$

Because

$$\mathbf{A} = \begin{pmatrix} 3 & 10 \\ 2 & 4 \end{pmatrix} \quad \therefore \quad \left\{ \begin{pmatrix} 3 & 10 \\ 2 & 4 \end{pmatrix} - (-1)\begin{pmatrix} 1 & 0 \\ 0 & 1 \end{pmatrix} \right\} \begin{pmatrix} x_1 \\ x_2 \end{pmatrix} = \begin{pmatrix} 0 \\ 0 \end{pmatrix}$$

$$\left\{ \begin{pmatrix} 3 & 10 \\ 2 & 4 \end{pmatrix} + \begin{pmatrix} 1 & 0 \\ 0 & 1 \end{pmatrix} \right\} \begin{pmatrix} x_1 \\ x_2 \end{pmatrix} = \begin{pmatrix} 0 \\ 0 \end{pmatrix}$$

$$\begin{pmatrix} 4 & 10 \\ 2 & 5 \end{pmatrix} \begin{pmatrix} x_1 \\ x_2 \end{pmatrix} = \begin{pmatrix} 0 \\ 0 \end{pmatrix}$$

$\therefore \quad 4x_1 + 10x_2 = 0 \quad \therefore \quad x_2 = -\dfrac{2}{5}x_1 \quad \mathbf{x_1} = \alpha\begin{pmatrix} 5 \\ -2 \end{pmatrix}$

$\therefore$ with $\alpha = 1 \quad \lambda_1 = -1$ and $\mathbf{x_1} = \begin{pmatrix} 5 \\ -2 \end{pmatrix}$

(b) In the same way the corresponding eigenvector $\mathbf{x_2}$ for $\lambda_2 = 8$ is

. . . .

61

$$\boxed{\mathbf{x_2} = \begin{pmatrix} 2 \\ 1 \end{pmatrix}}$$

Because

$$\left\{ \begin{pmatrix} 3 & 10 \\ 2 & 4 \end{pmatrix} - 8\begin{pmatrix} 1 & 0 \\ 0 & 1 \end{pmatrix} \right\} \begin{pmatrix} x_1 \\ x_2 \end{pmatrix} = \begin{pmatrix} 0 \\ 0 \end{pmatrix}$$

$$\left\{ \begin{pmatrix} 3 & 10 \\ 2 & 4 \end{pmatrix} - \begin{pmatrix} 8 & 0 \\ 0 & 8 \end{pmatrix} \right\} \begin{pmatrix} x_1 \\ x_2 \end{pmatrix} = \begin{pmatrix} 0 \\ 0 \end{pmatrix}$$

$$\begin{pmatrix} -5 & 10 \\ 2 & -4 \end{pmatrix} \begin{pmatrix} x_1 \\ x_2 \end{pmatrix} = \begin{pmatrix} 0 \\ 0 \end{pmatrix}$$

$\therefore \quad -5x_1 + 10x_2 = 0 \quad \therefore \quad x_2 = \dfrac{1}{2}x_1 \quad \mathbf{x_2} = \beta\begin{pmatrix} 2 \\ 1 \end{pmatrix}$

$\therefore$ with $\beta = 1, \quad \lambda_2 = 8$ and $\mathbf{x_2} = \begin{pmatrix} 2 \\ 1 \end{pmatrix}$

▶

The same basic method can similarly be applied to third-order sets of equations.

Example 3

Determine the eigenvalues and eigenvectors of $\mathbf{Ax} = \lambda\mathbf{x}$ where

$$\mathbf{A} = \begin{pmatrix} 1 & 0 & 4 \\ 0 & 2 & 0 \\ 3 & 1 & -3 \end{pmatrix}.$$

As before, we have $(\mathbf{A} - \lambda\mathbf{I})\mathbf{x} = \mathbf{0}$ with characteristic equation $|\mathbf{A} - \lambda\mathbf{I}| = 0$.

i.e. $\begin{vmatrix} 1-\lambda & 0 & 4 \\ 0 & 2-\lambda & 0 \\ 3 & 1 & -3-\lambda \end{vmatrix} = 0$

Expanding this we have

$$\lambda_1 = \ldots\ldots\ldots; \quad \lambda_2 = \ldots\ldots\ldots; \quad \lambda_3 = \ldots\ldots\ldots$$

$$\boxed{\lambda_1 = 2; \quad \lambda_2 = 3; \quad \lambda_3 = -5}$$

62

Because

$$(1-\lambda)\{(2-\lambda)(-3-\lambda) - 0\} + 4\{0 - 3(2-\lambda)\} = 0$$
$$(1-\lambda)(2-\lambda)(-3-\lambda) - 12(2-\lambda) = 0$$
$$\therefore \ (2-\lambda)\{(1-\lambda)(-3-\lambda) - 12\} = 0$$
$$\therefore \ \lambda = 2 \ \text{or} \ \lambda^2 + 2\lambda - 15 = 0 \ \therefore \ (\lambda-3)(\lambda+5) = 0$$
$$\therefore \ \lambda = 2, 3, \ \text{or} \ -5$$

(a) With $\lambda_1 = 2$, $\quad (\mathbf{A} - \lambda\mathbf{I})\mathbf{x} = \mathbf{0}$ becomes

$$\left\{\begin{pmatrix} 1 & 0 & 4 \\ 0 & 2 & 0 \\ 3 & 1 & -3 \end{pmatrix} - 2\begin{pmatrix} 1 & 0 & 0 \\ 0 & 1 & 0 \\ 0 & 0 & 1 \end{pmatrix}\right\}\begin{pmatrix} x_1 \\ x_2 \\ x_3 \end{pmatrix} = \begin{pmatrix} 0 \\ 0 \\ 0 \end{pmatrix}$$

$$\left\{\begin{pmatrix} 1 & 0 & 4 \\ 0 & 2 & 0 \\ 3 & 1 & -3 \end{pmatrix} - \begin{pmatrix} 2 & 0 & 0 \\ 0 & 2 & 0 \\ 0 & 0 & 2 \end{pmatrix}\right\}\begin{pmatrix} x_1 \\ x_2 \\ x_3 \end{pmatrix} = \begin{pmatrix} 0 \\ 0 \\ 0 \end{pmatrix}$$

$$\therefore \ \begin{pmatrix} -1 & 0 & 4 \\ 0 & 0 & 0 \\ 3 & 1 & -5 \end{pmatrix}\begin{pmatrix} x_1 \\ x_2 \\ x_3 \end{pmatrix} = \begin{pmatrix} 0 \\ 0 \\ 0 \end{pmatrix}$$

from which a corresponding eigenvector $\mathbf{x}_1$ is $\ldots\ldots\ldots$

63

$$\mathbf{x}_1 = \begin{pmatrix} 4 \\ -7 \\ 1 \end{pmatrix}$$

Because we have $-x_1 + 4x_3 = 0$ $\qquad \therefore \ x_3 = \frac{1}{4}x_1$

$3x_1 + x_2 - 5x_3 = 0 \quad \therefore \ 3x_1 + x_2 - \frac{5}{4}x_1 = 0 \quad \therefore \ x_2 = -\frac{7}{4}x_1$

$\therefore \ x_1, x_2, x_3$ are in the ratio $1 : -\dfrac{7}{4} : \dfrac{1}{4}$ i.e. $4 : -7 : 1$ $\ \therefore \ \mathbf{x_1} = \begin{pmatrix} 4 \\ -7 \\ 1 \end{pmatrix}$

(b) Similarly for $\lambda_2 = 3$, $(\mathbf{A} - \lambda\mathbf{I})\mathbf{x} = \mathbf{0}$

$$\left\{ \begin{pmatrix} 1 & 0 & 4 \\ 0 & 2 & 0 \\ 3 & 1 & -3 \end{pmatrix} - 3\begin{pmatrix} 1 & 0 & 0 \\ 0 & 1 & 0 \\ 0 & 0 & 1 \end{pmatrix} \right\} \begin{pmatrix} x_1 \\ x_2 \\ x_3 \end{pmatrix} = \begin{pmatrix} 0 \\ 0 \\ 0 \end{pmatrix}$$

from which a corresponding eigenvector is

$$\mathbf{x}_2 = \ldots\ldots\ldots\ldots$$

64

$$\mathbf{x}_2 = \begin{pmatrix} 2 \\ 0 \\ 1 \end{pmatrix}$$

Because

$$\left\{ \begin{pmatrix} 1 & 0 & 4 \\ 0 & 2 & 0 \\ 3 & 1 & -3 \end{pmatrix} - \begin{pmatrix} 3 & 0 & 0 \\ 0 & 3 & 0 \\ 0 & 0 & 3 \end{pmatrix} \right\} \begin{pmatrix} x_1 \\ x_2 \\ x_3 \end{pmatrix} = \begin{pmatrix} 0 \\ 0 \\ 0 \end{pmatrix}$$

$$\begin{pmatrix} -2 & 0 & 4 \\ 0 & -1 & 0 \\ 3 & 1 & -6 \end{pmatrix} \begin{pmatrix} x_1 \\ x_2 \\ x_3 \end{pmatrix} = \begin{pmatrix} 0 \\ 0 \\ 0 \end{pmatrix}$$

$$\therefore \ -2x_1 + 4x_3 = 0 \quad \therefore \ x_3 = \frac{1}{2}x_1$$

Also $\qquad -x_2 = 0 \quad \therefore \ x_2 = 0 \qquad \therefore \ \mathbf{x}_2 = \begin{pmatrix} 2 \\ 0 \\ 1 \end{pmatrix}$

(c) All that now remains is $\lambda_3 = -5$. A corresponding eigenvector $\mathbf{x}_3$ is

$$\mathbf{x}_3 = \ldots\ldots\ldots\ldots$$

Finish it on your own. Method just the same as before.

$$\mathbf{x_3} = \begin{pmatrix} 2 \\ 0 \\ -3 \end{pmatrix}$$

Check the working.

$$\mathbf{A} = \begin{pmatrix} 1 & 0 & 4 \\ 0 & 2 & 0 \\ 3 & 1 & -3 \end{pmatrix} \text{ and } \lambda_3 = -5 \text{ with } (\mathbf{A} - \lambda\mathbf{I})\mathbf{x} = \mathbf{0}.$$

$$\left\{ \begin{pmatrix} 1 & 0 & 4 \\ 0 & 2 & 0 \\ 3 & 1 & -3 \end{pmatrix} + 5 \begin{pmatrix} 1 & 0 & 0 \\ 0 & 1 & 0 \\ 0 & 0 & 1 \end{pmatrix} \right\} \begin{pmatrix} x_1 \\ x_2 \\ x_3 \end{pmatrix} = \begin{pmatrix} 0 \\ 0 \\ 0 \end{pmatrix}$$

$$\begin{pmatrix} 6 & 0 & 4 \\ 0 & 7 & 0 \\ 3 & 1 & 2 \end{pmatrix} \begin{pmatrix} x_1 \\ x_2 \\ x_3 \end{pmatrix} = \begin{pmatrix} 0 \\ 0 \\ 0 \end{pmatrix}$$

$$\therefore \ 6x_1 + 4x_3 = 0 \quad \therefore \ x_3 = -\tfrac{3}{2}x_1$$

$$7x_2 = 0 \quad \therefore \ x_2 = 0 \qquad \therefore \ \mathbf{x_3} = \begin{pmatrix} 2 \\ 0 \\ -3 \end{pmatrix}$$

Collecting the results together, we finally have

$$\lambda_1 = 2, \ \mathbf{x_1} = \begin{pmatrix} 4 \\ -7 \\ 1 \end{pmatrix}; \ \lambda_2 = 3, \ \mathbf{x_2} = \begin{pmatrix} 2 \\ 0 \\ 1 \end{pmatrix}; \ \lambda_3 = -5, \ \mathbf{x_3} = \begin{pmatrix} 2 \\ 0 \\ -3 \end{pmatrix}$$

Cayley–Hamilton theorem

66

The Cayley–Hamilton theorem states that every square matrix satisfies its characteristic equation. For example the matrix

$$\mathbf{A} = \begin{pmatrix} 2 & 3 \\ 4 & 1 \end{pmatrix}$$

of Frame 54 has the characteristic equation

$$\lambda^2 - 3\lambda - 10 = 0$$

and so the Cayley–Hamilton theorem tells us that

$$\mathbf{A}^2 - 3\mathbf{A} - 10\mathbf{I} = \mathbf{0}$$

▶

To verify this we note that

$$\mathbf{A}^2 = \begin{pmatrix} 2 & 3 \\ 4 & 1 \end{pmatrix} \begin{pmatrix} 2 & 3 \\ 4 & 1 \end{pmatrix} = \begin{pmatrix} 16 & 9 \\ 12 & 13 \end{pmatrix} \text{ so that}$$

$$\mathbf{A}^2 - 3\mathbf{A} - 10\mathbf{I} = \begin{pmatrix} 16 & 9 \\ 12 & 13 \end{pmatrix} - 3\begin{pmatrix} 2 & 3 \\ 4 & 1 \end{pmatrix} - 10\begin{pmatrix} 1 & 0 \\ 0 & 1 \end{pmatrix}$$

$$= \begin{pmatrix} 16 & 9 \\ 12 & 13 \end{pmatrix} - \begin{pmatrix} 6 & 9 \\ 12 & 3 \end{pmatrix} - \begin{pmatrix} 10 & 0 \\ 0 & 10 \end{pmatrix} = \begin{pmatrix} 0 & 0 \\ 0 & 0 \end{pmatrix}$$

You try one. Verify that the matrix $\mathbf{A} = \begin{pmatrix} 3 & 10 \\ 2 & 4 \end{pmatrix}$ of Frame 57 with the characteristic equation

$$\lambda^2 - 7\lambda - 8 = 0$$

satisfies the Cayley–Hamilton theorem, that is ....

67

$$\boxed{\mathbf{A}^2 - 7\mathbf{A} - 8\mathbf{I} = \mathbf{0}}$$

Because

$$\mathbf{A}^2 = \begin{pmatrix} 3 & 10 \\ 2 & 4 \end{pmatrix} \begin{pmatrix} 3 & 10 \\ 2 & 4 \end{pmatrix} = \begin{pmatrix} 29 & 70 \\ 14 & 36 \end{pmatrix} \text{ so that}$$

$$\mathbf{A}^2 - 7\mathbf{A} - 8\mathbf{I} = \begin{pmatrix} 29 & 70 \\ 14 & 36 \end{pmatrix} - 7\begin{pmatrix} 3 & 10 \\ 2 & 4 \end{pmatrix} - 8\begin{pmatrix} 1 & 0 \\ 0 & 1 \end{pmatrix}$$

$$= \begin{pmatrix} 29 & 70 \\ 14 & 36 \end{pmatrix} - \begin{pmatrix} 21 & 70 \\ 14 & 28 \end{pmatrix} - \begin{pmatrix} 8 & 0 \\ 0 & 8 \end{pmatrix} = \begin{pmatrix} 0 & 0 \\ 0 & 0 \end{pmatrix}$$

Now on to something different

Systems of first-order ordinary differential equations

68

Matrix methods involving eigenvalues and their associated eigenvectors can be used to solve systems of coupled differential equations, though we shall only consider cases where the relevant eigenvalues are distinct. We proceed by example.

Example 1

Consider the system of two coupled ordinary differential equations

$$\begin{aligned} f_1'(x) &= 2f_1(x) + 3f_2(x) \\ f_2'(x) &= 4f_1(x) + f_2(x) \end{aligned} \quad \text{where } f_1(0) = 2 \text{ and } f_2(0) = 1$$

These can be written in matrix form as ....

$$\begin{pmatrix} f_1'(x) \\ f_2'(x) \end{pmatrix} = \begin{pmatrix} 2 & 3 \\ 4 & 1 \end{pmatrix} \begin{pmatrix} f_1(x) \\ f_2(x) \end{pmatrix}$$

That is

$\mathbf{F}'(x) = \mathbf{A}\mathbf{F}(x)$

where $\mathbf{F}(x) = \begin{pmatrix} f_1(x) \\ f_2(x) \end{pmatrix}$, $\mathbf{F}'(x) = \begin{pmatrix} f_1'(x) \\ f_2'(x) \end{pmatrix}$ and $\mathbf{A} = \begin{pmatrix} 2 & 3 \\ 4 & 1 \end{pmatrix}$ and where

$\mathbf{F}(0) = \begin{pmatrix} f_1(0) \\ f_2(0) \end{pmatrix} = \begin{pmatrix} 2 \\ 1 \end{pmatrix}$ are the boundary conditions in matrix form.

The matrix differential equation $\mathbf{F}'(x) = \mathbf{A}\mathbf{F}(x)$ is similar in form to the single differential equation $f'(x) = af(x)$ (a constant) which has solution $f(x) = \alpha e^{ax}$ (α constant), so to solve the matrix equation we try a solution of the form

$\mathbf{F}(x) = \mathbf{C}e^{kx}$ where the number k and the constants c_1 and c_2 of the

matrix $\mathbf{C} = \begin{pmatrix} c_1 \\ c_2 \end{pmatrix}$ are to be determined.

Substituting $\mathbf{F}(x) = \mathbf{C}e^{kx}$ into the matrix equation $\mathbf{F}'(x) = \mathbf{A}\mathbf{F}(x)$ gives

. . . .

$$k\mathbf{C}e^{kx} = \mathbf{A}\mathbf{C}e^{kx}$$

Because

$\mathbf{F}(x) = \mathbf{C}e^{kx}$ so $\mathbf{F}'(x) = k\mathbf{C}e^{kx}$. Since $\mathbf{F}'(x) = \mathbf{A}\mathbf{F}(x)$ then $k\mathbf{C}e^{kx} = \mathbf{A}\mathbf{C}e^{kx}$

Dividing both sides by e^{kx} gives

$k\mathbf{C} = \mathbf{A}\mathbf{C}$ that is $\mathbf{A}\mathbf{C} = k\mathbf{C}$.

So, from Frame 53, k is an *eigenvalue* of $\mathbf{A}$ and $\mathbf{C}$ is the corresponding *eigenvector*. Therefore, we must first find the eigenvalues of $\mathbf{A}$ and for this matrix these have been found earlier in Frames 54 to 57. They are

$\lambda = -2$ (and so $k = -2$) with corresponding eigenvector $\begin{pmatrix} 3 \\ -4 \end{pmatrix}$

$\lambda = 5$ (and so $k = 5$) with corresponding eigenvector $\begin{pmatrix} 1 \\ 1 \end{pmatrix}$

To each eigenvalue the matrix $\mathbf{F}(x) = \mathbf{C}e^{kx}$ is a solution. The complete solution to $\mathbf{F}' = \mathbf{A}\mathbf{F}$ is then

$$\mathbf{F}_1(x) = \begin{pmatrix} \cdots \\ \cdots \end{pmatrix} a_1 e^{\cdots} \quad \text{and} \quad \mathbf{F}_2(x) = \begin{pmatrix} \cdots \\ \cdots \end{pmatrix} a_2 e^{\cdots}$$

71

$$\mathbf{F}_1(x) = \begin{pmatrix} 3 \\ -4 \end{pmatrix} a_1 e^{-2x} \quad \text{and} \quad \mathbf{F}_2(x) = \begin{pmatrix} 1 \\ 1 \end{pmatrix} a_2 e^{5x}$$

Because

$\mathbf{F}(x) = \mathbf{C}e^{kx}$ is the solution corresponding to the eigenvalue k with associated eigenvector $\mathbf{C}$.

The complete solution to the equation $\mathbf{F}'(x) = \mathbf{AF}(x)$ is then a combination of these two solutions in the form

$$\mathbf{F}(x) = A\begin{pmatrix} 3 \\ -4 \end{pmatrix} e^{-2x} + B\begin{pmatrix} 1 \\ 1 \end{pmatrix} e^{5x}$$

Applying the boundary conditions gives $\mathbf{F}(0) = \dots\dots\dots$

72

$$\mathbf{F}(0) = \begin{pmatrix} 3A + B \\ -4A + B \end{pmatrix} = \begin{pmatrix} 2 \\ 1 \end{pmatrix}$$

Because

$$\mathbf{F}(x) = A\begin{pmatrix} 3 \\ -4 \end{pmatrix} e^{-2x} + B\begin{pmatrix} 1 \\ 1 \end{pmatrix} e^{5x} \text{ and so } \mathbf{F}(0) = A\begin{pmatrix} 3 \\ -4 \end{pmatrix} + B\begin{pmatrix} 1 \\ 1 \end{pmatrix}$$
$$= \begin{pmatrix} 3A + B \\ -4A + B \end{pmatrix} = \begin{pmatrix} 2 \\ 1 \end{pmatrix}$$

Therefore

$\begin{matrix} 3A + B = 2 \\ -4A + B = 1 \end{matrix}$ with solution $A = 1/7$ and $B = 11/7$,

giving the final solution as $\mathbf{F}(x) = \dots\dots\dots$

73

$$\mathbf{F}(x) = \begin{pmatrix} 3/7 \\ -4/7 \end{pmatrix} e^{-2x} + \begin{pmatrix} 11/7 \\ 11/7 \end{pmatrix} e^{5x}$$

Summary

To solve an equation of the form

$\mathbf{F}'(x) = \mathbf{AF}(x)$

1 Find the eigenvalues $\lambda_1, \lambda_2, \dots, \lambda_n$ of $\mathbf{A}$ (assuming they are all distinct)
2 Find the associated eigenvectors $\mathbf{C}_1, \mathbf{C}_2, \dots, \mathbf{C}_n$
3 Write the solution of the equation as $\mathbf{F}(x) = \sum_{r=1}^{n}(A_r e^{\lambda_r x})\mathbf{C}_r$ and use the boundary conditions to find the values of a_r for $r = 1, 2, \dots, n$.

Now you try one.

Next frame

Example 2

74

The system of two coupled ordinary differential equations

$$f_1'(x) = 3f_1(x) + 10f_2(x)$$
$$f_2'(x) = 2f_1(x) + 4f_2(x)$$
where $f_1(0) = 0$ and $f_2(0) = 1$

has the solution (refer to Frames 57 to 61)

$$f_1(x) = \ldots\ldots\ldots\ldots$$
$$f_2(x) = \ldots\ldots\ldots\ldots$$

75

$$f_1(x) = -\frac{10}{9}e^{-x} + \frac{10}{9}e^{8x}$$
$$f_2(x) = \frac{4}{9}e^{-x} + \frac{5}{9}e^{8x}$$

Because

$$f_1'(x) = 3f_1(x) + 10f_2(x)$$
$$f_2'(x) = 2f_1(x) + 4f_2(x)$$

can be written in matrix form as

$$\ldots\ldots\ldots\ldots$$

76

$$\begin{pmatrix} f_1'(x) \\ f_2'(x) \end{pmatrix} = \begin{pmatrix} 3 & 10 \\ 2 & 4 \end{pmatrix} \begin{pmatrix} f_1(x) \\ f_2(x) \end{pmatrix}$$

That is

$$\mathbf{F}'(x) = \mathbf{A}\mathbf{F}(x)$$

where $\mathbf{F}(x) = \begin{pmatrix} f_1(x) \\ f_2(x) \end{pmatrix}$, $\mathbf{F}'(x) = \begin{pmatrix} f_1'(x) \\ f_2'(x) \end{pmatrix}$ and $\mathbf{A} = \begin{pmatrix} 3 & 10 \\ 2 & 4 \end{pmatrix}$ and where

$\mathbf{F}(0) = \begin{pmatrix} 0 \\ 1 \end{pmatrix}$.

To solve the matrix equation we first need the eigenvalues and associated eigenvectors of the matrix $\mathbf{A}$. These have already been found in Frames 57 to 61 and they are

$\lambda = -1$ with corresponding eigenvector $\begin{pmatrix} 5 \\ -2 \end{pmatrix}$

$\lambda = 8$ with corresponding eigenvector $\begin{pmatrix} 2 \\ 1 \end{pmatrix}$

The complete solution of $\mathbf{F}' = \mathbf{A}\mathbf{F}$ is then

$$\mathbf{F}(x) = A\begin{pmatrix} 5 \\ -2 \end{pmatrix}e^{-x} + B\begin{pmatrix} 2 \\ 1 \end{pmatrix}e^{8x}$$

$$\text{That is } f_1(x) = \ldots\ldots\ldots\ldots$$
$$f_2(x) = \ldots\ldots\ldots\ldots$$

77

$$f_1(x) = 5Ae^{-x} + 2Be^{8x}$$
$$f_2(x) = -2Ae^{-x} + Be^{8x}$$

Because

$$\mathbf{F}(x) = \begin{pmatrix} f_1(x) \\ f_2(x) \end{pmatrix} = A\begin{pmatrix} 5 \\ -2 \end{pmatrix}e^{-x} + B\begin{pmatrix} 2 \\ 1 \end{pmatrix}e^{8x}$$

and so

$$f_1(x) = 5Ae^{-x} + 2Be^{8x}$$
$$f_2(x) = -2Ae^{-x} + Be^{8x}$$

Applying the boundary conditions, we find

$$\mathbf{F}(0) = \begin{pmatrix} 0 \\ 1 \end{pmatrix} = \begin{pmatrix} \ldots A + \ldots B \\ \ldots A + \ldots B \end{pmatrix}$$

78

$$\mathbf{F}(0) = \begin{pmatrix} 0 \\ 1 \end{pmatrix} = \begin{pmatrix} f_1(0) \\ f_2(0) \end{pmatrix} = \begin{pmatrix} 5A + 2B \\ -2A + B \end{pmatrix}$$

Because

The boundary conditions are $f_1(0) = 0$ and $f_2(0) = 1$ therefore

$$\mathbf{F}(0) = \begin{pmatrix} 0 \\ 1 \end{pmatrix} = \begin{pmatrix} f_1(0) \\ f_2(0) \end{pmatrix} = \begin{pmatrix} 5A + 2B \\ -2A + B \end{pmatrix}$$

This gives the pair of simultaneous equations

$$5A + 2B = 0$$
$$-2A + B = 1$$ which have solution

$$A = \ldots\ldots\ldots\ldots \quad \text{and} \quad B = \ldots\ldots\ldots\ldots$$

79

$$A = -2/9 \quad \text{and} \quad B = 5/9$$

This gives the complete solution as

$$\mathbf{F}(x) = \begin{pmatrix} f_1(x) \\ f_2(x) \end{pmatrix} = \begin{pmatrix} -10/9 \\ 4/9 \end{pmatrix}e^{-x} + \begin{pmatrix} 10/9 \\ 5/9 \end{pmatrix}e^{8x}$$
$$f_1(x) = -\frac{10}{9}e^{-x} + \frac{10}{9}e^{8x}$$
$$f_2(x) = \frac{4}{9}e^{-x} + \frac{5}{9}e^{8x}$$

Diagonalization of a matrix

Modal matrix

80

We have already discussed the eigenvalues and eigenvectors of a matrix $\mathbf{A}$ of order n. In this section we shall assume that all the eigenvalues are distinct. If the n eigenvectors $\mathbf{x}_i$ are arranged as columns of a square matrix, the *modal matrix* of $\mathbf{A}$, denoted by $\mathbf{M}$, is formed

   i.e. $\mathbf{M} = (\mathbf{x}_1, \mathbf{x}_2, \mathbf{x}_3, \ldots, \mathbf{x}_n)$

For example, we have seen earlier that if

$$\mathbf{A} = \begin{pmatrix} 1 & 0 & 4 \\ 0 & 2 & 0 \\ 3 & 1 & -3 \end{pmatrix} \text{ then } \lambda_1 = 2, \lambda_2 = 3, \lambda_3 = -5$$

and the corresponding eigenvectors are

$$\mathbf{x}_1 = \begin{pmatrix} 4 \\ -7 \\ 1 \end{pmatrix}, \ \mathbf{x}_2 = \begin{pmatrix} 2 \\ 0 \\ 1 \end{pmatrix}, \ \mathbf{x}_3 = \begin{pmatrix} 2 \\ 0 \\ -3 \end{pmatrix}$$

Then the modal matrix $\mathbf{M} = \begin{pmatrix} 4 & 2 & 2 \\ -7 & 0 & 0 \\ 1 & 1 & -3 \end{pmatrix}$

Spectral matrix

Also, we define the *spectral matrix* of $\mathbf{A}$, i.e. $\mathbf{S}$, as a diagonal matrix with the eigenvalues only on the main diagonal

$$\text{i.e. } \mathbf{S} = \begin{pmatrix} \lambda_1 & 0 & 0 & \ldots & 0 \\ 0 & \lambda_2 & 0 & \ldots & 0 \\ \vdots & \vdots & \vdots & & \vdots \\ 0 & 0 & 0 & \ldots & \lambda_n \end{pmatrix}$$

So, in the example above, $\mathbf{S} = \ldots\ldots\ldots\ldots$

$$\mathbf{S} = \begin{pmatrix} 2 & 0 & 0 \\ 0 & 3 & 0 \\ 0 & 0 & -5 \end{pmatrix}$$

Note that the eigenvalues of $\mathbf{S}$ and $\mathbf{A}$ are the same.

So, if $\mathbf{A} = \begin{pmatrix} 5 & -6 & 1 \\ 1 & 1 & 0 \\ 3 & 0 & 1 \end{pmatrix}$ has eigenvalues $\lambda = 1, 2, 4$ and

corresponding eigenvectors $\begin{pmatrix} 0 \\ 1 \\ 6 \end{pmatrix}, \begin{pmatrix} 1 \\ 1 \\ 3 \end{pmatrix}, \begin{pmatrix} 3 \\ 1 \\ 3 \end{pmatrix}$

then $\mathbf{M} = \ldots\ldots\ldots\ldots$ and $\mathbf{S} = \ldots\ldots\ldots\ldots$

82

$$\mathbf{M} = \begin{pmatrix} 0 & 1 & 3 \\ 1 & 1 & 1 \\ 6 & 3 & 3 \end{pmatrix}; \quad \mathbf{S} = \begin{pmatrix} 1 & 0 & 0 \\ 0 & 2 & 0 \\ 0 & 0 & 4 \end{pmatrix}$$

Now how are these connected? Let us investigate.

The eigenvectors $\mathbf{x}$ arranged in the modal matrix satisfy the original equation

$$\mathbf{Ax} = \lambda\mathbf{x}$$

Also $\quad\quad \mathbf{M} = (\,\mathbf{x}_1 \quad \mathbf{x}_2 \quad \ldots \quad \mathbf{x}_n\,)$

Then $\quad\quad \mathbf{AM} = \mathbf{A}(\,\mathbf{x}_1 \quad \mathbf{x}_2 \quad \ldots \quad \mathbf{x}_n\,)$

$$= (\,\mathbf{Ax}_1 \quad \mathbf{Ax}_2 \quad \ldots \quad \mathbf{Ax}_n\,)$$

$$= (\,\lambda_1\mathbf{x}_1 \quad \lambda_2\mathbf{x}_2 \quad \ldots \quad \lambda_n\mathbf{x}_n\,) \quad \text{since } \mathbf{Ax} = \lambda\mathbf{x}$$

Now $\quad \mathbf{S} = \begin{pmatrix} \lambda_1 & 0 & \ldots & 0 \\ 0 & \lambda_2 & \ldots & 0 \\ \vdots & \vdots & & \vdots \\ 0 & 0 & \ldots & \lambda_n \end{pmatrix} \quad \therefore (\,\lambda_1\mathbf{x}_1 \quad \lambda_2\mathbf{x}_2 \quad \ldots \quad \lambda_n\mathbf{x}_n\,) = \mathbf{MS}$

$$\therefore \mathbf{AM} = \mathbf{MS}$$

If we now pre-multiply both sides by $\mathbf{M}^{-1}$ we have

$$\mathbf{M}^{-1}\mathbf{AM} = \mathbf{M}^{-1}\mathbf{MS} \quad\quad \text{But } \mathbf{M}^{-1}\mathbf{M} = \mathbf{I}$$

$$\therefore \mathbf{M}^{-1}\mathbf{AM} = \mathbf{S}$$

Make a note of this result. Then we will consider an example

83

Example 1

From the results of a previous example in Frame 65, if

$$\mathbf{A} = \begin{pmatrix} 1 & 0 & 4 \\ 0 & 2 & 0 \\ 3 & 1 & -3 \end{pmatrix} \text{ then } \lambda_1 = 2, \ \lambda_2 = 3, \ \lambda_3 = -5 \text{ and}$$

$$\mathbf{x}_1 = \begin{pmatrix} 4 \\ -7 \\ 1 \end{pmatrix}, \quad \mathbf{x}_2 = \begin{pmatrix} 2 \\ 0 \\ 1 \end{pmatrix}, \quad \mathbf{x}_3 = \begin{pmatrix} 2 \\ 0 \\ -3 \end{pmatrix}.$$

Also $\mathbf{M} = \begin{pmatrix} 4 & 2 & 2 \\ -7 & 0 & 0 \\ 1 & 1 & -3 \end{pmatrix}.$

We can find $\mathbf{M}^{-1}$ by any of the methods we have established previously.

$$\mathbf{M}^{-1} = \ldots\ldots\ldots\ldots$$

$$\mathbf{M}^{-1} = \begin{pmatrix} 0 & -1/7 & 0 \\ 3/8 & 1/4 & 1/4 \\ 1/8 & 1/28 & -1/4 \end{pmatrix}$$

Here is one way of determining the inverse. You may have done it by another.

$$\begin{pmatrix} 4 & 2 & 2 & \vdots & 1 & 0 & 0 \\ -7 & 0 & 0 & \vdots & 0 & 1 & 0 \\ 1 & 1 & -3 & \vdots & 0 & 0 & 1 \end{pmatrix} \sim \begin{pmatrix} 7 & 0 & 0 & \vdots & 0 & -1 & 0 \\ 1 & 1 & -3 & \vdots & 0 & 0 & 1 \\ 4 & 2 & 2 & \vdots & 1 & 0 & 0 \end{pmatrix}$$

$$\sim \begin{pmatrix} 1 & 0 & 0 & \vdots & 0 & -1/7 & 0 \\ 0 & 1 & -3 & \vdots & 0 & 1/7 & 1 \\ 0 & 2 & 2 & \vdots & 1 & 4/7 & 0 \end{pmatrix} \sim \begin{pmatrix} 1 & 0 & 0 & \vdots & 0 & -1/7 & 0 \\ 0 & 1 & -3 & \vdots & 0 & 1/7 & 1 \\ 0 & 0 & 8 & \vdots & 1 & 2/7 & -2 \end{pmatrix}$$

$$\sim \begin{pmatrix} 1 & 0 & 0 & \vdots & 0 & -1/7 & 0 \\ 0 & 1 & -3 & \vdots & 0 & 1/7 & 1 \\ 0 & 0 & 1 & \vdots & 1/8 & 1/28 & -1/4 \end{pmatrix}$$

$$\sim \begin{pmatrix} 1 & 0 & 0 & \vdots & 0 & -1/7 & 0 \\ 0 & 1 & 0 & \vdots & 3/8 & 7/28 & 1/4 \\ 0 & 0 & 1 & \vdots & 1/8 & 1/28 & -1/4 \end{pmatrix}$$

$$\therefore \mathbf{M}^{-1} = \begin{pmatrix} 0 & -1/7 & 0 \\ 3/8 & 1/4 & 1/4 \\ 1/8 & 1/28 & -1/4 \end{pmatrix}$$

So now $\mathbf{A} = \begin{pmatrix} 1 & 0 & 4 \\ 0 & 2 & 0 \\ 3 & 1 & -3 \end{pmatrix}$ and $\mathbf{M} = \begin{pmatrix} 4 & 2 & 2 \\ -7 & 0 & 0 \\ 1 & 1 & -3 \end{pmatrix}$

$$\therefore \mathbf{AM} = \begin{pmatrix} 1 & 0 & 4 \\ 0 & 2 & 0 \\ 3 & 1 & -3 \end{pmatrix} \begin{pmatrix} 4 & 2 & 2 \\ -7 & 0 & 0 \\ 1 & 1 & -3 \end{pmatrix} = \begin{pmatrix} 8 & 6 & -10 \\ -14 & 0 & 0 \\ 2 & 3 & 15 \end{pmatrix}$$

Then $\mathbf{M}^{-1}\mathbf{AM} = \begin{pmatrix} 0 & -1/7 & 0 \\ 3/8 & 1/4 & 1/4 \\ 1/8 & 1/28 & -1/4 \end{pmatrix} \begin{pmatrix} 8 & 6 & -10 \\ -14 & 0 & 0 \\ 2 & 3 & 15 \end{pmatrix}$

$$= \dots\dots\dots\dots$$

$$\mathbf{M}^{-1}\mathbf{AM} = \begin{pmatrix} 2 & 0 & 0 \\ 0 & 3 & 0 \\ 0 & 0 & -5 \end{pmatrix}$$

So we have transformed the original matrix **A** into a diagonal matrix and notice that the elements on the main diagonal are, in fact, the eigenvalues of **A**

   i.e. $\mathbf{M}^{-1}\mathbf{AM} = \mathbf{S}$

Therefore, let us list a few relevant facts

  1 $\mathbf{M}^{-1}\mathbf{AM}$ transforms the square matrix **A** into a diagonal matrix **S**.
  2 A square matrix **A** of order n can be so transformed if the matrix has n independent eigenvectors.
  3 A matrix **A** always has n linearly independent eigenvectors if it has n distinct eigenvalues or if it is a symmetric matrix.
  4 If the matrix has repeated eigenvalues and is not symmetric, it may or may not have n linearly independent eigenvectors.

Now here is one straightforward example with which to finish.

Example 2

If $\mathbf{A} = \begin{pmatrix} -6 & 5 \\ 4 & 2 \end{pmatrix}$, $\mathbf{M} = \ldots\ldots\ldots\ldots;$ $\mathbf{M}^{-1} = \ldots\ldots\ldots\ldots;$

and hence $\mathbf{M}^{-1}\mathbf{AM} = \ldots\ldots\ldots\ldots$

Work through it entirely on your own:

(1) Determine the eigenvalues and corresponding eigenvectors.

(2) Hence form the matrix **M**.

(3) Determine $\mathbf{M}^{-1}$, the inverse of **M**.

(4) Finally form the matrix products **AM** and $\mathbf{M}^{-1}(\mathbf{AM})$.

$$\mathbf{M} = \begin{pmatrix} 1 & 5 \\ 2 & -2 \end{pmatrix}; \quad \mathbf{M}^{-1} = \begin{pmatrix} 1/6 & 5/12 \\ 1/6 & -1/12 \end{pmatrix}; \quad \mathbf{M}^{-1}\mathbf{AM} = \begin{pmatrix} 4 & 0 \\ 0 & -8 \end{pmatrix}$$

Here is the working. See whether you agree.

$$\mathbf{A} = \begin{pmatrix} -6 & 5 \\ 4 & 2 \end{pmatrix} \qquad \therefore \begin{vmatrix} -6-\lambda & 5 \\ 4 & 2-\lambda \end{vmatrix} = 0$$

$$(-6-\lambda)(2-\lambda) - 20 = 0 \qquad \therefore \lambda^2 + 4\lambda - 32 = 0$$

$$(\lambda - 4)(\lambda + 8) = 0 \qquad \therefore \lambda = 4 \text{ or } -8$$

(a) $\lambda_1 = 4 \quad \left\{ \begin{pmatrix} -6 & 5 \\ 4 & 2 \end{pmatrix} - \begin{pmatrix} 4 & 0 \\ 0 & 4 \end{pmatrix} \right\} \begin{pmatrix} x_1 \\ x_2 \end{pmatrix} = \begin{pmatrix} 0 \\ 0 \end{pmatrix}$

$$\begin{pmatrix} -10 & 5 \\ 4 & -2 \end{pmatrix} \begin{pmatrix} x_1 \\ x_2 \end{pmatrix} = \begin{pmatrix} 0 \\ 0 \end{pmatrix}$$

$$\therefore -10x_1 + 5x_2 = 0 \quad \therefore x_2 = 2x_1 \quad \mathbf{x}_1 = \begin{pmatrix} 1 \\ 2 \end{pmatrix}$$

(b) $\lambda_2 = -8 \quad \left\{ \begin{pmatrix} -6 & 5 \\ 4 & 2 \end{pmatrix} + \begin{pmatrix} 8 & 0 \\ 0 & 8 \end{pmatrix} \right\} \begin{pmatrix} x_1 \\ x_2 \end{pmatrix} = \begin{pmatrix} 0 \\ 0 \end{pmatrix}$

$$\begin{pmatrix} 2 & 5 \\ 4 & 10 \end{pmatrix} \begin{pmatrix} x_1 \\ x_2 \end{pmatrix} = \begin{pmatrix} 0 \\ 0 \end{pmatrix}$$

$$\therefore 2x_1 + 5x_2 = 0 \quad \therefore x_2 = -\tfrac{2}{5}x_1 \quad \therefore \mathbf{x}_2 = \begin{pmatrix} 5 \\ -2 \end{pmatrix}$$

$$\therefore \mathbf{M} = \begin{pmatrix} 1 & 5 \\ 2 & -2 \end{pmatrix}$$

To find $\mathbf{M}^{-1}$ $\begin{pmatrix} 1 & 5 & | & 1 & 0 \\ 2 & -2 & | & 0 & 1 \end{pmatrix}$

Operating on rows, we have

$$\begin{pmatrix} 0 & 5 & | & 1 & 0 \\ 0 & -12 & | & -2 & 1 \end{pmatrix} = \begin{pmatrix} 1 & 5 & | & 1 & 0 \\ 0 & 1 & | & 1/6 & -1/12 \end{pmatrix}$$

$$= \begin{pmatrix} 1 & 0 & | & 1/6 & 5/12 \\ 0 & 1 & | & 1/6 & -1/12 \end{pmatrix}$$

$$\therefore \mathbf{M}^{-1} = \begin{pmatrix} 1/6 & 5/12 \\ 1/6 & -1/12 \end{pmatrix}$$

$$\therefore \mathbf{AM} = \begin{pmatrix} -6 & 5 \\ 4 & 2 \end{pmatrix} \begin{pmatrix} 1 & 5 \\ 2 & -2 \end{pmatrix} = \begin{pmatrix} 4 & -40 \\ 8 & 16 \end{pmatrix}$$

$$\therefore \mathbf{M}^{-1}\mathbf{AM} = \begin{pmatrix} 1/6 & 5/12 \\ 1/6 & -1/12 \end{pmatrix} \begin{pmatrix} 4 & -40 \\ 8 & 16 \end{pmatrix} = \begin{pmatrix} 4 & 0 \\ 0 & -8 \end{pmatrix}$$

$$\therefore \mathbf{M}^{-1}\mathbf{AM} = \begin{pmatrix} 4 & 0 \\ 0 & -8 \end{pmatrix}$$

Systems of second-order differential equations

87

The process of uncoupling a system of differential equations to obtain their solution can be achieved by diagonalizing the matrix of coefficients. For simplicity we shall only consider second-order equations and again, we proceed by example.

Example 1

Consider the system of coupled second-order differential equations

$$f_1''(x) = 2f_1(x) + 3f_2(x)$$
$$f_2''(x) = 4f_1(x) + f_2(x)$$

where $f_1(0) = 2$, $f_2(0) = 1$, $f_1'(0) = 4$ and $f_2'(0) = 3$

These can be written in matrix form as

. . . .

88

$$\boxed{\begin{pmatrix} f_1''(x) \\ f_2''(x) \end{pmatrix} = \begin{pmatrix} 2 & 3 \\ 4 & 1 \end{pmatrix} \begin{pmatrix} f_1(x) \\ f_2(x) \end{pmatrix}}$$

That is

$$\mathbf{F}''(x) = \mathbf{A}\mathbf{F}(x)$$

where $\mathbf{F}(x) = \begin{pmatrix} f_1(x) \\ f_2(x) \end{pmatrix}$, $\mathbf{F}''(x) = \begin{pmatrix} f_1''(x) \\ f_2''(x) \end{pmatrix}$ and $\mathbf{A} = \begin{pmatrix} 2 & 3 \\ 4 & 1 \end{pmatrix}$ and where

$\mathbf{F}(0) = \begin{pmatrix} f_1(0) \\ f_2(0) \end{pmatrix} = \begin{pmatrix} 2 \\ 1 \end{pmatrix}$ and $\mathbf{F}'(0) = \begin{pmatrix} f_1'(0) \\ f_2'(0) \end{pmatrix} = \begin{pmatrix} 4 \\ 3 \end{pmatrix}$ are the boundary conditions in matrix form.

The matrix differential equation $\mathbf{F}''(x) = \mathbf{A}\mathbf{F}(x)$ is similar in form to the single differential equation $f''(x) = af(x)$ (a constant) which has solution $f(x) = \alpha e^{\sqrt{a}x} + \beta e^{-\sqrt{a}x}$ (α, β constants), so to solve the matrix equation we try a solution of this form. We already know from Frames 54 to 57 that the eigenvalues and eigenvectors of matrix $\mathbf{A}$ are

$\lambda = -2$ with corresponding eigenvector $\begin{pmatrix} 3 \\ -4 \end{pmatrix}$

$\lambda = 5$ with corresponding eigenvector $\begin{pmatrix} 1 \\ 1 \end{pmatrix}$

The modal matrix of $\mathbf{A}$ is the matrix $\mathbf{M}$ and the spectral matrix of $\mathbf{A}$ is the matrix $\mathbf{S}$ where

$$\mathbf{M} = \begin{pmatrix} \cdots & \cdots \\ \cdots & \cdots \end{pmatrix} \text{ and } \mathbf{S} = \begin{pmatrix} \cdots & \cdots \\ \cdots & \cdots \end{pmatrix}$$

$$\mathbf{M} = \begin{pmatrix} 3 & 1 \\ -4 & 1 \end{pmatrix} \text{ and } \mathbf{S} = \begin{pmatrix} -2 & 0 \\ 0 & 5 \end{pmatrix}$$

Because

The modal matrix is formed from the eigenvectors of **A**. That is

$$\mathbf{M} = \begin{pmatrix} 3 & 1 \\ -4 & 1 \end{pmatrix} \text{ where the two eigenvectors are } \begin{pmatrix} 3 \\ -4 \end{pmatrix} \text{ and } \begin{pmatrix} 1 \\ 1 \end{pmatrix}$$

The spectral matrix is formed from the eigenvalues of **A**. That is

$$\mathbf{S} = \begin{pmatrix} -2 & 0 \\ 0 & 5 \end{pmatrix} \text{ where the two eigenvalues are } -2 \text{ and } 5$$

If we now define the matrix $\mathbf{G}(x)$ by the equation $\mathbf{F}(x) = \mathbf{MG}(x)$, then differentiating gives

$\mathbf{F}''(x) = [\mathbf{MG}(x)]'' = \mathbf{MG}''(x)$ where
$\mathbf{F}''(x) = \mathbf{AF}(x) = \mathbf{AMG}(x)$

and so, from Frame 85, $\mathbf{M}^{-1}\mathbf{MG}''(x) = \mathbf{G}''(x) = \mathbf{M}^{-1}\mathbf{AMG}(x) = \mathbf{SG}(x)$. That is

$\mathbf{G}''(x) = \mathbf{SG}(x)$

Therefore, in component terms

$$\mathbf{G}''(x) = \begin{pmatrix} g_1''(x) \\ g_2''(x) \end{pmatrix} = \mathbf{SG}(x) = \begin{pmatrix} -2 & 0 \\ 0 & 5 \end{pmatrix} \begin{pmatrix} g_1(x) \\ g_2(x) \end{pmatrix}$$

and so

$$g_1''(x) = \ldots g_1(x) \text{ with solution } g_1(x) = k_{11}e^{\ldots x} + k_{12}e^{-\ldots x}$$
$$g_2''(x) = \ldots g_2(x) \text{ with solution } g_2(x) = k_{21}e^{\ldots x} + k_{22}e^{-\ldots x}$$

$$g_1''(x) = -2g_1(x) \text{ with solution } g_1(x) = k_{11}e^{i\sqrt{2}x} + k_{12}e^{-i\sqrt{2}x}$$
$$g_2''(x) = 5g_2(x) \text{ with solution } g_2(x) = k_{21}e^{\sqrt{5}x} + k_{22}e^{-\sqrt{5}x}$$

Now, $\mathbf{F}(x) = \mathbf{MG}(x)$ so

$$\mathbf{F}(x) = \begin{pmatrix} f_1(x) \\ f_2(x) \end{pmatrix} = \begin{pmatrix} \cdots \\ \cdots \end{pmatrix}$$

91

$$\mathbf{F}(x) = \begin{pmatrix} f_1(x) \\ f_2(x) \end{pmatrix} = \begin{pmatrix} 3k_{11}e^{i\sqrt{2}x} + 3k_{12}e^{-i\sqrt{2}x} + k_{21}e^{\sqrt{5}x} + k_{22}e^{-\sqrt{5}x} \\ -4k_{11}e^{i\sqrt{2}x} - 4k_{12}e^{-i\sqrt{2}x} + k_{21}e^{\sqrt{5}x} + k_{22}e^{-\sqrt{5}x} \end{pmatrix}$$

Because

$$\mathbf{F}(x) = \begin{pmatrix} f_1(x) \\ f_2(x) \end{pmatrix} = \mathbf{MG}(x) = \begin{pmatrix} 3 & 1 \\ -4 & 1 \end{pmatrix}\begin{pmatrix} k_{11}e^{i\sqrt{2}x} + k_{12}e^{-i\sqrt{2}x} \\ k_{21}e^{\sqrt{5}x} + k_{22}e^{-\sqrt{5}x} \end{pmatrix}$$

and so

$$f_1(x) = 3k_{11}e^{i\sqrt{2}x} + 3k_{12}e^{-i\sqrt{2}x} + k_{21}e^{i\sqrt{5}x} + k_{22}e^{-\sqrt{5}x}$$

and

$$f_2(x) = -4k_{11}e^{i\sqrt{2}x} - 4k_{12}e^{-i\sqrt{2}x} + k_{21}e^{\sqrt{5}x} + k_{22}e^{-\sqrt{5}x}$$

This solution can be written in terms of circular and hyperbolic trigonometric expressions as

$$\mathbf{F}(x) = \begin{pmatrix} \cdots & \cdots \\ \cdots & \cdots \end{pmatrix}\begin{pmatrix} P\cos\ldots x + Q\sin\ldots x \\ R\cosh\ldots x + S\sinh\ldots x \end{pmatrix}$$

92

$$\mathbf{F}(x) = \begin{pmatrix} 3 & 1 \\ -4 & 1 \end{pmatrix}\begin{pmatrix} P\cos\sqrt{2}x + Q\sin\sqrt{2}x \\ R\cosh\sqrt{5}x + S\sinh\sqrt{5}x \end{pmatrix}$$

Because

$$3k_{11}e^{i\sqrt{2}x} + 3k_{12}e^{-i\sqrt{2}x}$$
$$= 3k_{11}\left(\cos\sqrt{2}x + i\sin\sqrt{2}x\right) + 3k_{12}\left(\cos\sqrt{2}x - i\sin\sqrt{2}x\right)$$
$$= P\cos\sqrt{2}x + Q\sin\sqrt{2}x$$

where $P = 3k_{11} + 3k_{12}$ and $Q = (3k_{11} - 3k_{12})i$

and

$$k_{21}e^{\sqrt{5}x} + k_{22}e^{-\sqrt{5}x}$$
$$= k_{21}\left(\cosh\sqrt{5}x + \sinh\sqrt{5}x\right) + k_{22}\left(\cosh\sqrt{5}x - \sinh\sqrt{5}x\right)$$
$$= R\cosh\sqrt{5}x + S\sinh\sqrt{5}x \text{ where } R = k_{21} + k_{22} \text{ and } S = k_{21} - k_{22}$$

Therefore

$$\mathbf{F}(x) = \begin{pmatrix} \cdots \\ \cdots \end{pmatrix}$$

93

$$\mathbf{F}(x) = \begin{pmatrix} 3P\cos\sqrt{2}x + 3Q\sin\sqrt{2}x + R\cosh\sqrt{5}x + S\sinh\sqrt{5}x \\ -4P\cos\sqrt{2}x - 4Q\sin\sqrt{2}x + R\cosh\sqrt{5}x + S\sinh\sqrt{5}x \end{pmatrix}$$

That is

$$f_1(x) = \ldots\ldots\ldots$$
$$f_2(x) = \ldots\ldots\ldots$$

94

$$f_1(x) = 3P\cos\sqrt{2}x + 3Q\sin\sqrt{2}x + R\cosh\sqrt{5}x + S\sinh\sqrt{5}x$$
$$f_2(x) = -4P\cos\sqrt{2}x - 4Q\sin\sqrt{2}x + R\cosh\sqrt{5}x + S\sinh\sqrt{5}x$$

Because

$$\mathbf{F}(x) = \begin{pmatrix} f_1(x) \\ f_2(x) \end{pmatrix}$$

and so

$$f_1(x) = 3P\cos\sqrt{2}x + 3Q\sin\sqrt{2}x + R\cosh\sqrt{5}x + S\sinh\sqrt{5}x$$
$$f_2(x) = -4P\cos\sqrt{2}x - 4Q\sin\sqrt{2}x + R\cosh\sqrt{5}x + S\sinh\sqrt{5}x$$

Applying the boundary conditions, we find

$$\mathbf{F}(0) = \begin{pmatrix} 2 \\ 1 \end{pmatrix} = \begin{pmatrix} \ldots P + \ldots R \\ \ldots P + \ldots R \end{pmatrix} \text{ and } \mathbf{F}'(0) = \begin{pmatrix} 4 \\ 3 \end{pmatrix} = \begin{pmatrix} \ldots Q + \ldots S \\ \ldots Q + \ldots S \end{pmatrix}$$

95

$$\mathbf{F}(0) = \begin{pmatrix} 2 \\ 1 \end{pmatrix} = \begin{pmatrix} 3P + R \\ -4P + R \end{pmatrix} \text{ and } \mathbf{F}'(0) = \begin{pmatrix} 4 \\ 3 \end{pmatrix} = \begin{pmatrix} 3\sqrt{2}Q + \sqrt{5}S \\ -4\sqrt{2}Q + \sqrt{5}S \end{pmatrix}$$

Because

$$f_1(0) = 2, f_2(0) = 1, f_1'(0) = 4 \text{ and } f_2'(0) = 3 \text{ and so}$$

$$\mathbf{F}(0) = \begin{pmatrix} 2 \\ 1 \end{pmatrix} = \begin{pmatrix} f_1(0) \\ f_2(0) \end{pmatrix} = \begin{pmatrix} 3P + R \\ -4P + R \end{pmatrix} \text{ and}$$

$$\mathbf{F}'(0) = \begin{pmatrix} 4 \\ 3 \end{pmatrix} = \begin{pmatrix} f_1'(0) \\ f_2'(0) \end{pmatrix} = \begin{pmatrix} 3\sqrt{2}Q + \sqrt{5}S \\ -4\sqrt{2}Q + \sqrt{5}S \end{pmatrix}$$

This gives the two sets of simultaneous equations

$$3P + R = 2 \qquad 3\sqrt{2}Q + \sqrt{5}S = 4$$

and which have solution

$$-4P + R = 1 \qquad -4\sqrt{2}Q + \sqrt{5}S = 3$$

$$P = \ldots\ldots\ldots, \quad R = \ldots\ldots\ldots,$$
$$Q = \ldots\ldots\ldots \text{ and } S = \ldots\ldots\ldots$$

96

$$\boxed{P = 1/7, \ R = 11/7, \ Q = 1/\left(7\sqrt{2}\right) \text{ and } S = 25/\left(7\sqrt{5}\right)}$$

This gives the complete solution as

$$f_1(x) = \frac{3}{7}\cos\sqrt{2}x + \frac{3}{7\sqrt{2}}\sin\sqrt{2}x + \frac{11}{7}\cosh\sqrt{5}x + \frac{25}{7\sqrt{5}}\sinh\sqrt{5}x$$

$$f_2(x) = -\frac{4}{7}\cos\sqrt{2}x - \frac{4}{7\sqrt{2}}\sin\sqrt{2}x + \frac{11}{7}\cosh\sqrt{5}x + \frac{25}{7\sqrt{5}}\sinh\sqrt{5}x$$

This method is quite straightforwardly extended to three or more such coupled differential equations.

Summary

To solve the system of coupled second-order differential equations

$$\mathbf{F}''(x) = \mathbf{A}\mathbf{F}(x)$$

1 Find the eigenvalues and eigenvectors of matrix $\mathbf{A}$ and construct the modal matrix $\mathbf{M}$ and the diagonal spectral matrix $\mathbf{S}$
2 Solve the equation $\mathbf{G}'(x) = \mathbf{S}\mathbf{G}(x)$
   (note that even though $\mathbf{M}^{-1}$ is used there was no need to calculate it)
3 Apply $\mathbf{F}(x) = \mathbf{M}\mathbf{G}(x)$ to find $\mathbf{F}(x)$.

Try one yourself.

Next frame

97

Example 2

The system of coupled second-order differential equations (refer to Frames 57 to 61)

$$f_1''(x) = 3f_1(x) + 10f_2(x)$$

$$f_2''(x) = 2f_1(x) + 4f_2(x)$$

where $f_1(0) = 0$, $f_2(0) = 1$, $f_1'(0) = 1$ and $f_2'(0) = 0$

has the solution (refer to Frames 57 to 61)

$$f_1(x) = \ldots\ldots\ldots\ldots$$
$$f_2(x) = \ldots\ldots\ldots\ldots$$

$$f_1(x) = 10\cos x + \frac{5}{9}\sin x + 10\cosh 2\sqrt{2}x + \frac{2}{9\sqrt{2}}\sinh 2\sqrt{2}x$$
$$f_2(x) = -4\cos x - \frac{2}{9}\sin x + 5\cosh 2\sqrt{2}x + \frac{1}{9\sqrt{2}}\sinh 2\sqrt{2}x$$

Because

$$f_1''(x) = 3f_1(x) + 10f_2(x)$$
$$f_2''(x) = 2f_1(x) + 4f_2(x)$$

can be written in matrix form as

. . . .

$$\begin{pmatrix} f_1''(x) \\ f_2''(x) \end{pmatrix} = \begin{pmatrix} 3 & 10 \\ 2 & 4 \end{pmatrix} \begin{pmatrix} f_1(x) \\ f_2(x) \end{pmatrix}$$

That is

$$\mathbf{F}''(x) = \mathbf{A}\mathbf{F}(x)$$

where $\mathbf{F}(x) = \begin{pmatrix} f_1(x) \\ f_2(x) \end{pmatrix}$, $\mathbf{F}''(x) = \begin{pmatrix} f_1''(x) \\ f_2''(x) \end{pmatrix}$ and $\mathbf{A} = \begin{pmatrix} 3 & 10 \\ 2 & 4 \end{pmatrix}$

and where $\mathbf{F}(0) = \begin{pmatrix} 0 \\ 1 \end{pmatrix}$ and $\mathbf{F}'(0) = \begin{pmatrix} f_1'(0) \\ f_2'(0) \end{pmatrix} = \begin{pmatrix} 1 \\ 0 \end{pmatrix}$.

To solve the matrix equation we first need the eigenvalues and associated eigenvectors of the matrix $\mathbf{A}$. These have already been found in Frames 57 to 61 and they are

$\lambda = -1$ with corresponding eigenvector $\begin{pmatrix} 5 \\ -2 \end{pmatrix}$

$\lambda = 8$ with corresponding eigenvector $\begin{pmatrix} 2 \\ 1 \end{pmatrix}$

The complete solution of $\mathbf{F}'' = \mathbf{A}\mathbf{F}$ is then

$$\mathbf{F}(x) = (P\cos x + Q\sin x)\begin{pmatrix} 5 \\ -2 \end{pmatrix} + \left(R\cosh 2\sqrt{2}x + S\sinh 2\sqrt{2}x\right)\begin{pmatrix} 2 \\ 1 \end{pmatrix}$$

$$= \begin{pmatrix} 5P\cos x + 5Q\sin x + 2R\cosh 2\sqrt{2}x + 2S\sinh 2\sqrt{2}x \\ -2P\cos x - 2Q\sin x + R\cosh 2\sqrt{2}x + S\sinh 2\sqrt{2}x \end{pmatrix}$$

That is

$$f_1(x) = \ldots\ldots\ldots\ldots$$
$$f_2(x) = \ldots\ldots\ldots\ldots$$

100

$$f_1(x) = 5P\cos x + 5Q\sin x + 2R\cosh 2\sqrt{2}x + 2S\sinh 2\sqrt{2}x$$
$$f_2(x) = -2P\cos x - 2Q\sin x + R\cosh 2\sqrt{2}x + S\sinh 2\sqrt{2}x$$

Because

$$\mathbf{F}(x) = \begin{pmatrix} f_1(x) \\ f_2(x) \end{pmatrix}$$

and so

$$f_1(x) = 5P\cos x + 5Q\sin x + 2R\cosh 2\sqrt{2}x + 2S\sinh 2\sqrt{2}x$$
$$f_2(x) = -2P\cos x - 2Q\sin x + R\cosh 2\sqrt{2}x + S\sinh 2\sqrt{2}x$$

Applying the boundary conditions, we find

$$\mathbf{F}(0) = \begin{pmatrix} 0 \\ 1 \end{pmatrix} = \begin{pmatrix} \ldots P + \ldots R \\ \ldots P + \ldots R \end{pmatrix} \text{ and } \mathbf{F}'(0) = \begin{pmatrix} 1 \\ 0 \end{pmatrix} = \begin{pmatrix} \ldots Q + \ldots S \\ \ldots Q + \ldots S \end{pmatrix}$$

101

$$\mathbf{F}(0) = \begin{pmatrix} 0 \\ 1 \end{pmatrix} = \begin{pmatrix} 5P + 2R \\ -2P + R \end{pmatrix} \text{ and } \mathbf{F}'(0) = \begin{pmatrix} 1 \\ 0 \end{pmatrix} = \begin{pmatrix} 5Q + 4\sqrt{2}S \\ -2Q + 2\sqrt{2}S \end{pmatrix}$$

Because

The boundary conditions are $f_1(0) = 0$, $f_2(0) = 1$, $f_1'(0) = 1$ and $f_2'(0) = 0$, therefore

$$\mathbf{F}(0) = \begin{pmatrix} 0 \\ 1 \end{pmatrix} = \begin{pmatrix} f_1(0) \\ f_2(0) \end{pmatrix} = \begin{pmatrix} 5P + 2R \\ -2P + R \end{pmatrix} \text{ and }$$

$$\mathbf{F}'(0) = \begin{pmatrix} 1 \\ 0 \end{pmatrix} = \begin{pmatrix} f_1'(0) \\ f_2'(0) \end{pmatrix} = \begin{pmatrix} 5Q + 4\sqrt{2}S \\ -2Q + 2\sqrt{2}S \end{pmatrix}$$

This gives the two sets of simultaneous equations

$$\begin{aligned} 5P + 2R &= 0 \\ -2P + R &= 1 \end{aligned} \quad \text{and} \quad \begin{aligned} 5Q + 4\sqrt{2}S &= 1 \\ -2Q + 2\sqrt{2}S &= 0 \end{aligned} \quad \text{which have solution}$$

$$P = \ldots\ldots\ldots, \quad R = \ldots\ldots\ldots,$$
$$Q = \ldots\ldots\ldots \text{ and } S = \ldots\ldots\ldots$$

102

$$P = -2/9, \quad R = 5/9, \quad Q = 1/9 \text{ and } S = 1/\left(9\sqrt{2}\right)$$

This gives the complete solution as

$$f_1(x) = -\frac{10}{9}\cos x + \frac{5}{9}\sin x + \frac{10}{9}\cosh 2\sqrt{2}x + \frac{2}{9\sqrt{2}}\sinh 2\sqrt{2}x$$

$$f_2(x) = \frac{4}{9}\cos x - \frac{2}{9}\sin x + \frac{5}{9}\cosh 2\sqrt{2}x + \frac{1}{9\sqrt{2}}\sinh 2\sqrt{2}x$$

Matrix transformation

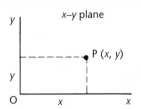 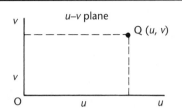

103

If for every point Q (u, v) in the u–v plane there is a corresponding point P (x, y) in the x–y plane, then there is a relationship between the two sets of coordinates. In the simple case of scaling the coordinate where

$$u = ax \text{ and } v = by$$

we have a *linear transformation* and we can combine these in matrix form

$$\begin{pmatrix} u \\ v \end{pmatrix} = \begin{pmatrix} a & 0 \\ 0 & b \end{pmatrix} \begin{pmatrix} x \\ y \end{pmatrix}$$

The matrix $\begin{pmatrix} a & 0 \\ 0 & b \end{pmatrix}$ then provides the transformation between the vector $\begin{pmatrix} x \\ y \end{pmatrix}$ in one set of coordinates and the vector $\begin{pmatrix} u \\ v \end{pmatrix}$ in the other set of coordinates.

Similarly, if we solve the two equations for x and y, we have

$$x = \frac{1}{a}u \quad \text{and} \quad y = \frac{1}{b}v$$

$$\therefore \begin{pmatrix} x \\ y \end{pmatrix} = \begin{pmatrix} 1/a & 0 \\ 0 & 1/b \end{pmatrix} \begin{pmatrix} u \\ v \end{pmatrix}$$

which allows us to transform back from the u–v plane coordinates to the x–y plane coordinates.

Now for an example.

▶

Example

If $\mathbf{X} = \begin{pmatrix} x \\ y \end{pmatrix} = \begin{pmatrix} 2 \\ 1 \end{pmatrix}$ with the transformation $\mathbf{T} = \begin{pmatrix} -2 & 0 \\ 2 & 1 \end{pmatrix}$ determine

$\mathbf{U} = \begin{pmatrix} u \\ v \end{pmatrix} = \mathbf{TX}$ and show the positions on the x–y and u–v planes.

In this case

$$\begin{pmatrix} u \\ v \end{pmatrix} = \begin{pmatrix} -2 & 0 \\ 2 & 1 \end{pmatrix} \begin{pmatrix} 2 \\ 1 \end{pmatrix} = \begin{pmatrix} -4 \\ 5 \end{pmatrix}$$

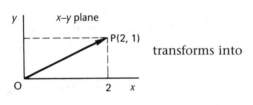   transforms into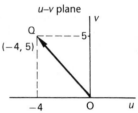

If $\mathbf{T}$ is non-singular and $\mathbf{U} = \mathbf{TX}$ then $\mathbf{X} = \mathbf{T}^{-1}\mathbf{U}$ and since

$$\mathbf{T} = \begin{pmatrix} -2 & 0 \\ 2 & 1 \end{pmatrix} \text{ then } \mathbf{T}^{-1} = \dots\dots\dots$$

104

$$\boxed{\mathbf{T}^{-1} = \begin{pmatrix} -1/2 & 0 \\ 1 & 1 \end{pmatrix}}$$

There are several ways of finding the inverse of a matrix. One method is as follows.

$$\mathbf{T} = \begin{pmatrix} -2 & 0 \\ 2 & 1 \end{pmatrix}$$

$$\begin{pmatrix} -2 & 0 & | & 1 & 0 \\ 2 & 1 & | & 0 & 1 \end{pmatrix} \sim \begin{pmatrix} -2 & 0 & | & 1 & 0 \\ 0 & 1 & | & 1 & 1 \end{pmatrix}$$

$$\sim \begin{pmatrix} 1 & 0 & | & -1/2 & 0 \\ 0 & 1 & | & 1 & 1 \end{pmatrix}$$

$$\therefore \mathbf{T}^{-1} = \begin{pmatrix} -1/2 & 0 \\ 1 & 1 \end{pmatrix}$$

So we have $\mathbf{U} = \mathbf{TX}$ $\therefore \mathbf{X} = \mathbf{T}^{-1}\mathbf{U}$

$$\therefore \begin{pmatrix} x \\ y \end{pmatrix} = \begin{pmatrix} -1/2 & 0 \\ 1 & 1 \end{pmatrix} \begin{pmatrix} u \\ v \end{pmatrix}$$

Hence a vector $\begin{pmatrix} 1 \\ 4 \end{pmatrix}$ in the u–v plane transforms into $\begin{pmatrix} x \\ y \end{pmatrix}$ in the x–y

plane where $\begin{pmatrix} x \\ y \end{pmatrix} = \dots\dots\dots$

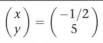

$$\begin{pmatrix} x \\ y \end{pmatrix} = \begin{pmatrix} -1/2 \\ 5 \end{pmatrix}$$

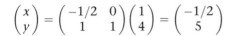

$$\begin{pmatrix} x \\ y \end{pmatrix} = \begin{pmatrix} -1/2 & 0 \\ 1 & 1 \end{pmatrix}\begin{pmatrix} 1 \\ 4 \end{pmatrix} = \begin{pmatrix} -1/2 \\ 5 \end{pmatrix}$$

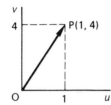

transforms into

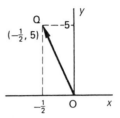

Rotation of axes

A more interesting case occurs with a degree of rotation between the two sets of coordinate axes.

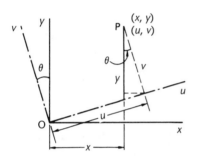

Let P be the point (x, y) in the x–y plane and the point (u, v) in the u–v plane.

Let θ be the angle of rotation between the two systems. From the diagram we can see that

$$\left. \begin{array}{l} x = u\cos\theta - v\sin\theta \\ y = u\sin\theta + v\cos\theta \end{array} \right\} \tag{1}$$

In matrix form, this becomes $\begin{pmatrix} x \\ y \end{pmatrix} = \begin{pmatrix} \cos\theta & -\sin\theta \\ \sin\theta & \cos\theta \end{pmatrix}\begin{pmatrix} u \\ v \end{pmatrix}$

which enables us to transform from the u–v plane coordinates to the corresponding x–y plane coordinates.

Make a note of this and then move on

106

If we solve equations (1) for u and v, we have

$$x \sin \theta = u \sin \theta \cos \theta - v \sin^2 \theta$$
$$y \cos \theta = u \sin \theta \cos \theta + v \cos^2 \theta$$
$$\therefore \ y \cos \theta - x \sin \theta = v(\cos^2 \theta + \sin^2 \theta) = v$$

Also
$$x \cos \theta = u \cos^2 \theta - v \sin \theta \cos \theta$$
$$y \sin \theta = u \sin^2 \theta + v \sin \theta \cos \theta$$
$$\therefore \ x \cos \theta + y \sin \theta = u(\cos^2 \theta + \sin^2 \theta) = u$$

So
$$u = x \cos \theta + y \sin \theta$$
$$v = -x \sin \theta + y \cos \theta$$

and written in matrix form, this is ....

107

$$\boxed{\begin{pmatrix} u \\ v \end{pmatrix} = \begin{pmatrix} \cos \theta & \sin \theta \\ -\sin \theta & \cos \theta \end{pmatrix} \begin{pmatrix} x \\ y \end{pmatrix}}$$

So we have

$$\begin{pmatrix} x \\ y \end{pmatrix} = \begin{pmatrix} \cos \theta & -\sin \theta \\ \sin \theta & \cos \theta \end{pmatrix} \begin{pmatrix} u \\ v \end{pmatrix}$$

and
$$\begin{pmatrix} u \\ v \end{pmatrix} = \begin{pmatrix} \cos \theta & \sin \theta \\ -\sin \theta & \cos \theta \end{pmatrix} \begin{pmatrix} x \\ y \end{pmatrix}$$

i.e. $\mathbf{X} = \mathbf{T}\mathbf{U}$ and $\mathbf{U} = \mathbf{T}^{-1}\mathbf{X}$

where $\mathbf{T}$ is the matrix of transformation and the equations provide a linear transformation between the two sets of coordinates.

Example

If the u–v plane axes rotate through $30°$ in an anticlockwise manner from the x–y plane axes, determine the (u, v) coordinates of a point whose (x, y) coordinates are $x = 2$, $y = 3$ in the x–y plane.

This is a straightforward application of the results above.

So
$$\begin{pmatrix} u \\ v \end{pmatrix} = \ ....$$

$$\begin{pmatrix} u \\ v \end{pmatrix} = \begin{pmatrix} \sqrt{3} + 3/2 \\ -1 + 3\sqrt{3}/2 \end{pmatrix} = \begin{pmatrix} 3 \cdot 23 \\ 1 \cdot 60 \end{pmatrix}$$

Because

$$\begin{pmatrix} u \\ v \end{pmatrix} = \begin{pmatrix} \cos\theta & \sin\theta \\ -\sin\theta & \cos\theta \end{pmatrix} \begin{pmatrix} 2 \\ 3 \end{pmatrix} \qquad \begin{array}{l} \cos\theta = \sqrt{3}/2 \\ \sin\theta = 1/2 \end{array}$$

$$= \begin{pmatrix} \sqrt{3}/2 & 1/2 \\ -1/2 & \sqrt{3}/2 \end{pmatrix} \begin{pmatrix} 2 \\ 3 \end{pmatrix}$$

$$= \begin{pmatrix} \sqrt{3} + 3/2 \\ -1 + 3\sqrt{3}/2 \end{pmatrix} = \begin{pmatrix} 3 \cdot 23 \\ 1 \cdot 60 \end{pmatrix}$$

As usual, the Program ends with the **Review summary**, to be read in conjunction with the **Can You?** checklist. Go back to the relevant part of the Program for any points on which you are unsure. The **Test exercise** should then be straightforward and the **Further problems** give valuable additional practice.

Review summary

109

1 *Singular* square matrix: $|\mathbf{A}| = 0$
   Non-singular square matrix $|\mathbf{A}| \neq 0$.

2 *Rank of a matrix* – order of the largest non-zero determinant that can be formed from the elements of the matrix.

3 *Elementary operations and equivalent matrices*
   Each of the following row operations on matrix **A** produces a *row equivalent matrix* **B** where the order and rank of **B** are the same as those of **A**. We write $\mathbf{A} \sim \mathbf{B}$.

   (1) Interchanging two rows

   (2) Multiplying each element of a row by the same non-zero scalar quantity

   (3) Adding or subtracting corresponding elements from those of another row.

   These operations are called *elementary row operations*. There is a corresponding set of three *elementary column operations* that can be used to form *column equivalent matrices*.

4 *Consistency* of a set of n equations in n unknowns with coefficient matrix **A** and augmented matrix $\mathbf{A}_b$.

   (a) Consistent if rank of **A** = rank of $\mathbf{A}_b$

   (b) Inconsistent if rank of **A** < rank of $\mathbf{A}_b$.

▶

5   *Uniqueness of solutions* – n equations with n unknowns.
  (a) rank of $\mathbf{A}$ = rank of $\mathbf{A}_b = n$ *unique solutions*
  (b) rank of $\mathbf{A}$ = rank of $\mathbf{A}_b = m < n$ *infinite number of solutions*
  (c) rank of $\mathbf{A} <$ rank of $\mathbf{A}_b$ *no solution*

6   *Solution of sets of equations*
  (a) *Inverse matrix method* $\mathbf{Ax} = \mathbf{b}$; $\mathbf{x} = \mathbf{A}^{-1}\mathbf{b}$
    To find $\mathbf{A}^{-1}$
    (1) evaluate $|\mathbf{A}|$
    (2) form $\mathbf{C}$, the matrix of cofactors of $\mathbf{A}$
    (3) write $\mathbf{C}^{\mathrm{T}}$, the transpose of $\mathbf{A}$
    (4) $\mathbf{A}^{-1} = \dfrac{1}{|\mathbf{A}|} \times \mathbf{C}^{\mathrm{T}}$.

  (b) *Row transformation method* $\mathbf{Ax} = \mathbf{b}$; $\mathbf{Ax} = \mathbf{Ib}$
    (1) form the combined coefficient matrix $[\mathbf{A}|\mathbf{I}]$
    (2) row transformations to convert to $[\mathbf{I}|\mathbf{A}^{-1}]$
    (3) then solve $\mathbf{x} = \mathbf{A}^{-1}\mathbf{b}$.

  (c) *Gaussian elimination method* $\mathbf{Ax} = \mathbf{b}$
    (1) form augmented matrix $[\mathbf{A}|\mathbf{b}]$
    (2) operate on rows to convert to $[\mathbf{U}|\mathbf{b}']$ where $\mathbf{U}$ is the upper-triangular matrix.
    (3) expand from bottom row to obtain $\mathbf{x}$.

  (d) *Triangular decomposition method* $\mathbf{Ax} = \mathbf{b}$
    Write $\mathbf{A}$ as the product of upper and lower triangular matrices.
      $\mathbf{A} = \mathbf{LU}$, $\mathbf{L}(\mathbf{Ux}) = \mathbf{b}$. Put $\mathbf{Ux} = \mathbf{y}$ $\therefore$ $\mathbf{Ly} = \mathbf{b}$
    (1) solve $\mathbf{Ly} = \mathbf{b}$ to obtain $\mathbf{y}$
    (2) solve $\mathbf{Ux} = \mathbf{y}$ to obtain $\mathbf{x}$.

7   *Eigenvalues and eigenvectors* $\mathbf{Ax} = \lambda\mathbf{x}$
  Sets of equations of form $\mathbf{Ax} = \lambda\mathbf{x}$, where $\mathbf{A}$ = coefficient matrix, $\mathbf{x}$ = column matrix, λ = scalar quantity.
  Equations become $(\mathbf{A} - \lambda\mathbf{I})\mathbf{x} = \mathbf{0}$.
  For non-trivial solutions, $|\mathbf{A} - \lambda\mathbf{I}| = 0$ is the *characteristic equation* and gives values of λ i.e. the *eigenvalues*.
  Substitution of each eigenvalue gives a corresponding *eigenvector*.

8   *Cayley–Hamilton theorem*
  Every square matrix satisfies its own characteristic equation.

9   *Solving systems of first-order ordinary differential equations*
  To solve the system of coupled first-order differential equations
    $\mathbf{F}'(x) = \mathbf{AF}(x)$

▶

(a) Find the eigenvalues and eigenvectors of matrix **A** and construct the modal matrix **M** and the diagonal spectral matrix **S**

(b) Solve the equation $\mathbf{G}'(x) = \mathbf{S}\mathbf{G}(x)$

(c) Apply $\mathbf{F}(x) = \mathbf{M}\mathbf{G}(x)$ to find $\mathbf{F}(x)$.

10 *Diagonalization of a matrix*

Modal matrix of **A**

If **A** has distinct eigenvalues $\mathbf{M} = (\mathbf{x}_1, \mathbf{x}_2, \ldots, \mathbf{x}_n)$, where $\mathbf{x}_1, \mathbf{x}_2, \ldots, \mathbf{x}_n$ are eigenvectors of **A**, then $\mathbf{M}^{-1}\mathbf{A}\mathbf{M} = \mathbf{S}$ where **S** is the *spectral matrix* of **A**

$$\text{and} \quad \mathbf{S} = \begin{pmatrix} \lambda_1 & 0 & \ldots & 0 \\ 0 & \lambda_2 & \ldots & 0 \\ \cdot & \cdot & & \cdot \\ \cdot & \cdot & & \cdot \\ \cdot & \cdot & & \cdot \\ 0 & 0 & \ldots & \lambda_n \end{pmatrix}$$

$\lambda_1, \lambda_2, \ldots, \lambda_n$ are the eigenvalues of **A**.

11 *Solving systems of second-order ordinary differential equations*

To solve an equation of the form

$$\mathbf{F}''(x) = \mathbf{A}\mathbf{F}(x)$$

(a) Find the eigenvalues $\lambda_1, \lambda_2, \ldots, \lambda_n$ of **A**

(b) Assuming the eigenvectors are all distinct, find the associated eigenvectors $\mathbf{C}_1, \mathbf{C}_2, \ldots, \mathbf{C}_n$

(c) Write the solution of the equation as

$$\mathbf{F}(x) = \sum_{r=1}^{n} \left(a_r e^{\sqrt{\lambda_r} x} + b_r e^{-\sqrt{\lambda_r} x} \right) \mathbf{C}_r$$

and use the boundary conditions to find the values of a_r and b_r for $r = 1, 2, \ldots, n$.

12 *Matrix transformation*

(a) $\mathbf{U} = \mathbf{T}\mathbf{X}$, where **T** is a transformation matrix, transforms a vector in the x–y plane to a corresponding vector in the u–v plane. Similarly, $\mathbf{X} = \mathbf{T}^{-1}\mathbf{U}$ converts a vector in the u–v plane to a corresponding vector in the x–y plane.

(b) *Rotation of axes*

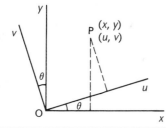

$$\begin{pmatrix} u \\ v \end{pmatrix} = \begin{pmatrix} \cos\theta & \sin\theta \\ -\sin\theta & \cos\theta \end{pmatrix} \begin{pmatrix} x \\ y \end{pmatrix}$$

$$\begin{pmatrix} x \\ y \end{pmatrix} = \begin{pmatrix} \cos\theta & -\sin\theta \\ \sin\theta & \cos\theta \end{pmatrix} \begin{pmatrix} u \\ v \end{pmatrix}$$

✅ **Can You?**

110 **Checklist 10**

Check this list before and after you try the end of Program test.

On a scale of 1 to 5 how confident are you that you can: Frames

- Determine whether a matrix is singular or non-singular?
  Yes ☐ ☐ ☐ ☐ ☐ *No*   **1** to **3**

- Determine the rank of a matrix?
  Yes ☐ ☐ ☐ ☐ ☐ *No*   **3** to **13**

- Determine the consistency of a set of linear equations and
  hence demonstrate the uniqueness of their solution?
  Yes ☐ ☐ ☐ ☐ ☐ *No*   **14** to **23**

- Obtain the solution of a set of simultaneous linear equations
  by using matrix inversion, by row transformation, by Gaussian
  elimination and by triangular decomposition?
  Yes ☐ ☐ ☐ ☐ ☐ *No*   **24** to **52**

- Obtain the eigenvalues and corresponding eigenvectors of a
  square matrix?
  Yes ☐ ☐ ☐ ☐ ☐ *No*   **53** to **65**

- Demonstrate the validity of the Cayley–Hamilton theorem?   **66** and **67**
  Yes ☐ ☐ ☐ ☐ ☐ *No*

- Solve systems of first-order ordinary differential equations
  using eigenvalue and eigenvector methods?
  Yes ☐ ☐ ☐ ☐ ☐ *No*   **68** to **79**

- Construct the modal matrix from the eigenvectors of a matrix
  and the spectral matrix from the eigenvalues?
  Yes ☐ ☐ ☐ ☐ ☐ *No*   **80** to **86**

- Solve systems of second-order ordinary differential equations
  using diagonalization?
  Yes ☐ ☐ ☐ ☐ ☐ *No*   **87** to **102**

- Use matrices to represent transformations between coordinate
  systems?
  Yes ☐ ☐ ☐ ☐ ☐ *No*   **103** to **108**

🚲 **Test exercise 10**

---

1 Determine the rank of **A** and of $\mathbf{A}_b$ for the following sets of equations and hence determine the nature of the solutions. Do *not* solve the equations. ⎡111⎤

(a) $x_1 + 3x_2 - 2x_3 = 6$ (b) $x_1 + 2x_2 - 4x_3 = 3$
  $4x_1 + 5x_2 + 2x_3 = 3$ $x_1 + 2x_2 + 3x_3 = -4$
  $x_1 + 3x_2 + 4x_3 = 7$ $2x_1 + 4x_2 + x_3 = -3.$

2 If $\mathbf{Ax} = \mathbf{b}$ where $\mathbf{A} = \begin{pmatrix} 2 & 3 & -2 \\ 3 & 5 & -4 \\ 1 & 2 & -3 \end{pmatrix}$ and $\mathbf{b} = \begin{pmatrix} 4 \\ 10 \\ 9 \end{pmatrix}$, determine $\mathbf{A}^{-1}$

and hence solve the set of equations.

3 Given that $3x_1 + 2x_2 + x_3 = 1$
  $x_1 - x_2 + 3x_3 = 5$
  $2x_1 + 5x_2 - 2x_3 = 0$

apply the method of row transformation to obtain the value of x_1, x_2, x_3.

4 By the method of Gaussian elimination, solve the equations $\mathbf{Ax} = \mathbf{b}$,

where $\mathbf{A} = \begin{pmatrix} 1 & -2 & -4 \\ 2 & 1 & -3 \\ 1 & 3 & 2 \end{pmatrix}$ and $\mathbf{b} = \begin{pmatrix} -3 \\ 4 \\ 5 \end{pmatrix}$.

5 If $\mathbf{Ax} = \mathbf{b}$ where $\mathbf{A} = \begin{pmatrix} 1 & -2 & 1 \\ 3 & 1 & -2 \\ 5 & 3 & 3 \end{pmatrix}$ and $\mathbf{b} = \begin{pmatrix} 7 \\ -3 \\ 5 \end{pmatrix}$, express **A** as the

product $\mathbf{A} = \mathbf{LU}$ where **L** and **U** are lower and upper-triangular matrices and hence determine the values of x_1, x_2, x_3.

6 Determine the eigenvalues and corresponding eigenvectors of $\mathbf{Ax} = \lambda\mathbf{x}$

where $\mathbf{A} = \begin{pmatrix} 1 & 3 & 0 \\ 1 & 2 & 1 \\ -2 & 1 & -1 \end{pmatrix}$.

7 If $\mathbf{x_1}$ and $\mathbf{x_2}$ are eigenvectors of $\mathbf{Ax} = \lambda\mathbf{x}$ where $\mathbf{A} = \begin{pmatrix} 3 & 2 \\ 4 & 1 \end{pmatrix}$ determine

(a) $\mathbf{M} = (\mathbf{x}_1 \mathbf{x}_2)$

(b) $\mathbf{M}^{-1}$

(c) $\mathbf{M}^{-1}\mathbf{AM}$.

8 Solve the system of first-order differential equations
$f_1'(x) = 5f_1(x) - 2f_2(x)$
$f_2'(x) = -f_1(x) + 4f_2(x)$ where $f_1(0) = -3$ and $f_2(0) = 2$.

▶

9 Solve the system of second-order differential equations

$$f_1''(x) = f_1(x) + 6f_2(x)$$
$$f_2''(x) = 3f_1(x) - 2f_2(x)$$
where $f_1(0) = 1$, $f_2(0) = 0$, $f_1'(0) = 2$, $f_2'(0) = -1$.

10 (a) Determine the vector in the u–v plane formed by $\mathbf{U} = \mathbf{TX}$, where the transformation matrix is $\mathbf{T} = \begin{pmatrix} -2 & 1 \\ 3 & 4 \end{pmatrix}$ and $\mathbf{X} = \begin{pmatrix} 3 \\ -2 \end{pmatrix}$ is a vector in the x–y plane.

(b) The coordinate axes in the x–y plane and in the u–v plane have the same origin O, but OU is inclined to OX at an angle of $60°$ in an anticlockwise manner. Transform a vector $\mathbf{X} = \begin{pmatrix} 4 \\ 6 \end{pmatrix}$ in the x–y plane into the corresponding vector in the u–v plane.

🚴 Further problems 10

112

1 If $\mathbf{Ax} = \mathbf{b}$ where $\mathbf{A} = \begin{pmatrix} 5 & 2 & 3 \\ 3 & -2 & -2 \\ 4 & 3 & 1 \end{pmatrix}$ and $\mathbf{b} = \begin{pmatrix} 6 \\ 5 \\ -5 \end{pmatrix}$, determine $\mathbf{A}^{-1}$ and hence solve the set of equations.

2 Apply the method of row transformation to solve the following sets of equations.

(a) $x_1 - 3x_2 - 2x_3 = 8$
$2x_1 + 2x_2 + x_3 = 4$
$3x_1 - 4x_2 + 2x_3 = -3$

(b) $x_1 - 3x_2 + 2x_3 = 8$
$2x_1 - x_2 + x_3 = 9$
$3x_1 + 2x_2 + 3x_3 = 5.$

3 Solve the following sets of equations by Gaussian elimination.

(a) $x_1 - 2x_2 - x_3 + 3x_4 = 4$
$2x_1 + x_2 + x_3 - 4x_4 = 3$
$3x_1 - x_2 - 2x_3 + 2x_4 = 6$
$x_1 + 3x_2 - x_3 + x_4 = 8$

(b) $2x_1 + 3x_2 - 2x_3 + 2x_4 = 2$
$4x_1 + 2x_2 - 3x_3 - x_4 = 6$
$x_1 - x_2 + 4x_3 - 2x_4 = 7$
$3x_1 + 2x_2 + x_3 - x_4 = 5$

(c) $x_1 + 2x_2 + 5x_3 + x_4 = 4$
$3x_1 - 4x_2 + 3x_3 - 2x_4 = 7$
$4x_1 + 3x_2 + 2x_3 - x_4 = 1$
$x_1 - 2x_2 - 4x_3 - x_4 = 2.$

▶

4 Using the method of triangular decomposition, solve the following sets of equations.

(a) $\begin{pmatrix} 1 & 4 & -1 \\ 4 & 2 & 3 \\ 7 & -3 & 2 \end{pmatrix} \begin{pmatrix} x_1 \\ x_2 \\ x_3 \end{pmatrix} = \begin{pmatrix} -2 \\ -1 \\ -18 \end{pmatrix}$

(b) $\begin{pmatrix} 1 & -2 & 3 \\ 2 & 1 & -5 \\ 6 & -3 & 2 \end{pmatrix} \begin{pmatrix} x_1 \\ x_2 \\ x_3 \end{pmatrix} = \begin{pmatrix} -2 \\ 17 \\ 22 \end{pmatrix}$

(c) $\begin{pmatrix} 1 & -2 & 3 & -1 \\ 3 & 1 & -3 & 2 \\ 5 & 3 & 2 & 3 \\ 2 & -4 & -2 & 4 \end{pmatrix} \begin{pmatrix} x_1 \\ x_2 \\ x_3 \\ x_4 \end{pmatrix} = \begin{pmatrix} -3 \\ 14 \\ 21 \\ -10 \end{pmatrix}.$

5 If $\mathbf{Ax} = \lambda\mathbf{x}$, determine the eigenvalues and corresponding eigenvectors in each of the following cases.

(a) $\mathbf{A} = \begin{pmatrix} 4 & 3 \\ 2 & 5 \end{pmatrix}$

(b) $\mathbf{A} = \begin{pmatrix} 2 & -5 \\ 1 & -4 \end{pmatrix}$

(c) $\mathbf{A} = \begin{pmatrix} -6 & 5 \\ 4 & 2 \end{pmatrix}$

(d) $\mathbf{A} = \begin{pmatrix} -5 & 9 \\ 1 & 3 \end{pmatrix}$

(e) $\mathbf{A} = \begin{pmatrix} 2 & 7 & 0 \\ 1 & 3 & 1 \\ 5 & 0 & 8 \end{pmatrix}$

(f) $\mathbf{A} = \begin{pmatrix} 5 & -6 & 1 \\ 1 & 1 & 0 \\ 3 & 0 & 1 \end{pmatrix}$

(g) $\mathbf{A} = \begin{pmatrix} -3 & 0 & 6 \\ 4 & 5 & 3 \\ 1 & 2 & 1 \end{pmatrix}$

(h) $\mathbf{A} = \begin{pmatrix} 4 & 10 & -8 \\ 1 & 2 & 1 \\ -1 & 2 & 3 \end{pmatrix}.$

6 Solve each of the following systems of first-order differential equations.

(a) $f_1'(x) = 2f_1(x) - 5f_2(x)$
$f_2'(x) = f_1(x) - 4f_2(x)$
where $f_1(0) = 1$ and $f_2(0) = 0$

(b) $f_1'(x) = -5f_1(x) + 9f_2(x)$
$f_2'(x) = f_1(x) + 3f_2(x)$
where $f_1(0) = 0$ and $f_2(0) = -2$

(c) $f_1'(x) = 5f_1(x) - 6f_2(x) + f_3(x)$
$f_2'(x) = f_1(x) + f_2(x)$
$f_3'(x) = 3f_1(x) + f_3(x)$
where $f_1(0) = 1$, $f_2(0) = 0$ and $f_3(0) = 2$

(d) $f_1'(x) = 4f_1(x) + 10f_2(x) - 8f_3(x)$
$f_2'(x) = f_1(x) + 2f_2(x) + f_3(x)$
$f_3'(x) = -f_1(x) + 2f_2(x) + 3f_3(x)$
where $f_1(0) = 4$, $f_2(0) = -2$ and $f_3(0) = -1$.

▶

7 If $\mathbf{A} = \begin{pmatrix} 1 & 3 & 0 \\ 3 & 10 & -3 \\ 0 & -3 & 9 \end{pmatrix}$, determine the three eigenvalues λ_1, λ_2, λ_3 of $\mathbf{A}$ and

verify that if $\mathbf{M} = \begin{pmatrix} -9 & 1 & 1 \\ 3 & 2 & 4 \\ 1 & 3 & -3 \end{pmatrix}$ then $\mathbf{M}^{-1}\mathbf{AM} = \mathbf{S}$, where $\mathbf{S}$ is a diagonal

matrix with elements λ_1, λ_2, λ_3.

8 Invert the matrix $\mathbf{A} = \begin{pmatrix} 8 & 10 & 7 \\ 5 & 9 & 4 \\ 9 & 11 & 8 \end{pmatrix}$ and hence solve the equations

$$8I_1 + 10I_2 + 7I_3 = 0$$
$$5I_1 + 9I_2 + 4I_3 = -9$$
$$9I_1 + 11I_2 + 8I_3 = 1.$$

9 If $\mathbf{A} = \begin{pmatrix} 1 & 2 & 3 \\ 4 & 6 & 7 \\ 5 & 8 & 9 \end{pmatrix}$ and $\mathbf{B} = \begin{pmatrix} -2 & 6 & -4 \\ -1 & -6 & 5 \\ 2 & 2 & -2 \end{pmatrix}$, verify that $\mathbf{AB} = k\mathbf{I}$ where

$\mathbf{I}$ is a unit matrix and k is a constant. Hence solve the equations
$$x_1 + 2x_2 + 3x_3 = 2$$
$$4x_1 + 6x_2 + 7x_3 = 2$$
$$5x_1 + 8x_2 + 9x_3 = 3.$$

10 Solve each of the following systems of second-order differential equations.
(a) $f_1''(x) = 4f_1(x) + 3f_2(x)$
$f_2''(x) = 2f_1(x) + 5f_2(x)$
where $f_1(0) = 0$, $f_2(0) = 1$, $f_1'(0) = 4$ and $f_2'(0) = 1$

(b) $f_1''(x) = -6f_1(x) + 5f_2(x)$
$f_2''(x) = 4f_1(x) + 2f_2(x)$
where $f_1(0) = 0$, $f_2(0) = 1$, $f_1'(0) = 1$ and $f_2'(0) = 0$

(c) $f_1''(x) = 2f_1(x) + 7f_2(x)$
$f_2''(x) = f_1(x) + 3f_2(x) + f_3(x)$
$f_3''(x) = 5f_1(x) + 8f_3(x)$
where $f_1(0) = 1$, $f_2(0) = 1$, $f_3(0) = 0$, $f_1'(0) = 0$, $f_2'(0) = 0$
and $f_3'(0) = 1$

(d) $f_1''(x) = -3f_1(x) + 6f_3(x)$
$f_2''(x) = 4f_1(x) + 5f_2(x) + 3f_3(x)$
$f_3''(x) = f_1(x) + 2f_2(x) + f_3(x)$
where $f_1(0) = 1$, $f_2(0) = 1$, $f_3(0) = 0$, $f_1'(0) = 0$, $f_2'(0) = 0$, $f_3'(0) = 1$.

Numerical solutions of ordinary differential equations

Frames
1 to 69

Learning outcomes

When you have completed this Program you will be able to:

- Derive a form of Taylor's series from Maclaurin's series and from it describe a function increment as a series of first and higher-order derivatives of the function

- Describe and apply by means of a spreadsheet the Euler method, the Euler–Cauchy method and the Runge–Kutta method for first-order differential equations

- Describe and apply by means of a spreadsheet the Euler second-order method and the Runge–Kutta method for second-order ordinary differential equations

- Describe and apply by means of a spreadsheet a simple predictor–corrector method.

Introduction

1

The range of differential equations that can be solved by straightforward analytical methods is relatively restricted. Even solution in series may not always be satisfactory, either because of the slow convergence of the resulting series or because of the involved manipulation in repeated stages of differentiation.

In such cases, where a differential equation and known boundary conditions are given, an approximate solution is often obtainable by the application of numerical methods, where a numerical solution is obtained at discrete values of the independent variable.

The solution of differential equations by numerical methods is a wide subject. The present Program introduces some of the simpler methods, which nevertheless are of practical use.

Taylor's series

Let us start off by briefly revising the fundamentals of Maclaurin's and Taylor's series.

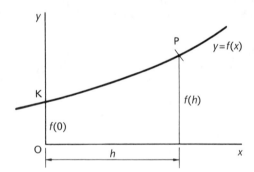

Maclaurin's series for $f(x)$ is

$$f(x) = f(0) + xf'(0) + \frac{x^2}{2!}f''(0) + \ldots + \frac{x^n}{n!}f^n(0) + \ldots \tag{1}$$

and expresses the function $f(x)$ in terms of its successive derivatives at $x = 0$, i.e. at the point K.

Therefore, at P, $f(h) = \ldots\ldots\ldots\ldots$

$$f(h) = f(0) + hf'(0) + \frac{h^2}{2!}f''(0) + \dots + \frac{h^n}{n!}f^n(0) + \dots \qquad (2)$$

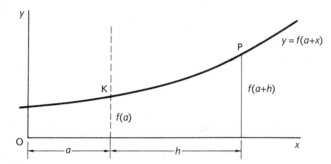

If the y-axis and origin are moved a units to the left, the equation of the same curve relative to the new axes becomes $y = f(a + x)$ and the function value at K is $f(a)$.

At P, $\quad f(a + h) = f(a) + hf'(a) + \dfrac{h^2}{2!}f''(a) + \dots + \dfrac{h^n}{n!}f^n(a) + \dots$

This is one common form of Taylor's series.

Make a note of it and then move on

Function increment

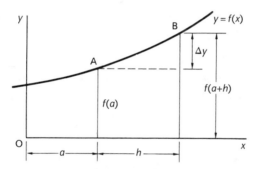

If we know the function value $f(a)$ at A, i.e. at $x = a$, we can apply Taylor's series to determine the function value at a neighbouring point B, i.e. at $x = a + h$.

$$f(a + h) = f(a) + hf'(a) + \frac{h^2}{2!}f''(a) + \frac{h^3}{3!}f'''(a) + \dots \qquad (3)$$

The *function increment* from A to B $= \Delta y = f(a + h) - f(a)$

i.e. $\quad f(a + h) = f(a) + \Delta y$

where $\Delta y = hf'(a) + \dfrac{h^2}{2!}f''(a) + \dfrac{h^3}{3!}f'''(a) + \dots$

This entails evaluation of an infinite number of derivatives at $x = a$: in practice an approximation is accepted by restricting the number of terms that are used in the series.

This approximation of Taylor's series forms the basis of several numerical methods, some of which we shall now introduce. It should be noted that these early examples have been selected because exact solutions can also be found. The purpose of this is to enable a comparison between the results obtained by a particular method with those obtained from an exact solution, and so to demonstrate the accuracy of the method.

On then to the next frame

First-order differential equations

4

Numerical solution of $\dfrac{dy}{dx} = f(x, y)$ with the initial condition that, at $x = x_0$, $y = y_0$.

Euler's method

The simplest of the numerical methods for solving first-order differential equations is *Euler's method*, in which the Taylor's series

$$f(a + h) = f(a) + hf'(a) \;\left| \; + \frac{h^2}{2!}f''(a) + \frac{h^3}{3!}f'''(a) + \ldots \right.$$

is truncated after the second term to give

$$f(a + h) \approx f(a) + hf'(a) \tag{4}$$

This is a severe approximation, but in practice the 'approximately equals' sign is replaced by the normal 'equals' sign, in the knowledge that the result we obtain will necessarily differ to some extent from the function value we seek. With this in mind, we write

$$f(a + h) = f(a) + hf'(a)$$

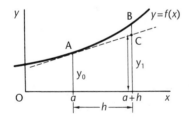

If h is the interval between two near ordinates and if we denote $f(a)$ by y_0, then the relationship

$$f(a + h) = f(a) + hf'(a)$$

becomes

$$y_1 = y_0 + h(y')_0 \tag{5}$$

Hence, knowing y_0, h and $(y')_0$, we can compute y_1, an approximate value for the function value at B.

Make a note of result (5): we shall be using it quite a lot.

Then move on for an example

Example 1

Given that $\dfrac{dy}{dx} = 2(1+x) - y$ with the initial condition that at $x = 2$, $y = 5$, we can find an approximate value of y at $x = 2.2$, as follows.

We have $y' = 2(1+x) - y$ with $x_0 = 2,\ y_0 = 5$

$$\therefore\ (y')_0 = \ldots\ldots\ldots\ldots$$

$$\boxed{(y')_0 = 1}$$

We obtain this by substituting x_0 and y_0 in the given equation:

$$(y')_0 = 2(1 + x_0) - y_0 = 2(1 + 2) - 5 \quad \therefore\ (y')_0 = 1$$

So we have $x_0 = 2$; $y_0 = 5$; $(y')_0 = 1$; $x_1 = 2.2$; $h = 0.2$.

By Euler's relationship:

$$y_1 = y_0 + h(y')_0 \quad \therefore\ y_1 = \ldots\ldots\ldots\ldots$$

$$\boxed{y_1 = 5.2}$$

Because

$$y_1 = y_0 + h(y')_0 = 5 + (0.2)1 = 5.2$$

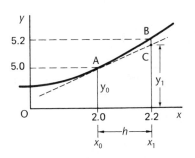

At B, $x_1 = 2.2$; $y_1 = 5.2$; and

$$(y')_1 = \ldots\ldots\ldots\ldots$$

8

$$(y')_1 = 1.2$$

$$(y')_1 = 2(1 + x_1) - y_1 = 2(1 + 2.2) - 5.2 = 1.2$$

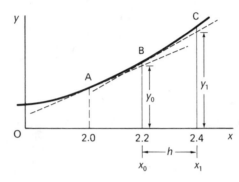

If we take the values of x, y and y' that we have just found for the point B and treat these as new starter values x_0, y_0, $(y')_0$, we can repeat the process and find values corresponding to the point C.

At B, $x_0 = 2.2$; $y_0 = 5.2$; $(y')_0 = 1.2$; $x_1 = 2.4$.

Then at C: $y_1 = \dots\dots\dots$; $(y')_1 = \dots\dots\dots$

9

$$y_1 = 5.44; \quad (y')_1 = 1.36$$

$$y_1 = y_0 + h(y')_0 = 5.2 + (0.2)1.2 = 5.44$$
$$(y')_1 = 2(1 + x_1) - y_1 = 2(1 + 2.4) - 5.44 = 1.36$$

So we could continue in a step-by-step method. At each stage, the determined values of x_1, y_1 and $(y')_1$ become the new starter values x_0, y_0 and $(y')_0$ for the next stage.

Our results so far can be tabulated thus

x_0	y_0	$(y')_0$	x_1	y_1	$(y')_1$
2.0	5.0	1.0	2.2	5.2	1.2
2.2	5.2	1.2	2.4	5.44	1.36
2.4	5.44	1.36			

Continue the table with a constant interval of $h = 0.2$. The third row can be completed to give

$$x_1 = \dots\dots\dots; \quad y_1 = \dots\dots\dots; \quad (y')_1 = \dots\dots\dots$$

$$x_1 = 2.6; \quad y_1 = 5.712; \quad (y')_1 = 1.488$$

Because

$$x_1 = x_0 + h = 2.4 + 0.2 = 2.6$$
$$y_1 = y_0 + h(y')_0 = 5.44 + (0.2)1.36 = 5.712$$
$$(y')_1 = 2(1 + x_1) - y_1 = 2(1 + 2.6) - 5.712 = 1.488$$

Now you can continue in the same way and complete the table for

$$x = 2.0, \ 2.2, \ 2.4, \ 2.6, \ 2.8, \ 3.0$$

Finish it off and compare results with the next frame

Here is the result.

x_0	y_0	$(y')_0$	x_1	y_1	$(y')_1$
2.0	5.0	1.0	2.2	5.2	1.2
2.2	5.2	1.2	2.4	5.44	1.36
2.4	5.44	1.36	2.6	5.712	1.488
2.6	5.712	1.488	2.8	6.009 6	1.590 4
2.8	6.009 6	1.590 4	3.0	6.327 68	1.672 32
3.0	6.327 68	1.672 32			

In practice, we do not, in fact, enter the values in the right-hand half of the table, but write them in directly as new starter values in the left-hand section of the table.

x_0	y_0	$(y')_0$
2.0	5.0	1.0
2.2	5.2	1.2
2.4	5.44	1.36
2.6	5.712	1.488
2.8	6.009 6	1.590 4
3.0	6.327 68	1.672 32

The particular solution is given by the values of y against x and a graph of the function can be drawn.

Draw the graph of the function carefully on graph paper.

12

Graph of the solution of $\dfrac{dy}{dx} = 2(1+x) - y$ with $y = 5$ at $x = 2$.

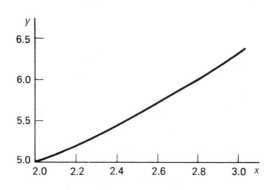

13

It is an advantage to plot the points step-by-step as the results are built up. In that way, one can check that there is a smooth progression and that no apparent errors in the calculations occur at any one stage.

The differential equation $\dfrac{dy}{dx} = 2(1+x) - y$ can be solved by the integration factor method (see Program 1) to give the solution

$$y = 2x + e^{2-x}$$

and in the following table we compare our results with the actual values to determine the errors.

x	y (Euler)	y (actual)	Absolute error
2.0	5.0	5.0	0
2.2	5.2	5.218731	0.018731
2.4	5.44	5.470320	0.030320
2.6	5.712	5.748812	0.036812
2.8	6.0096	6.049329	0.039729
3.0	6.32768	6.367879	0.040199

The errors involved in the process are shown. These errors are due mainly to
....

> the fact that Taylor's series was truncated after the second term

By now you will appreciate the amount of arithmetic manipulation involved in solving these differential equations – a large amount of which is repetitive. To avoid the tedium and to make the computations more efficient we shall resort to the use of a spreadsheet. If you have even a limited knowledge of spreadsheets, then you will be able to follow the text from here. The spreadsheet we shall be using here is Microsoft Excel, though all commercial spreadsheets possess the equivalent functionality. Alternatively, an iteration process can be used in any computer algebra package such as *Derive*, *Maple* or *Mathematica*.

Open your spreadsheet and in cell A1 enter the letter *n* and press **Enter**. In this first column we are going to enter the iteration numbers. In cell A2 enter the number 0 and press **Enter**. Place the cell highlight in cell A2 and highlight the block of cells A2 to A12 by holding down the mouse button and wiping the highlight down to cell A12. Click the **Edit** command on the Command bar and point at **Fill** from the drop-down menu. Select **Series** from the next drop-down menu and accept the default **Step value** of 1 by clicking **OK** in the Series window.

The cells A3 to A12 fill with . . . .

> The numbers 1 to 10

In cell B1 enter the letter *x* – this column is going to contain the successive *x*-values for which the *y*-value is going to be enumerated. In cell B2 enter the number 2 – the initial *x*-value. We now could fill the column in much the same way as we filled the first column, but we have a better way.

Place the cell highlight in cell F1 and enter the number 0.2 – this is the value of *h*, the increment in *x*. Now place the cell highlight in cell B3 and enter the formula

$= B2 + \$F\$1$ followed by **Enter**   (uppercase or lowercase, it does not matter)

The number 2.2 appears in cell B3. Place the cell highlight in cell B3, click the **Edit** command and select **Copy** from the drop-down menu. You have now copied the contents of cell B3 to the clipboard. Now place the cell highlight in cell B4 and highlight the block of cells from B4 to B12. Click the **Edit** command again but this time select **Paste** from the drop-down menu.

The cells B4 to B12 fill with the numbers . . . .

16

The numbers 2.4 to 4.0 in intervals of 0.2

How has this happened? When you typed in the cell reference B2 into the formula in cell B3, the spreadsheet understood this to mean *the contents of the cell immediately above current cell B3*. When the formula is copied into cell B4 it means *the contents of the cell immediately above current cell B4*. Entered in this way the address B2 is a *relative address*. On the other hand, when you typed in F1 the spreadsheet understood this to mean the contents of cell F1 and that meaning remains when it is copied – the dollar signs indicate an *absolute address*. So as you move down the column the contents of a cell contain the contents of the cell immediately above it plus the contents of cell F1. You will shortly see the advantages of all this.

For now, place the cell highlight in cell C1 and enter the letter y – this column is going to contain the computed y-values against the corresponding x-values in column B. Place the cell highlight in cell C2 and enter the number 5 – the initial y-value. Before we can compute the y-values in column C we need to be able to tabulate the values of y' – the derivatives of y. Place the cell highlight in cell D1 and enter y' – this column will contain the values of the derivatives of y against the corresponding x-values. Cell D2 will contain the initial value of y' which can be computed from the equation

$$y' = 2(1 + x) - y$$

When $x = x_0 = 2$ and $y = y_0 = 5$ then

$$y'_0 = 2(1 + x_0) - y_0 = 2(1 + 2) - 5 = 1$$

so place the cell highlight in cell D2 and enter the formula

= 2 (1 + B2) – C2   (B2 contains x_0 and C2 contains y_0)

The number 1 appears in cell D2. We need to copy this formula down the y' column. Place the cell highlight in cell D2, click **Edit** and select **Copy**. Now place the cell highlight in cell D3 and highlight the block of cells D3 to D12. Click the **Edit** command again and select **Paste**.

The cells D3 to D12 fill with . . . .

> The numbers 6.4 to 10.0 in intervals of 0.4

Because the cells in the C2 column are currently empty, these values are just $2(1 + B2) - 0$.

Now, to compute the y-values we use the equation $y_1 = y_0 + h(y')_0$. Place the cell highlight in cell C3 and enter the formula

$= C2 + \$F\$1 \ D2$ (C2 contains y_0, F1 contains h and D2 contains $(y')_0$)

and the number 5.2 appears. That is, $y_1 = 5 + (0.2)(1) = 5.2$. This now completes the sequence of operations required to find y_1. To find the values of $y_2 = y(x_2) = y(2.4)$ this sequence is repeated and, to ensure this, all that remains is to copy the formula in cell C3 into cells C4 to C12. So do this to reveal the following display

n	x	y	y'	0.2
0	2	5	1	
1	2.2	5.2	1.2	
2	2.4	5.44	1.36	
3	2.6	5.712	1.488	
4	2.8	6.0096	1.5904	
5	3	6.32768	1.67232	
6	3.2	6.662144	1.737856	
7	3.4	7.0097152	1.7902848	
8	3.6	7.36777216	1.83222784	
9	3.8	7.734217728	1.865782272	
10	4	8.107374182	1.892625818	

Now that was a lot easier than all that arithmetic manipulation by hand, wasn't it? We can tidy this display up by using the **Format** command and by using the various options on the tool bars to change the column widths and to display the numbers in a regular format of 10 decimal places to produce a display that is easier to read.

Next frame

18

n	x	y	y′	*h*=0.2
0	2.0	5.0000000000	1.0000000000	
1	2.2	5.2000000000	1.2000000000	
2	2.4	5.4400000000	1.3600000000	
3	2.6	5.7120000000	1.4880000000	
4	2.8	6.0096000000	1.5904000000	
5	3.0	6.3276800000	1.6723200000	
6	3.2	6.6621440000	1.7378560000	
7	3.4	7.0097152000	1.7902848000	
8	3.6	7.3677721600	1.8322278400	
9	3.8	7.7342177280	1.8657822720	
10	4.0	8.1073741824	1.8926258176	

Notice that we have added **h=** in cell E1 and justified it to the right and then justified the number 0.2 in F2 to the left so that together they read as an equation. The advantage of isolating the step value 0.2 in cell F1, as we have done, is that we can change the value and immediately see the effects on the calculations. For example, if the contents of F1 are changed to 0.1 the display changes automatically to

n	x	y	y′	*h*=0.1
0	2.0	5.0000000000	1.0000000000	
1	2.1	5.1000000000	1.1000000000	
2	2.2	5.2100000000	1.1900000000	
3	2.3	5.3290000000	1.2710000000	
4	2.4	5.4561000000	1.3439000000	
5	2.5	5.5904900000	1.4095100000	
6	2.6	5.7314410000	1.4685590000	
7	2.7	5.8782969000	1.5217031000	
8	2.8	6.0304672100	1.5695327900	
9	2.9	6.1874204890	1.6125795110	
10	3.0	6.3486784401	1.6513215599	

Notice that the different values of h produce different corresponding values in the tables. For example, for $h = 0.2$ we find that $y(3.0) = 6.3276800000$ whereas for $h = 0.1$ we have $y(3.0) = 6.3486784401$. The smaller the value of h then, the smaller the errors in the calculation – we shall see this demonstrated explicitly in the next frame.

Go to the next frame

The exact value and the errors

The differential equation

$$y' = 2(1 + x) - y$$

can be solved using the integration factor method (see Program 1) to give the solution

$$y = 2x + e^{2-x}$$

We can program this into the spreadsheet to compare the exact solution with the solution obtained numerically and compute the actual errors. Place the cell highlight in cell E1 and highlight cells E1 and F1. Click **Insert** on the Command bar and select **Columns**. Immediately two new columns appear. Notice that the numbers in the display do not change despite the fact that the *h*-value of 0.2 has moved from F1 to H1 – all the formulas in the spreadsheet will have automatically adjusted themselves. You can check this by high-lighting a cell with a formula in it to see the change.

In cell E1 enter the word **Exact** and in cell F1 enter **Errors (%)**. In cell E2 enter the right-hand side of the equation $y = 2x + e^{2-x}$ by using the formula

= 2 B2 + EXP(2 – B2)   (the EXP stands for the exponential function)

and copy this into the block of cells E3 to E12. In cell F2 enter the formula for the error

= (E2 – C2) 100/E2   (the error as a percentage of the exact value)

and copy this into the block of cells F3 to F12 to produce the following display

n	x	y	y'	Exact	Errors (%)	h=0.2
0	2.0	5.0000000000	1.0000000000	5.0000000000	0.00	
1	2.2	5.2000000000	1.2000000000	5.2187307531	0.36	
2	2.4	5.4400000000	1.3600000000	5.4703200460	0.55	
3	2.6	5.7120000000	1.4880000000	5.7488116361	0.64	
4	2.8	6.0096000000	1.5904000000	6.0493289641	0.66	
5	3.0	6.3276800000	1.6723200000	6.3678794412	0.63	
6	3.2	6.6621440000	1.7378560000	6.7011942119	0.58	
7	3.4	7.0097152000	1.7902848000	7.0465969639	0.52	
8	3.6	7.3677721600	1.8322278400	7.4018965180	0.46	
9	3.8	7.7342177280	1.8657822720	7.7652988882	0.40	
10	4.0	8.1073741824	1.8926258176	8.1353352832	0.34	

▶

Change the value of h to 0.1 and produce the following display

n	x	y	y′	Exact	Errors (%)	h=0.1
0	2.0	5.0000000000	1.0000000000	5.0000000000	0.00	
1	2.1	5.1000000000	1.1000000000	5.1048374180	0.09	
2	2.2	5.2100000000	1.1900000000	5.2187307531	0.17	
3	2.3	5.3290000000	1.2710000000	5.3408182207	0.22	
4	2.4	5.4561000000	1.3439000000	5.4703200460	0.26	
5	2.5	5.5904900000	1.4095100000	5.6065306597	0.29	
6	2.6	5.7314410000	1.4685590000	5.7488116361	0.30	
7	2.7	5.8782969000	1.5217031000	5.8965853038	0.31	
8	2.8	6.0304672100	1.5695327900	6.0493289641	0.31	
9	2.9	6.1874204890	1.6125795110	6.2065696597	0.31	
10	3.0	6.3486784401	1.6513215599	6.3678794412	0.30	

When $h = 0.2$ the error in $y(3.0)$ is 0.63% whereas when $h = 0.1$ the error in $y(3.0)$ is 0.30%.

The smaller the value of h the ....

20

smaller the error

Having completed your first spreadsheet you can now use it as a template for similar problems.

To avoid losing the work that you have already done, save your spreadsheet under some suitable name. When that is complete, highlight all the cells from A1 to G12 and copy them onto the clipboard using the **Edit-Copy** sequence of commands. Now click the **Sheet 2** tab at the bottom of your spreadsheet to reveal a blank worksheet. Place the cell highlight in cell A1, click **Edit** and select **Paste**. The entire contents of **Sheet 1** are now copied to **Sheet 2** in readiness for editing to accommodate a new problem.

So let's look at another example.

Example 2

Obtain a numerical solution of the equation

$$\frac{dy}{dx} = 1 + x - y$$

with the initial condition that $y = 2$ at $x = 1$, for the range $x = 1.0(0.2)3.0$, that is from $x = 1.0$ to $x = 3.0$ with step length $x = 0.2$.

As initial conditions, we have

$$x_0 = \text{............ and } y_0 = \text{............}$$

21

$$x_0 = 1, \quad y_0 = 2$$

Because

$x_0 = 1$ and $y_0 = 2$ are given initial conditions.

These values can now be inserted into the
spreadsheet in cells ....

22

$$x_1 = 1 \text{ in B2}, \quad y_0 = 2 \text{ in C2}$$

Notice how the numbers in column B have changed to accommodate the new
sequence of x-values. The contents of the cells in column C do not need to be
changed as they refer to the equation

$$y_1 = y_0 + h(y')_0$$

which is the same in this spreadsheet as it was in the previous spreadsheet.
The contents of column D do have to be changed because they currently refer
to the equation to be solved in the previous problem. The equation to be
solved here is

$$y' = 1 + x - y$$

so in cell D3 the contents need to be changed to ....

23

$$= 1 + \text{B2} - \text{C2}$$

This formula must then be copied into cells C3 to C12. Finally, the **Exact**
column needs to be amended to reflect the exact solution to this equation,
which is again found by using the integration factor method as

$$y = x + e^{1-x}$$

So, in E2, enter the formula ....

24

$$= \text{B2} + \text{EXP}(1 - \text{B2})$$

This formula needs to be copied into cells E3 to E12. This completes the editing of the spreadsheet to reflect the new problem to give the display

n	x	y	y′	Exact	Errors (%)	h=0.2
0	1.0	2.0000000000	0.0000000000	2.0000000000	0.00	
1	1.2	2.0000000000	0.2000000000	2.0187307531	0.93	
2	1.4	2.0400000000	0.3600000000	2.0703200460	1.46	
3	1.6	2.1120000000	0.4880000000	2.1488116361	1.71	
4	1.8	2.2096000000	0.5904000000	2.2493289641	1.77	
5	2.0	2.3276800000	0.6723200000	2.3678794412	1.70	
6	2.2	2.4621440000	0.7378560000	2.5011942119	1.56	
7	2.4	2.6097152000	0.7902848000	2.6465969639	1.39	
8	2.6	2.7677721600	0.8322278400	2.8018965180	1.22	
9	2.8	2.9342177280	0.8657822720	2.9652988882	1.05	
10	3.0	3.1073741824	0.8926258176	3.1353352832	0.89	

A plot of the graph of y against x for both the computed value and the exact value looks as follows

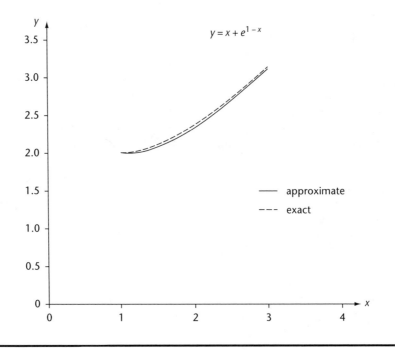

Graphical interpretation of Euler's method

25

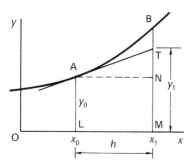

If AT is the tangent to the curve at A,

$$\text{then } \frac{\text{NT}}{\text{AN}} = \left[\frac{dy}{dx}\right]_{x=x_0} = (y')_0$$

$$\frac{\text{NT}}{h} = (y')_0 \qquad \therefore \text{ NT} = h(y')_0$$

$$\therefore \text{ At } x = x_1, \text{ MT} = y_0 + h(y')_0$$

By Euler's relationship, $y_1 = y_0 + h(y')_0$ i.e. MT.

The difference between the calculated value of y, i.e. MT, and the actual value of the function y, i.e. MB, at $x = x_1$, is indicated by TB. This error can be considerable, depending on the curvature of the graph and the size of the interval h. It is inherent to the method and corresponds to the truncation of the Taylor's series after the second term.

Euler's method, then

(a) is simple in procedure

(b) is lacking in accuracy, especially away from the starter values of the initial conditions

(c) is of use only for very small values of the interval h.

In spite of its practical limitations, it is the foundation of several more sophisticated methods and hence it is worthy of note.

Here is one more example to work on your own.

Example 3

Obtain the solution of $\dfrac{dy}{dx} = x + y$ with the initial condition that $y = 1$ at $x = 0$, for the range $x = 0(0.1)0.5$.

By using a previously constructed spreadsheet as a template, the solution is

. . . .

The function values are given in the next frame

26

n	x	y	y'	Exact	Errors (%)	h = 0.1
0	0.0	1.0000000000	1.0000000000	1.0000000000	0.00	
1	0.1	1.1000000000	1.2000000000	1.1103418362	0.93	
2	0.2	1.2200000000	1.4200000000	1.2428055163	1.84	
3	0.3	1.3620000000	1.6620000000	1.3997176152	2.69	
4	0.4	1.5282000000	1.9282000000	1.5836493953	3.50	
5	0.5	1.7210200000	2.2210200000	1.7974425414	4.25	
6	0.6	1.9431220000	2.5431220000	2.0442376008	4.95	
7	0.7	2.1974342000	2.8974342000	2.3275054149	5.59	
8	0.8	2.4871776200	3.2871776200	2.6510818570	6.18	
9	0.9	2.8158953820	3.7158953820	3.0192062223	6.73	
10	1.0	3.1874849202	4.1874849202	3.4365636569	7.25	

Because

The initial conditions are entered as

0 in cell B2   (the initial *x*-value)
1 in cell C2   (the initial *y*-value)
0.1 in cell H1   (the *x* step length)

The formulas are entered as

= B2 + C2 in cell D2, copied into cells D3 to D12
            (the successive *y'*-values)
= C2 + \$H\$1 D2 in cell C3 copied into cells C4 to C12
            (the successive *y*-values)

The exact solution found by using the integration factor method is $y = 2e^x - x - 1$ and so

= 2 EXP(B2) – B2 – 1 is entered into cell E2 and copied into cells E3 to E12

Notice how the errrors here are significant, which is very evident from the graphs of the computed values and the exact values.

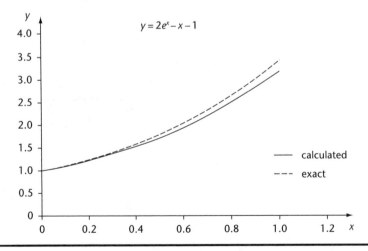

The Euler–Cauchy method – or the improved Euler method

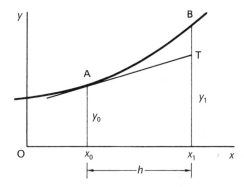

In Euler's method, we use the slope $(y')_0$ at A (x_0, y_0) across the whole interval h to obtain an approximate value of y_1 at B. TB is the resulting error in the result.

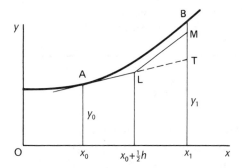

In the Euler–Cauchy method, we use the slope at A (x_0, y_0) across half the interval and then continue with a line whose slope approximates to the slope of the curve at x_1.

Let $\overline{\overline{y}}_1$ be the y-value of the point at T.

The error (MB) in the result is now considerably less than the error (TB) associated with the basic Euler method and the calculated results will accordingly be of greater accuracy.

28 Euler–Cauchy calculations

The steps in the Euler–Cauchy method are as follows.

1 We start with the given equation $y' = f(x, y)$ with the initial condition that at $x = x_0$, $y = y_0$. We have to determine function values for $x = x_0(h)x_n$.

2 From the equation and the initial condition we obtain $(y')_0 = f(x_0, y_0)$.

3 Knowing x_0, y_0, $(y')_0$ and h, we then evaluate

(a) $x_1 = x_0 + h$

(b) the auxiliary value of y, denoted by $\bar{\bar{y}}$ where
$\bar{\bar{y}}_1 = y_0 + h(y')_0$. This is the same step as in Euler's method.

(c) Then $y_1 = y_0 + \frac{1}{2}h\{(y')_0 + f(x_1, \bar{\bar{y}}_1)\}$

Note that $f(x_1, y_1)$ is the right-hand side of the given equation with x and y replaced by the calculated values of x_1 and $\bar{\bar{y}}_1$.

(d) Finally $(y')_1 = f(x_1, y_1)$.

We have thus evaluated x_1, y_1 and $(y')_1$.

The whole process is then repeated, the calculated values of x_1, y_1 and $(y')_1$ becoming the starter values x_0, y_0, $(y')_0$ for the next stage.

Make a note of the relationships above. We shall be using them quite often.

Then on to the next frame for an example of their use

29 Example 1

Apply the Euler–Cauchy method to solve the equation
$$y' = x + y$$
with the initial condition that at $x = 0$, $y = 1$, for the range $x = 0(0.1)1.0$.

We proceed as before by copying our template solution to a new worksheet. Before we continue we need to decide what the entries are going to be in our spreadsheet.

1 We are going to have to enter new initial conditions, so

Enter 0 in cell B2   that is $x_0 = 0$
Enter 1 in cell C2   that is $y_0 = 1$
Enter 0.1 in cell H1   this is the x step length

2 The equation to be solved is $y' = x + y$, so enter the formula

$= \text{B2} + \text{C2}$ in cell D2 and copy the contents of D2 into cells D3 to D12

3 The Euler–Cauchy method tells us that
$$y_1 = y_0 + \frac{1}{2}h\{(y')_0 + f(x, \bar{\bar{y}}_1)\}$$
where $\bar{\bar{y}}_1 = y_0 + h(y')_0$ so that
$$f(x_1, \bar{\bar{y}}_1) = x_1 + \bar{\bar{y}}_1 = x_1 + y_0 + h(y')_0$$

Therefore $y_1 = \ldots\ldots\ldots$

$$y_1 = y_0 + \frac{1}{2}h\{x_1 + y_0 + (1+h)(y')_0\}$$

Because

By replacing $f(x_1, \bar{\bar{y}}_1)$ with $x_1 + y_0 + h(y')_0$ in the expression

$$y_1 = y_0 + \tfrac{1}{2}h\{(y')_0 + f(x_1, \bar{\bar{y}}_1)\}$$

we find that

$$y_1 = y_0 + \tfrac{1}{2}h\{(y')_0 + x_1 + y_0 + h(y')_0\}$$
$$= y_0 + \tfrac{1}{2}h\{x_1 + y_0 + (1+h)(y')_0\}$$

In cell C3 enter the formula ....

31

$$= C2 + (0.5)\,\$H\$1\,(B3 + C2 + (1 + \$H\$1)\,D2)$$

Because

y_0 is in cell C2, h is in cell H1, x_1 is in cell B3 and $(y')_0$ is in cell D2.

Copy the contents of cell C3 into cells C4 to C12.

4 Finally, for comparison purposes, the exact solution of this equation is $y = 2e^x - x - 1$ and this is

entered into E2 by the formula ....
and copied into cells ....

$$= 2\,EXP(B2) - B2 - 1 \text{ and copied into cells E3 to E12}$$

The resulting display looks as follows

n	x	y	y'	Exact	Errors (%)	$h=0.1$
0	0.0	1.0000000000	1.0000000000	1.0000000000	0.00	
1	0.1	1.1100000000	1.2100000000	1.1103418362	0.03	
2	0.2	1.2420500000	1.4420500000	1.2428055163	0.06	
3	0.3	1.3984652500	1.6984652500	1.3997176152	0.09	
4	0.4	1.5818041013	1.9818041013	1.5836493953	0.12	
5	0.5	1.7948935319	2.2948935319	1.7974425414	0.14	
6	0.6	2.0408573527	2.6408573527	2.0442376008	0.17	
7	0.7	2.3231473748	3.0231473748	2.3275054149	0.19	
8	0.8	2.6455778491	3.4455778491	2.6510818570	0.21	
9	0.9	3.0123635233	3.9123635233	3.0192062223	0.23	
10	1.0	3.4281616932	4.4281616932	3.4365636569	0.24	

▶

Comparing these results with the same equation being solved by the Euler method demonstrates how much more accurate the Euler–Cauchy method is, as can be seen from the following table of comparative errors

x	Euler	Euler–Cauchy
0.0	0.00	0.00
0.1	0.93	0.03
0.2	1.84	0.06
0.3	2.69	0.09
0.4	3.50	0.12
0.5	4.25	0.14
0.6	4.95	0.17
0.7	5.59	0.19
0.8	6.18	0.21
0.9	6.73	0.23
1.0	7.25	0.24

Next frame

33

Now for another example, but before that, complete the following without reference to your notes – if possible. In the Euler–Cauchy method the relevant relationships are

$$x_1 = \ldots\ldots\ldots$$
$$\bar{\bar{y}}_1 = \ldots\ldots\ldots$$
$$y_1 = \ldots\ldots\ldots$$
$$(y')_1 = \ldots\ldots\ldots$$

Next frame

34

$$
\begin{aligned}
x_1 &= x_0 + h \\
\bar{\bar{y}}_1 &= y_0 + h(y')_0 \\
y_1 &= y_0 + \frac{1}{2}h\{(y')_0 + f(x_1, \bar{\bar{y}}_1)\} \\
(y')_1 &= f(x_1, y_1)
\end{aligned}
$$

Example 2

Determine a numerical solution of the equation $y' = 2(1+x) - y$ with the initial condition that $y = 5$ when $x = 2$, for the range 2.0(0.2)4.0. Try this one yourself.

The exact solution is given as $y = 2x + e^{-2x}$
and the final display of results is $\ldots\ldots\ldots$

n	x	y	y′	Exact	Errors (%) $h = 0.2$
0	2.0	5.0000000000	1.0000000000	5.0000000000	0.00
1	2.2	5.2200000000	1.1800000000	5.2187307531	−0.02
2	2.4	5.4724000000	1.3276000000	5.4703200460	−0.04
3	2.6	5.7513680000	1.4486320000	5.7488116361	−0.04
4	2.8	6.0521217600	1.5478782400	6.0493289641	−0.05
5	3.0	6.3707398432	1.6292601568	6.3678794412	−0.04
6	3.2	6.7040066714	1.6959933286	6.7011942119	−0.04
7	3.4	7.0492854706	1.7507145294	7.0465969639	−0.04
8	3.6	7.4044140859	1.7955859141	7.4018965180	−0.03
9	3.8	7.7676195504	1.8323804496	7.7652988882	−0.03
10	4.0	8.1374480313	1.8625519687	8.1353352832	−0.03

Because

1 The initial conditions are entered as

Enter 2 in cell B2 (that is $x_0 = 2$); enter 5 in cell C2 (that is $y_0 = 5$)
Enter 0.2 in cell H1 (this is the x step length)

2 The equation to be solved is $y' = 2(1 + x) - y$, so enter the formula

$= 2 (1 + B2) - C2$ in cell D2 and copy the contents of D2 into cells D3 to D12

3 The Euler–Cauchy method tells us that

$$y_1 = y_0 + \frac{1}{2}h\{(y')_0 + f(x_1, \bar{\bar{y}}_1)\} \quad \text{where } \bar{\bar{y}}_1 = y_0 + h(y')_0 \text{ so that}$$

$$f(x_1, \bar{\bar{y}}_1) = 2(1 + x_1) - \bar{\bar{y}}_1 = 2(1 + x_1) - y_0 - h(y')_0 \text{ therefore}$$

$$y_1 = y_0 + \frac{1}{2}h\{(y')_0 + 2(1 + x_1) - y_0 - h(y')_0\} \text{ that is}$$

$$y_1 = y_0 + \frac{1}{2}h\{2(1 + x_1) - y_0 + (1 - h)(y')_0\}$$

This is accommodated by the formula in C3 (copied into cells C4 to C12)

$= C2 + (0.5) \$H\$1 (2 (1 + B3) - C2 + (1 - \$H\$1) D2)$

4 Finally the exact solution $y = 2x + e^{-2x}$ is entered into cell E2 as $= 2 B2 + EXP(-2 B2)$ and copied into cells E3 to E12.

Refer to Frame 19 for a comparison of errors between this method and the Euler method. Then another example for you to try just to make sure you are clear about the processes involved.

Next frame

36

Example 3

Solve the equation $y' = y^2 + xy$ with initial condition that at $x = 1$, $y = 1$, for the range $x = 1.0(0.1)1.7$. Use the Euler–Cauchy method and work to 6 places of decimals.

The solution is ....

37

n	x	y	y'	$h = 0.1$
0	1.0	1.000000	2.000000	
1	1.1	1.238000	2.894444	
2	1.2	1.591023	4.440583	
3	1.3	2.152410	7.431004	
4	1.4	3.145846	14.300528	
5	1.5	5.251007	35.449581	
6	1.6	11.595613	153.011211	
7	1.7	57.704110	3427.861242	

Because

1 The initial conditions are entered as

Enter 1 in cell B2 (that is $x_0 = 1$); enter 1 in cell C2 (that is $y_0 = 1$)
Enter 0.1 in cell H1 (this is the x step length)

2 The equation to be solved is $y' = y^2 + xy$, so

Enter the formula $= C2^2 + B2\,C2$ in cell D2 and copy the contents of D2 into cells D3 to D9. Note that C2^2 = C2 C2 – the 'hat' indicates raising to a power.

3 The Euler-Cauchey method tell us that

$$y_1 = y_0 + \frac{1}{2}h\{(y')_0 + f(x_1, \bar{\bar{y}}_1)\} \quad \text{where } \bar{\bar{y}}_1 = y_0 + h(y')_0 \text{ so that}$$

$$f(x_1, \bar{\bar{y}}_1) = \bar{\bar{y}}_1^2 + x_1\bar{\bar{y}}_1 = (y_0 + h(y')_0)^2 + x_1(y_0 + h(y')_0) \text{ therefore}$$

$$y_1 = y_0 + \frac{1}{2}h\{(y')_0 + (y_0 + h(y')_0)^2 + x_1(y_0 + h(y')_0)\}$$

This is accommodated by the formula in C3 (copied into cells C4 to C9)

$$= C2 + (0.5)\,\$F\$1\,(D2 + (C2 + \$F\$1\,D2))^2 + B3\,(C2 + \$F\$1\,D2))$$

The table shows that as x increases, the computed values of y and its derivative increase dramatically. This is an indication that the exact solution increases without bound near to the larger values of x considered, so bringing the accuracy of these computed values into question. This emphasises the importance of checking every method against a known solution so as to form some idea of the method's accuracy. However, all numerical methods produce significant accuracies whenever the exact solution diverges in this way.

Runge–Kutta method

38

The Runge–Kutta method for solving first-order differential equations is widely used and affords a high degree of accuracy. It is a further step-by-step process where a table of function values for a range of values of x is accumulated. Several intermediate calculations are required at each stage, but these are straightforward and present little difficulty.

In general terms, the method is as follows.

To solve $y' = f(x, y)$ with initial condition $y = y_0$ at $x = x_0$, for a range of values of $x = x_0(h)x_n$.

Starting as usual with $x = x_0$, $y = y_0$, $y' = (y')_0$ and h, we have

$$x_1 = x_0 + h$$

Finding y_1 requires four intermediate calculations

$$k_1 = hf(x_0, y_0) = h(y')_0$$
$$k_2 = hf(x_0 + \tfrac{1}{2}h, \ y_0 + \tfrac{1}{2}k_1)$$
$$k_3 = hf(x_0 + \tfrac{1}{2}h, \ y_0 + \tfrac{1}{2}k_2)$$
$$k_4 = hf(x_0 + h, \ y_0 + k_3)$$

The increment Δy_0 in the y-values from $x = x_0$ to $x = x_1$ is then

$$\Delta y_0 = \tfrac{1}{6}\{k_1 + 2k_2 + 2k_3 + k_4\}$$

and finally $\qquad y_1 = y_0 + \Delta y_0.$

We shall be using these repeatedly, so make a note of them for future reference. Then let us see an example

39

Example 1

Find the numerical solution of $y' = x + y$ using the Runge–Kutta method with $y = 1$ and $x = 0$ for values in the range $x = 0(0.1)1.0$.

We shall proceed with the solution of this differential equation using a spreadsheet in much the same manner as before. However, we are going to require a different structure in order to accommodate the four variables k_i for $i = 1, 2, 3, 4$. The structure we shall use is headed by

	A	B	C	D	E	F	G	H	I
1	**n**	**x**	**k1**	**k2**	**k3**	**k4**	**y**	**y'**	**h =**

where the value of h is held in cell J1.

1 Enter the values 0 to 10 in column A from A2 to A12 using the **Edit-Fill-Series** sequence of commands. These are the iteration numbers.

2 Enter the x step value of 0.1 in cell J1.

3 Enter the initial value of x in cell B2 as 0 and in B3 enter the formula = B2 + J1. Now copy the contents of B3 into cells B4 to B12.

4 Enter the initial value of y in cell G2 as 1.

We can now progressively enter the table of values from the left.

5 $k_1 = hf(x_0, y_0) = h(y')_0$ – the y'-values are in column H, so in cell C2 enter the formula = J1 H2. Copy the contents of C2 into cells C3 to C12.

6 $k_2 = hf\left(x_0 + \frac{1}{2}h, y_0 + \frac{1}{2}k_1\right) = h\left(x_0 + \frac{1}{2}h + y_0 + \frac{1}{2}k_1\right)$, so in cell D2 enter the formula = J1 (B2 + 0.5 J1 + G2 + 0.5 C2). Copy the contents of D2 into cells D3 to D12.

7 $k_3 = hf(x_0 + \frac{1}{2}h, y_0 + \frac{1}{2}k_2) = h(x_0 + \frac{1}{2}h + y_0 + \frac{1}{2}k_2)$, so in cell E2 enter the formula = J1 (B2 + 0.5 J1 + G2 + 0.5 D2). Copy the contents of E2 into cells E3 to E12.

8 $k_4 = hf(x_0 + h, y_0 + k_3) = h(x_0 + h + y_0 + k_3)$, so in cell F2 enter the formula = J1 (B2 + J1 + G2 + E2). Copy the contents of F2 into cells F3 to F12.

9 $y_1 = y_0 + \frac{1}{6}\{k_1 + 2k_2 + 2k_3 + k_4\}$, so in cell G3 enter the formula = G2+(1/6) (C2 + 2 D2 + 2 E2 + F2). Copy the contents of G3 into cells G4 to G12.

10 $y' = x + y$, so in H2 enter the formula = B2 + G2. Copy the contents of H2 into cells H3 to H12.

The results are displayed in the next frame

n	x	k1	k2	k3	k4	y	y'	$h = 0.1$
0	0.0	0.1000000	0.1100000	0.1105000	0.1210500	1.0000000	1.0000000	
1	0.1	0.1210342	0.1320859	0.1326385	0.1442980	1.1103417	1.2103417	
2	0.2	0.1442805	0.1564945	0.1571052	0.1699910	1.2428051	1.4428051	
3	0.3	0.1699717	0.1834703	0.1841452	0.1983862	1.3997170	1.6997170	
4	0.4	0.1983648	0.2132831	0.2140290	0.2297677	1.5836485	1.9836485	
5	0.5	0.2297441	0.2462313	0.2470557	0.2644497	1.7974413	2.2974413	
6	0.6	0.2644236	0.2826448	0.2835558	0.3027792	2.0442359	2.6442359	
7	0.7	0.3027503	0.3228878	0.3238947	0.3451398	2.3275033	3.0275033	
8	0.8	0.3451079	0.3673633	0.3684761	0.3919555	2.6510791	3.4510791	
9	0.9	0.3919203	0.4165163	0.4177461	0.4436949	3.0192028	3.9192028	
10	1.0	0.4436559	0.4708387	0.4721979	0.5008757	3.4365595	4.4365595	

with the following errors

n	x	Exact	Error (%)	Error (%)
0	0.0	1.0000000	0.0000000	0.00
1	0.1	1.1103418	0.0000153	0.93
2	0.2	1.2428055	0.0000301	1.84
3	0.3	1.3997176	0.0000444	2.69
4	0.4	1.5836494	0.0000578	3.50
5	0.5	1.7974425	0.0000703	4.25
6	0.6	2.0442376	0.0000820	4.95
7	0.7	2.3275054	0.0000929	5.59
8	0.8	2.6510819	0.0001030	6.18
9	0.9	3.0192062	0.0001124	6.73
10	1.0	3.4365637	0.0001213	7.25

The column to the far right contains the errors using the Euler method and, as you can see, the Runge–Kutta method provides a significant improvement in accuracy.

Now, without reference to your notes, complete the following expressions for

$$k_1 = \ldots\ldots\ldots$$
$$k_2 = \ldots\ldots\ldots$$
$$k_3 = \ldots\ldots\ldots$$
$$k_4 = \ldots\ldots\ldots$$
$$\Delta y_0 = \ldots\ldots\ldots$$
$$y_1 = \ldots\ldots\ldots$$

It speeds up your working if you can remember them.

41

$$
\begin{aligned}
k_1 &= h(y')_0 \\
k_2 &= hf\left(x_0 + \tfrac{1}{2}h,\ y_0 + \tfrac{1}{2}k_1\right) \\
k_3 &= hf\left(x_0 + \tfrac{1}{2}h,\ y_0 + \tfrac{1}{2}k_2\right) \\
k_4 &= hf(x_0 + h,\ y_0 + k_3) \\
\Delta y_0 &= \tfrac{1}{6}(k_1 + 2k_2 + 2k_3 + k_4) \\
y_1 &= y_0 + \Delta y_0
\end{aligned}
$$

With those in mind, let us move on to a further example. Next frame

42

Example 2

Solve $y' = \sqrt{x^2 + y}$ for $x = 0(0.2)2.0$ given that at $x = 0$, $y = 0.8$.

Using the spreadsheet for the previous example as a template for this example. The solution is ....

43

n	x	k1	k2	k3	k4	y	y′	h=0.2
0	0.0	0.1788854	0.1896779	0.1902460	0.2030021	0.8000000	0.8944272	
1	0.2	0.2030063	0.2174206	0.2180825	0.2339548	0.9902892	1.0150316	
2	0.4	0.2339473	0.2510185	0.2516977	0.2698134	1.2082838	1.1697366	
3	0.6	0.2698011	0.2887709	0.2894271	0.3091435	1.4598160	1.3490055	
4	0.8	0.3091304	0.3294604	0.3300769	0.3509482	1.7490394	1.5456518	
5	1.0	0.3509358	0.3722562	0.3728285	0.3945492	2.0788983	1.7546790	
6	1.2	0.3945381	0.4165946	0.4171237	0.4394829	2.4515074	1.9726904	
7	1.4	0.4394732	0.4620889	0.4625781	0.4854274	2.8684170	2.1973659	
8	1.6	0.4854190	0.5084682	0.5089213	0.5321545	3.3307894	2.4270948	
9	1.8	0.5321472	0.5555390	0.5559599	0.5794989	3.8395148	2.6607358	
10	2.0	0.5794925	0.6031595	0.6035518	0.6273385	4.3952888	2.8974625	

Because

1 The initial conditions are entered as $x_0 = 0$ and $y_0 = 0.8$. The x step length is entered as 0.2
2 The formula for the variable k_1 remains the same as $= \$J\$1\ H2$
3 The formula for the variable k_2 is changed to
   $= \$J\$1\ (((B2+0.5\ \$J\$1)^2+G2+0.5\ C2)^0.5)$
4 The formula for the variable k_3 is changed to
   $= \$J\$1\ (((B2+0.5\ \$J\$1)^2+G2+0.5\ D2)^0.5)$
5 The formula for the variable k_4 is changed to
   $= \$J\$1\ (((B2+\$J\$1)^2+G2+E2)^0.5)$
6 The formula for y remains the same as
   $= G2+\{1/6\}\ (C2+2\ D2+2\ E2+F2)$
7 The formula for y' is changed to $= (B2^2+G2)^0.5$

That is it. Now move on to the next frame where we make a new start and apply similar methods to the solution of second-order differential equations by numerical methods.

Second-order differential equations

Euler second-order method

44

The first method we will deal with is really an extension of the Euler method for the first-order equations and is a direct application of a truncated form of Taylor's series. We anticipate, therefore, that the method will be relatively easy, but the results will not be accurate to a high degree.

Taylor's series:

$$f(x + h) = f(x) + hf'(x) + \frac{h^2}{2!}f''(x) + \frac{h^3}{3!}f'''(x) + \ldots$$

Differentiating term by term with respect to x, we obtain

$$f'(x + h) = f'(x) + hf''(x) + \frac{h^2}{2!}f'''(x) + \frac{h^3}{3!}f''''(x) + \ldots$$

If we neglect terms in $f'''(x)$ and subsequent terms in each of these two series, we have the approximations

$$f(x + h) \approx \ldots\ldots\ldots\ldots$$
$$f'(x + h) \approx \ldots\ldots\ldots\ldots$$

45

$$\boxed{\begin{array}{l} f(x + h) \approx f(x) + hf'(x) + \frac{h^2}{2!}f''(x) \\ f'(x + h) \approx f'(x) + hf''(x) \end{array}}$$

Although these are approximations, in practice we tend to write them with the 'equals' sign. Therefore, at $x = a$, these become

$$\ldots\ldots\ldots\ldots\ldots\ldots\ldots\ldots$$

and

$$\ldots\ldots\ldots\ldots\ldots\ldots\ldots\ldots$$

46

$$\boxed{\begin{array}{l} f(a + h) = f(a) + hf'(a) + \frac{h^2}{2!}f''(a) \\ f'(a + h) = f'(a) + hf''(a) \end{array}}$$

and these, with the notation we have previously used, can be written

$$y_1 = y_0 + h(y')_0 + \frac{h^2}{2!}(y'')_0$$

$$(y')_1 = (y')_0 + h(y'')_0$$

Thus, if x_0, y_0, $(y')_0$ and $(y'')_0$ are known, we can find an approximate value of y_1 at $x_1 = x_0 + h$.

Make a note of these two relationships: then we can apply them.

47

Example

Solve the equation $y'' = xy' + y$ for $x = 0(0.2)2.0$ given that at $x = 0$, $y = 1$ and $y' = 0$.

We shall set about finding the numerical solution to this equation as we have done previously by using a spreadsheet. The headings for the sheet will be

	A	B	C	D	E	F	G	H
1	**n**	**x**	**y**	**y'**	**y''**	**Exact**	**Errors (%)**	**h=**

The entries will then be

1 Column A contains the iteration number from 0 in A2 to 10 in A12.
2 Cell I1 contains the x step length which is 0.2.
3 Column B contains the successive x-values from 0.0 to 2.0 in steps of 0.2. The initial value of $x_0 = 0$ is entered into cell B2 and the formula $= B2 + \$I\$1$ is entered into cell B3 and copied into cells B4 to B12.
4 Column C contains the computed y-values. The initial value of $y_0 = 1$ is entered into cell C2 and the equation

$$y_1 = y_0 + h(y')_0 + \frac{h^2}{2!}(y'')_0$$

is represented in cell C3 by the formula

$$= C2 + \$I\$1\ D2 + (\$I\$1\verb|^|2)\ E2/2$$

copied into cells C4 to C12.
5 Column D contains the computed y'-values. The initial value of $(y')_0 = 0$ is entered into cell D2 and the equation

$$(y')_1 = (y')_0 + h(y'')_0$$

is represented in cell D3 by the formula $= D2 + \$I\$1\ E2$ copied into cells D4 to D12.
6 Column E contains the y''-values which are obtained from the equation $y'' = xy' + y$ which is represented in cell E2 by the formula $= B2\ D2 + C2$ copied into cells E3 to E12.
7 Column F contains the values obtained from the exact solution which can be shown to be $y = e^{x^2/2}$. This is represented in cell F2 by the formula $= \mathrm{EXP}((B2\verb|^|2)/2)$ copied into cells F3 to F12.
8 Column G contains the percentage errors. In cell G2 enter the formula $= (F2 - C2)\ 100/F2$ copied into cells G3 to G12.

Your spreadsheet should now look like the one on the next page (with the appropriate formatting to make it easier to read).

▶

n	x	y	y′	y″	Exact	Errors (%) h=0.2
0	0.0	1.0000000	0.0000000	1.0000000	1.0000000	0.00
1	0.2	1.0200000	0.2000000	1.0600000	1.0202013	0.02
2	0.4	1.0812000	0.4120000	1.2460000	1.0832871	0.19
3	0.6	1.1885200	0.6612000	1.5852400	1.1972174	0.73
4	0.8	1.3524648	0.9782480	2.1350632	1.3771278	1.79
5	1.0	1.5908157	1.4052606	2.9960763	1.6487213	3.51
6	1.2	1.9317893	2.0044759	4.3371604	2.0544332	5.97
7	1.4	2.4194277	2.8719080	6.4400989	2.6644562	9.20
8	1.6	3.1226113	4.1599278	9.7784957	3.5966397	13.18
9	1.8	4.1501667	6.1156269	15.1582952	5.0530903	17.87
10	2.0	5.6764580	9.1472859	23.9710299	7.3890561	23.18

You will notice that the errors are significant and grow dramatically as the value of x increases. The main cause of errors is ....

48

the truncation of the Taylor's series on which the method is based

A greater degree of accuracy can be obtained by using the Runge–Kutta method for second-order differential equations, which is an extension of the method we have already used for first-order equations. As before, more intermediate calculations are required, but the reliability of results reflects the extra work involved.

Runge–Kutta method for second-order differential equations

Starting with the given equation $y'' = f(x, y, y')$ and initial conditions that at $x = x_0$, $y = y_0$ and $y' = (y')_0$, we can obtain the value of y_1 at $x_1 = x_0 + h$ as follows.

(a) We evaluate

$$k_1 = \tfrac{1}{2}h^2 f\{x_0,\ y_0,\ (y')_0\} = \tfrac{1}{2}h^2(y'')_0$$

$$k_2 = \tfrac{1}{2}h^2 f\left\{x_0 + \tfrac{1}{2}h,\ y_0 + \tfrac{1}{2}h(y')_0 + \tfrac{1}{4}k_1,\ (y')_0 + \frac{k_1}{h}\right\}$$

$$k_3 = \tfrac{1}{2}h^2 f\left\{x_0 + \tfrac{1}{2}h,\ y_0 + \tfrac{1}{2}h(y')_0 + \tfrac{1}{4}k_1,\ (y')_0 + \frac{k_2}{h}\right\}$$

$$k_4 = \tfrac{1}{2}h^2 f\left\{x_0 + h,\ y_0 + h(y')_0 + k_3,\ (y')_0 + \frac{2k_3}{h}\right\}$$

(b) From these results, we then determine

$$P = \tfrac{1}{3}\{k_1 + k_2 + k_3\}$$
$$Q = \tfrac{1}{3}\{k_1 + 2k_2 + 2k_3 + k_4\}$$

▶

(c) Finally, we have

$$x_1 = x_0 + h$$
$$y_1 = y_0 + h(y')_0 + P$$
$$(y')_1 = (y')_0 + \frac{Q}{h}$$

It is not as complicated as it looks at first sight. Copy down this list of relationships for reference when dealing with some examples that follow.

Then move on

49 Note the following

1   Four evaluations for k are required to determine a single new point on the solution curve.
2   The method is self-starting in that no preliminary calculations are required. The equation and initial conditions are sufficient to provide the next point on the curve.
3   As with the Runge–Kutta method for first-order equations, the method contains no self-correcting element or indication of any error involved.

Example 1

Use the Runge–Kutta method to solve the equation $y'' = xy' + y$ for $x = 0.0(0.2)2.0$ given that at $x = 0$, $y = 1$ and $y' = 0$.

This is the same problem that we have just encountered and in due course we shall compare results. As expected, we shall use a spreadsheet to derive the solution. The headings for the sheet this time will be

	A	B	C	D	E	F	G	H	I	J	K	L
1	n	x	k1	k2	k3	k4	P	Q	y	y'	y''	h=

The entries will then be

1   Column A contains the iteration number from 0 in A2 to 10 in A12.
2   Cell M1 contains the x step length which is 0.2.
3   Column B contains the successive x-values from 0.0 to 2.0 in steps of 0.2. The initial value of $x_0 = 0$ is entered into cell B2 and the formula $= B2 + \$M\$1$ is entered into cell B3 and copied into cells B4 to B12.
4   Column C contains the computed k_1-values and the equation $k_1 = \frac{1}{2}h^2(y'')_0$ is represented in cell C2 by the formula . . . .

50

$$= (0.5)\ (\$M\$1^2)\ K2$$

The contents of cell C2 are then copied into cells C3 to C12.

5 Column D contains the computed k_2-values and the equation

$$k_2 = \tfrac{1}{2}h^2 f\left(x_0 + \tfrac{1}{2}h,\ y_0 + \tfrac{1}{2}h(y')_0 + \tfrac{1}{4}k_1,\ (y')_0 + k_1/h\right)$$
$$= \tfrac{1}{2}h^2\left((x_0 + \tfrac{1}{2}h)((y')_0 + k_1/h) + y_0 + \tfrac{1}{2}h(y')_0 + \tfrac{1}{4}k_1\right)$$

is represented in cell D2 by the formula ....

51

$$=(0.5)\ (\$M\$1^2)\ ((B2+0.5\ \$M\$1)\ (J2+C2/\$M\$1)$$
$$+\,I2+0.5\ \$M\$1\ J2+0.25\ C2)$$

The contents of cell D2 are then copied into cells D3 to D12.

6 Column E contains the computed k_3-values and the equation

$$k_3 = \tfrac{1}{2}h^2 f\left(x_0 + \tfrac{1}{2}h,\ y_0 + \tfrac{1}{2}h(y')_0 + \tfrac{1}{4}k_1,\ (y')_0 + k_2/h\right)$$
$$= \tfrac{1}{2}h^2\left((x_0 + \tfrac{1}{2}h)((y')_0 + k_2/h) + y_0 + \tfrac{1}{2}h(y')_0 + \tfrac{1}{4}k_1\right)$$

is represented in cell E2 by the formula ....

52

$$=(0.5)\ (\$M\$1^2)\ ((B2+0.5\ \$M\$1)\ (J2+D2/\$M\$1)$$
$$+\,I2+0.5\ \$M\$1\ J2+0.25\ C2)$$

The contents of cell E2 are then copied into cells E3 to E12.

7 Column F contains the computed k_4-values and the equation

$$k_4 = \tfrac{1}{2}h^2 f\left(x_0 + h,\ y_0 + h(y')_0 + k_3,\ (y')_0 + 2k_3/h\right)$$
$$= \tfrac{1}{2}h^2\left((x_0 + h)((y')_0 + 2k_3/h) + y_0 + h(y')_0 + k_3\right)$$

is represented in cell F2 by the formula ....

53

$$=(0.5)\ (\$M\$1^2)\ ((B2+\$M\$1)\ (J2+2\ E2/\$M\$1)$$
$$+\,I2+\$M\$1\ J2+E2)$$

The contents of cell F2 are then copied into cells F3 to F12.

8 Column G contains the computed *P*-values and the equation $P = \tfrac{1}{3}(k_1 + k_2 + k_3)$ is represented in cell G2 by the formula ....

54

$$=(1/3)\,(C2+D2+E2)$$

The contents of cell G2 are then copied into cells G3 to G12.

9 Column H contains the computed Q-values and the equation $Q = \frac{1}{3}(k_1 + 2k_2 + 2k_3 + k_4)$ is represented in cell H2 by the formula
. . . .

55

$$=(1/3)\,(C2+2\,D2+2\,E2+F2)$$

The contents of cell H2 are then copied into cells H3 to H12.

10 Column I contains the computed y-values. The initial value of $y_0 = 1$ is entered into cell I2 and the equation

$$y_1 = y_0 + h(y')_0 + P$$

is represented in cell I3 by the formula . . . .

56

$$=I2+\$M\$1\,J2+G2$$

The contents of cell I3 are then copied into cells I4 to I12.

11 Column J contains the computed y'-values. The initial value of $(y')_0 = 0$ is entered into cell J2 and the equation $(y')_1 = (y')_0 + Q/h$ is represented in cell J3 by the formula . . . .

57

$$=J2+H2/\$M\$1$$

The contents of cell J3 are then copied into cells J4 to J12.

12 Column K contains the y''-values which are obtained from the equation $y'' = xy' + y$ which is represented in cell K2 by the formula . . . .

58

$$= B2 \ J2 + I2$$

The contents of cell K2 are then copied into cells K3 to K12 and the final spreadsheet looks like the following

n	x	k1	k2	k3	k4	P	Q	y	y'	y''	h = 0.2
0	0.0	0.0200000	0.0203000	0.0203030	0.0212182	0.0202010	0.0408081	1.0000000	0.0000000	1.0000000	
1	0.2	0.0212202	0.0227790	0.0228258	0.0251351	0.0222750	0.0458550	1.0202010	0.2040403	1.0610091	
2	0.4	0.0251322	0.0282477	0.0284035	0.0325752	0.0272612	0.0570033	1.0832841	0.4333153	1.2566102	
3	0.6	0.0325641	0.0378798	0.0382519	0.0451961	0.0362319	0.0766745	1.1972083	0.7183318	1.6282074	
4	0.8	0.0451694	0.0538673	0.0546501	0.0660061	0.0512289	0.1094035	1.3771066	1.1017045	2.2584702	
5	1.0	0.0659480	0.0801269	0.0816865	0.1003762	0.0759205	0.1633170	1.6486764	1.6487218	3.2973982	
6	1.2	0.1002542	0.1236497	0.1266912	0.1579840	0.1168650	0.2529733	2.0543413	2.4653068	5.0127095	
7	1.4	0.1577302	0.1970991	0.2030044	0.2565931	0.1859446	0.4048434	2.6642677	3.7301734	7.8865105	
8	1.6	0.2560654	0.3238945	0.3354254	0.4295622	0.3051284	0.6680891	3.5962469	5.7543906	12.8032719	
9	1.8	0.4284592	0.5483881	0.5711745	0.7411112	0.5160073	1.1362318	5.0522535	9.0948361	21.4229585	
10	2.0	0.7387844	0.9567270	1.0024949	1.3181400	0.8993354	1.9917894	7.3872279	14.7759954	36.9392186	

The errors have been dramatically reduced, as can be seen from the following table in comparison with those in Frame 47.

n	x	Exact	Error (%)
0	0.0	1.0000000	0.00
1	0.2	1.0202013	0.00
2	0.4	1.0832871	0.00
3	0.6	1.1972174	0.00
4	0.8	1.3771278	0.00
5	1.0	1.6487213	0.00
6	1.2	2.0544332	0.00
7	1.4	2.6644562	0.01
8	1.6	3.5966397	0.01
9	1.8	5.0530903	0.02
10	2.0	7.3890561	0.02

Next frame

59

Now here is one for you to do entirely on your own. The method is exactly the same as before and there are no snags. Use the spreadsheet that you created for the previous example as a template for this one.

Example 2

Solve the equation

$$y'' = x - y^2$$

for $x = 0.0(0.2)2.0$ where at $x = 0$, $y = 0$ and $y' = 0$.

When you have finished, check the results with the next frame

60

n	x	k1	k2	k3	k4	P	Q	y	y'	y''	h=0.2
0	0.0	0.0000000	0.0020000	0.0020000	0.0039999	0.0013333	0.0040000	0.0000000	0.0000000	0.0000000	
1	0.2	0.0040000	0.0059996	0.0059996	0.0079974	0.0053331	0.0119986	0.0013333	0.0199999	0.1999982	
2	0.4	0.0079977	0.0099915	0.0099915	0.0119731	0.0093269	0.0199789	0.0106664	0.0799930	0.3998862	
3	0.6	0.0119741	0.0139351	0.0139351	0.0158524	0.0132814	0.0278556	0.0359919	0.1798875	0.5987046	
4	0.8	0.0158546	0.0177065	0.0177065	0.0194436	0.0170892	0.0353748	0.0852508	0.3191655	0.7927323	
5	1.0	0.0194477	0.0210264	0.0210264	0.0223594	0.0205002	0.0419709	0.1661731	0.4960396	0.9723865	
6	1.2	0.0223654	0.0233782	0.0233782	0.0239421	0.0230406	0.0466068	0.2858812	0.7058940	1.1182719	
7	1.4	0.0239482	0.0239504	0.0239504	0.0232394	0.0239497	0.0476631	0.4501006	0.9389280	1.1974094	
8	1.6	0.0232395	0.0216639	0.0216639	0.0191107	0.0221891	0.0430019	0.6618359	1.1772436	1.1619732	
9	1.8	0.0190914	0.0153806	0.0153806	0.0105578	0.0166175	0.0303905	0.9194737	1.3922530	0.9545681	
10	2.0	0.0104978	0.0043750	0.0043750	−0.0026809	0.0064159	0.0084390	1.2145418	1.5442053	0.5248882	

Because

The only items that need amending from the previous spreadsheet are the references to the actual differential equation. Consequently

The formula in D2 for k_2 now reads as

$$= (0.5)\ (\$M\$1\,{}^{\wedge}2)\ (B2 + 0.5\ \$M\$1 -$$
$$(I2 + 0.5\ \$M\$1\ J2 + 0.25\ C2)^{\wedge}2)$$

The formula in E2 for k_3 now reads as

$$= 0.5\ (\$M\$1\,{}^{\wedge}2)\ (B2 + 0.5\ \$M\$1 -$$
$$(I2 + 0.5\ \$M\$1\ J2 + 0.25\ C2)^{\wedge}2)$$

The formula in F2 for k_4 now reads as

$$= 0.5\ (\$M\$1\,{}^{\wedge}2)\ (B2 + \$M\$1 - (I2 + \$M\$1\ J2 + E2)^{\wedge}2)$$

The formula in K2 for y'' now reads as

$$= B2 - I2\,{}^{\wedge}2$$

Predictor–corrector methods

61

So far, all the methods that we have used for the numerical solution of differential equations have been *single-step* methods. By this is meant that, given the differential equation $y' = f(x, y)$, a set of starting values (x_0 and y_0) and a step length (h), we can then find the value of y_1. The values of x_1 and y_1 become the starting values for the next iteration and so the procedure goes on, one step at a time. More accurate methods employ a *multi-step* procedure where, instead of starting with just a single set of initial values, we use a collection of previously calculated values.

A very simple multi-step method is given by the equations

$$\bar{y}_1 = y_0 + hf(x_0, y_0)$$
$$y_1 = y_0 + \tfrac{1}{2}h\big(f(x_0, y_0) + f(x_1, \bar{y}_1)\big)$$

Here we calculate $\bar{y}_1$ first from the given initial conditions x_0 and y_0. We call this equation the *predictor* because it gives $\bar{y}_1$ as a first estimate of y_1. Using $\bar{y}_1$ in the second equation then gives a more accurate value for y_1. We call this equation the *corrector*.

▶

An even better pair of predictor–corrector equations is given by

$$\bar{y}_{i+1} = y_i + \tfrac{1}{2}h(3f(x_i, y_i) - f(x_{i-1}, y_{i-1}))$$
$$y_{i+1} = y_i + \tfrac{1}{2}h(f(x_i, y_i) + f(x_{i+1}, \bar{y}_{i+1})) \quad \text{for } i = 0, 1, 2, 3, \ldots$$

Here, in order to use the predictor for the first time when $i = 0$ we need to know the value of $f(x_{0-1}, y_{0-1}) = f(x_{-1}, y_{-1})$, which we do not. Instead we shall use the equation $\bar{y}_1 = y_0 + hf(x_0, y_0)$ when $i = 0$.

In the next frame we shall look at an example

Example　　　　　　　　　　　　　　　　　　　　　　　　　　**62**

Solve the equation $y' = x + y$ for $x = 0.0(0.1)1.0$ where $y = 1$ when $x = 0$.

We have solved this equation before in Frame 32 using the Euler–Cauchy method and have viewed the accuracy of this method when compared with the exact solution. Here we shall see that this predictor–corrector method is even more accurate. Set up the following heading on your spreadsheet

	A	B	C	D	E	F	G
1	**n**	**x**	**y***	**y**	**Exact**	**Errors (%)**	**h=**

As usual, column A contains the iteration numbers 0 to 10 in cells A2 to A12 and column B contains the x-values stepped according to the step length $h = 0.1$ which is in cell H1. The initial value of $y = 1$ must be entered into cell D2.

Column C contains the predictor values given by the equations

$$\bar{y}_1 = y_0 + hf(x_0, y_0)$$
$$\bar{y}_{i+1} = y_i + \tfrac{1}{2}h(3f(x_i, y_i) - f(x_{i-1}, y_{i-1})) \quad \text{for } i > 0$$

To accommodate these equations in cell C3 enter the formula . . . .

63

$$= D2 + \$H\$1\,(B2 + D2)$$

And in cell C4 enter the formula . . . .

64

$$= D3 + 0.5\,\$H\$1\,(3\ B3 + 3\ D3 - B2 - D2)$$

And copy into cells C5 to C12.

Column D contains the corrector values given by the equation

$$y_{i+1} = y_i + \tfrac{1}{2}h(f(x_i, y_i) + f(x_{i+1}, \bar{y}_{i+1}))$$

To accommodate this equation in cell D3 enter the formula . . . .

$$= D2 + 0.5 \ \$H\$1 \ (B2 + D2 + B3 + C3)$$

And copy into cells D4 to D12.

We have seen that the exact solution to this equation is $2e^x - x - 1$, so this can be programd into the sheet entering the formula

$$= 2 \ EXP(B2) - B2 - 1 \quad \text{in cell E2 and then copying it into cells E3 to E12.}$$

The final table looks as follows

n	x	y*	y	Exact	Error (%)	h = 0.1
0	0.0		1.0000000	1.0000000	0.00	
1	0.1	1.1000000	1.1100000	1.1103418	0.03	
2	0.2	1.2415000	1.2425750	1.2428055	0.02	
3	0.3	1.3984613	1.3996268	1.3997176	0.01	
4	0.4	1.5824421	1.5837303	1.5836494	−0.01	
5	0.5	1.7963085	1.7977322	1.7974425	−0.02	
6	0.6	2.0432055	2.0447791	2.0442376	−0.03	
7	0.7	2.3266093	2.3283485	2.3275054	−0.04	
8	0.8	2.6503618	2.6522840	2.6510819	−0.05	
9	0.9	3.0187092	3.0208337	3.0192062	−0.05	
10	1.0	3.4363445	3.4386926	3.4365637	−0.06	

Here the errors are significantly reduced, as seen from the comparisons below.

1	2	3
0.00	0.00	0.00
0.93	0.03	0.03
1.84	0.06	0.02
2.69	0.09	0.01
3.50	0.12	−0.01
4.25	0.14	−0.02
4.95	0.17	−0.03
5.59	0.19	−0.04
6.18	0.21	−0.05
6.73	0.23	−0.05
7.25	0.24	−0.06

Here **1** refers to Euler, **2** refers to Euler–Cauchy and **3** refers to the predictor–corrector method just used.

And that is it. There are many other more sophisticated methods for the solution of ordinary differential equations by numerical methods and a detailed study of these is a course in itself. The methods we have used give an introduction to the processes and are practical in application.

The **Review summary** and **Can You?** checklist now follow as usual. Check them carefully and refer back to the Program for any points that may need further brushing up. Then you will be ready for the **Test exercise**, and the **Further problems** provide further practice.

Review summary

1 *Taylor's series*

$$f(a+h) = f(a) + hf'(a) + \frac{h}{2!}f''(a) + \frac{h}{3!}f'''(a) + \ldots$$

2 *Solution of first-order differential equations*
Equation $y' = f(x,y)$ with $y = y_0$ at $x = x_0$ for $x_0(h)x_n$.

(a) *Euler's method*
$$y_1 = y_0 + h(y')_0.$$

(b) *Euler–Cauchy method*
$$x_1 = x_0 + h$$
$$\bar{\bar{y}}_1 = y_0 + h(y')_0$$
$$y_1 = y_0 + \tfrac{1}{2}h\{(y')_0 + f(x_1, \bar{\bar{y}}_1)\}$$
$$(y')_1 = f(x_1, y_1).$$

(c) *Runge–Kutta method*
$$x_1 = x_0 + h$$
$$k_1 = hf(x_0, y_0) = h(y')_0$$
$$k_2 = hf(x_0 + \tfrac{1}{2}h, y_0 + \tfrac{1}{2}k_1)$$
$$k_3 = hf(x_0 + \tfrac{1}{2}h, y_0 + \tfrac{1}{2}k_2)$$
$$k_4 = hf(x_0 + h, y_0 + k_3)$$
$$\Delta y_0 = \tfrac{1}{6}(k_1 + 2k_2 + 2k_3 + k_4)$$
$$y_1 = y_0 + \Delta y_0$$
$$(y')_1 = f(x_1, y_1).$$

3 *Solution of second-order differential equations*
Equation $y'' = f(x,y,y')$ with $y = y_0$ and $y' = (y')_0$ at $x = x_0$ for $x = x_0(h)x_n$.

(a) *Euler's second-order method*
$$y_1 = y_0 + h(y')_0 + \frac{h^2}{2!}(y'')_0$$
$$(y')_1 = (y')_0 + h(y'')_0.$$

▶

(b) *Runge–Kutta method*

$$x_1 = x_0 + h$$

$$k_1 = \tfrac{1}{2}h^2 f\{x_0, y_0, (y')_0\} = \tfrac{1}{2}h^2(y'')_0$$

$$k_2 = \tfrac{1}{2}h^2 f\left\{x_0 + \tfrac{1}{2}h, \; y_0 + \tfrac{1}{2}h(y')_0 + \tfrac{1}{4}k_1, \; (y')_0 + \frac{k_1}{h}\right\}$$

$$k_3 = \tfrac{1}{2}h^2 f\left\{x_0 + \tfrac{1}{2}h, \; y_0 + \tfrac{1}{2}h(y')_0 + \tfrac{1}{4}k_1, \; (y')_0 + \frac{k_2}{h}\right\}$$

$$k_4 = \tfrac{1}{2}h^2 f\left\{x_0 + h, \; y_0 + h(y')_0 + k_3, \; (y')_0 + \frac{2k_3}{h}\right\}$$

$$P = \tfrac{1}{3}(k_1 + k_2 + k_3)$$

$$Q = \tfrac{1}{3}(k_1 + 2k_2 + 2k_3 + k_4)$$

$$y_1 = y_0 + h(y')_0 + P$$

$$(y')_1 = (y')_0 + \frac{Q}{h}$$

$$(y'')_1 = f\{x_1, y_1, (y')_1\}.$$

4 *Predictor–corrector*

Equation $y' = f(x, y)$ with $y = y_0$ and $y' = (y')_0$ at $x = x_0$ for $x = x_0(h)x_n$, then

Predictor

$$\bar{y}_{i+1} = y_i + \tfrac{1}{2}h(3f(x_i, y_i) - f(x_{i-1}, y_{i-1})) \quad \text{for } i = 1, 2, 3, \ldots$$

$$\bar{y}_1 = y_0 + hf(x_0, y_0) \quad \text{for } i = 0$$

Corrector

$$y_{i+1} = y_i + \tfrac{1}{2}h\big(f(x_i, y_i) + f(x_{i+1}, \bar{y}_{i+1})\big) \quad \text{for } i = 0, 1, 2, 3, \ldots$$

☑ Can You?

Checklist 11

67

Check this list before and after you try the end of Program test.

On a scale of 1 to 5 how confident are you that you can: Frames

• Derive a form of Taylor's series from Maclaurin's series and from it describe a function increment as a series of first and higher-order derivatives of the function?

Yes ☐ ☐ ☐ ☐ ☐ *No*   1 to 3

• Describe and apply by means of a spreadsheet the Euler method, the Euler–Cauchy method and the Runge–Kutta method for first-order differential equations?

Yes ☐ ☐ ☐ ☐ ☐ *No*   4 to 43

• Describe and apply by means of a spreadsheet the Euler second-order method and the Runge–Kutta method for second-order ordinary differential equations?

Yes ☐ ☐ ☐ ☐ ☐ *No*   44 to 60

• Describe and apply by means of a spreadsheet a simple predictor–corrector method?

Yes ☐ ☐ ☐ ☐ ☐ *No*   61 to 65

🚴 Test exercise 11

68

1   Apply Euler's method to solve the equation

$$\frac{dy}{dx} = 1 + xy \quad \text{for} \quad x = 0(0.1)0.5$$

given that at $x = 0$, $y = 1$.

2   The equation $\dfrac{dy}{dx} = x^2 - 2y$ is subject to the initial condition $y = 0$ at $x = 1$. Use the Euler–Cauchy method to obtain function values for $x = 1.0(0.2)2.0$.

3   Using the Runge–Kutta method, solve the equation

$$\frac{dy}{dx} = 1 + y - x \quad \text{for} \quad x = 0(0.1)0.5$$

given that $y = 1$ when $x = 0$.

▶

4   Apply Euler's second-order method to solve the equation

$$y'' = y - x \quad \text{for} \quad x = 2.0(0.1)2.5$$

given that at $x = 2$, $y = 3$ and $y' = 0$.

5   Use the Runge–Kutta method to solve the equation

$$y'' = (y'/x) + y \quad \text{for} \quad x = 1.0(0.1)1.5$$

given the initial conditions that at $x = 1.0$, $y = 0$ and $y' = 1.0$.

6   Use the predictor–corrector method in the text to solve the equation

$$y' = 1 + xy \quad \text{for } x = 0(0.1)1$$

given that $x = 0$ when $y = 0$.

🚲 **Further problems 11**

69   Solve the following differential equations by the methods indicated.

Euler's method

1	$y' = 2x - y$	$x = 0, y = 1$	$x = 0(0.2)1.0$
2	$y' = 2x + y^2$	$x = 0, y = 1.4$	$x = 0(0.1)0.5$

Euler–Cauchy method

3	$y' = 2 - y/x$	$x = 1, y = 2$	$x = 1.0(0.2)2.0$
4	$y' = x^2 - 2x + y$	$x = 0, y = 0.5$	$x = 0(0.1)0.5$
5	$y' = (y - x^2)^{\frac{1}{2}}$	$x = 0, y = 1$	$x = 0(0.1)0.5$
6	$y' = \dfrac{x + y}{xy}$	$x = 1, y = 1$	$x = 1.0(0.1)1.5$
7	$y' = y \sin x + \cos x$	$x = 0, y = 0$	$x = 0(0.1)0.5$

Runge Kutta method

8	$y' = 2x - y$	$x = 0, y = 1$	$x = 0(0.2)1.0$
9	$y' = x - y^2$	$x = 0, y = 1$	$x = 0(0.1)0.5$
10	$y' = y^2 - xy$	$x = 0, y = 0.4$	$x = 0(0.2)1.0$
11	$y' = \sqrt{2x + y}$	$x = 1, y = 2$	$x = 1.0(0.2)2.0$
12	$y' = 1 - x^3/y$	$x = 0, y = 1$	$x = 0(0.2)1.0$
13	$y' = \dfrac{y - x}{y + x}$	$x = 0, y = 1$	$x = 0(0.2)1.0$

▶

Euler second-order method

14 $y'' = (x+1)y' + y$ $\quad x = 0,\, y = 1,\, y' = 1$ $\quad x = 0(0.1)0.5$

15 $y'' = 2(xy' - 4y)$ $\quad x = 0,\, y = 3,\, y' = 0$ $\quad x = 0(0.1)0.5$

Runge–Kutta second-order method

16 $y'' = x - y - xy'$ $\quad x = 0,\, y = 0,\, y' = 1$ $\quad x = 0(0.2)1.0$

17 $y'' = (1-x)y' - y$ $\quad x = 0,\, y = 1,\, y' = 1$ $\quad x = 0(0.2)1.0$

18 $y'' = 1 + x - y^2$ $\quad x = 0,\, y = 2,\, y' = 1$ $\quad x = 0(0.1)0.5$

19 $y'' = (x+2)y - 2y'$ $\quad x = 0,\, y = 1,\, y' = 0$ $\quad x = 0(0.2)1.0$

20 $y'' = \dfrac{y - xy'}{x^2}$ $\quad x = 1,\, y = 0,\, y' = 1$ $\quad x = 1.0(0.2)2.0$

Predictor–corrector

21 $y' = 2 - y/x$ $\qquad\quad x = 1,\, y = 2$ $\qquad\quad x = 1.0(0.2)2.0$

22 $y' = 2x - y$ $\qquad\qquad x = 0,\, y = 1$ $\qquad\quad x = 0.0(0.2)1.0$

23 $y' = \sqrt{2x + y}$ $\qquad\quad x = 1,\, y = 2$ $\qquad\quad x = 1.0(0.2)2.0$

Answers

Test exercise 1 (page 37)

1 $y = \dfrac{x^2}{2} + 2x - 3\ln x + C$ **2** $\tan^{-1} y = C - \dfrac{1}{1+x}$ **3** $y = \dfrac{e^{3x}}{5} + Ce^{-2x}$

4 $y = x^2 + Cx$ **5** $y = -\dfrac{x\cos 3x}{3} + \dfrac{\sin 3x}{9} - \dfrac{4}{x} + C$ **6** $\sin y = Ax$ **7** $y^2 - x^2 = Ax^2 y$

8 $y(x^2 - 1) = \dfrac{x^2}{2} + C$ **9** $y = \cosh x + \dfrac{C}{\cosh x}$ **10** $y = x^2(\sin x + C)$

11 $xy^2(Cx + 2) = 1$ **12** $y = 1/(Cx^3 + x^2)$

Further problems 1 (page 38)

1 $x^4 y^3 = Ae^y$ **2** $y^3 = 4(1 + x^3)$ **3** $3x^4 + 4(y+1)^3 = A$ **4** $(1 + e^x)\sec y = 2\sqrt{2}$

5 $x^2 + y^2 + 2x - 2y + 2\ln(x-1) + 2\ln(y+1) = A$ **6** $y^2 - xy - x^2 + 1 = 0$

7 $xy = Ae^{y/x}$ **8** $x^3 - 2y^3 = Ax$ **9** $A(x - 2y)^5(3x + 2y)^3 = 1$ **10** $(x^2 - y^2)^2 = Axy$

11 $2y = x^3 + 6x^2 - 4x\ln x + Ax$ **12** $y = \cos x(A + \ln\sec x)$

13 $y = x(1 + x\sin x + \cos x)$ **14** $(3y - 5)(1 + x^2)^{3/2} = 2\sqrt{2}$ **15** $y\sin x + 5e^{\cos x} = 1$

16 $x + 3y + 2\ln(x + y - 2) = A$ **17** $x = Aye^{xy}$

18 $\ln\left\{4y^2 + (x-1)^2\right\} + \tan^{-1}\left\{\dfrac{2y}{x-1}\right\} = A$ **19** $(y - x + 1)^2(y + x - 1)^5 = A$

20 $2x^2 y^2 \ln y - 2xy - 1 = Ax^2 y^2$ **21** $\dfrac{2}{y^2} = 2x + 1 + Ce^{2x}$ **22** $\dfrac{1}{y^3} = \dfrac{3e^x}{2} + Ce^{3x}$

23 $y^2(x + Ce^x) = 1$ **24** $\dfrac{\sec^2 x}{y} = C - \dfrac{\tan^3 x}{3}$ **25** $\cos^2 x = y^2(C - 2\tan x)$

26 $y\sqrt{1 - x^2} = A + \sin^{-1} x$ **27** $x + \ln Ax = \sqrt{y^2 - 1}$

28 $\ln(x - y) = A + \dfrac{2x^2}{(x - y)^2} - \dfrac{4x}{(x - y)}$ **29** $y = \dfrac{\sqrt{2}\sin 2x}{2(\cos x - \sqrt{2})}$ **30** $(x - 4)y^4 = Ax$

31 $y = x\cos x - \dfrac{\pi}{8}\sec x$ **32** $(x - y)^3 - Axy = 0$ **33** $2\tan^{-1} y = \ln(1 + x^2) + A$

34 $2x^2 y = 2x^3 - x^2 - 4$ **35** $y = e^{\frac{x-y}{x}}$ **36** $3e^{2y} = 2e^{3x} + 1$

37 $4xy = \sin 2x - 2x\cos 2x + 2\pi - 1$ **38** $y = Ae^{y/x}$ **39** $x^3 - 3xy^2 = A$

40 $x^2 - 4xy + 4y^2 + 2x - 3 = 0$ **41** $y(1 - x^3)^{-1/3} = -\dfrac{1}{2}(1 - x^3)^{2/3} + C$

42 $xy + x\cos x - \sin x + 1 = 0$ **43** $2\tan^{-1} y = 1 - x^2$ **44** $y = \dfrac{x^2 + C}{2x(1 - x^2)}$

45 $y\sqrt{1 + x^2} = x + \dfrac{x^3}{3} + C$ **46** $1 + y^2 = 5(1 + x^2)$ **47** $\sin^2\theta(a^2 - r^2) = \dfrac{a^2}{2}$

48 $y = \dfrac{1}{2}\sin x$ **49** $y = \dfrac{1}{x(A - x)}$

Test exercise 2 (page 63)

1 $y = Ae^{-x} + Be^{2x} - 4$ **2** $y = Ae^{2x} + Be^{-2x} + 2e^{3x}$ **3** $y = e^{-x}(A + Bx) + e^{-2x}$

4 $y = A\cos 5x + B\sin 5x + \dfrac{1}{125}(25x^2 + 5x - 2)$ **5** $y = e^x(A + Bx) + 2\cos x$

6 $y = e^{-2x}(2 - \cos x)$ **7** $y = Ae^x + Be^{-x/3} - 2x + 7$ **8** $y = Ae^{2x} + Be^{4x} + 4xe^{4x}$

Further problems 2 (page 64)

1 $y = Ae^{4x} + Be^{-x/2} - \dfrac{e^{3x}}{7}$ **2** $y = e^{3x}(A + Bx) + 6x + 6$

3 $y = 4\cos 4x - 2\sin 4x + Ae^{2x} + Be^{3x}$ **4** $y = e^{-x}(Ax + B) + \dfrac{e^x}{2} - x^2 e^{-x}$

5 $y = Ae^x + Be^{-2x} + \dfrac{e^{2x}}{4} - \dfrac{xe^{-2x}}{3}$ **6** $y = e^{3x}(A\cos x + B\sin x) + 2 - \dfrac{e^{2x}}{2}$

7 $y = e^{-2x}(A + Bx) + \dfrac{1}{4} + \dfrac{1}{8}\sin 2x$ **8** $y = Ae^x + Be^{3x} + \dfrac{1}{9}(3x + 4) - e^{2x}$

9 $y = e^x(A\cos 2x + B\sin 2x) + \dfrac{x^2}{3} + \dfrac{4x}{9} - \dfrac{7}{27}$ **10** $y = Ae^{3x} + Be^{-3x} - \dfrac{1}{18}\sin 3x + \dfrac{1}{6}xe^{3x}$

11 $y = \dfrac{wx^2}{24EI}\{x^2 - 4lx + 6l^2\}; \quad y = \dfrac{wl^4}{8EI}$ **12** $x = \dfrac{1}{2}(1 - t)e^{-3t}$

13 $y = e^{-2t}(A\cos t + B\sin t) - \dfrac{3}{4}(\cos t - \sin t);$ amplitude $\dfrac{3\sqrt{2}}{4}$, frequency $\dfrac{1}{2\pi}$

14 $x = -\dfrac{1}{2}e^t + \dfrac{1}{5}e^{2t} + \dfrac{1}{10}(\sin t + 3\cos t)$ **15** $y = e^{-2x} - e^{-x} + \dfrac{3}{10}(\sin x - 3\cos x)$

16 $y = e^{-3x}(A\cos x + B\sin x) + 5x - 3$ **17** $x = e^{-t}(6\cos t + 7\sin t) - 6\cos 3t - 7\sin 3t$

18 $y = \sin x - \dfrac{1}{2}\sin 2x; \quad y_{max} = 1.299$ at $x = \dfrac{2\pi}{3}$ **19** $T = \dfrac{\pi}{2\sqrt{6}} = 0.641s; A = \dfrac{1}{6}$

20 $x = \dfrac{1}{10}\{e^{-3t} - e^{-2t} + \cos t + \sin t\};$ steady state: $x = \dfrac{\sqrt{2}}{10}\sin\left(t + \dfrac{\pi}{4}\right)$

Test exercise 3 (page 91)

1 230 **2** 2.488, 25.958 **3** 1812 **4** (a) convergent (b) divergent
(c) divergent (d) convergent **5** (a) convergent for all values of x
(b) convergent for $-1 \leq x \leq 1$ (c) convergent for $-1 \leq x \leq 1$

Further problems 3 (page 92)

1 $\dfrac{n}{3}(4n^2 - 1)$ **2** $\dfrac{n(3n + 1)}{4(n + 1)(n + 2)}$ **3** $\dfrac{n}{4}(n + 1)(n + 4)(n + 5)$ **4** (a) $\dfrac{n}{3}(n + 1)(n + 5)$

(b) $\dfrac{1}{4}(n^2 + 3n)(n^2 + 3n + 4)$ **5** 2 **6** $S_n = \dfrac{10}{3}\left\{1 + \dfrac{(-1)^{n+1}}{2^n}\right\}; S_\infty = \dfrac{10}{3}$

7 (a) 0.6 (b) 0.5 **8** (a) diverges (b) diverges (c) converges (d) converges
9 $-1 \leq x \leq 1$ **11** $-1 \leq x \leq 1$ **12** All values of x **13** $0 < x \leq 1$
16 (a) convergent (b) divergent (c) divergent (d) divergent **18** (a) convergent
(b) convergent **19** $1 \leq x \leq 3$ **20** $\dfrac{n}{6}(n + 1)(4n + 5) + 2^{n+2} - 4$

Test exercise 4 (page 119)

1 $f(x) = f(0) + xf'(0) + \dfrac{x^2}{2!}f''(0) + \ldots$ **2** $1 - x^2 + \dfrac{x^4}{3} - \dfrac{2x^6}{45} + \ldots$

3 $1 + \dfrac{x^2}{2} + \dfrac{5x^4}{24} + \ldots$ **5** $x + x^2 + \dfrac{5x^3}{6} + \dfrac{x^4}{2} + \ldots$ **6** 1.0247

7 (a) $-\dfrac{1}{10}$ (b) $\dfrac{2}{9}$ (c) $-\dfrac{1}{2}$ **8** 0.85719

Further problems 4 (page 120)

3 (a) $-\dfrac{1}{10}$ (b) $\dfrac{1}{3}$ (c) $\dfrac{1}{2}$ (d) $-\dfrac{1}{6}$ (e) 2 **6** $-\dfrac{1}{4}$ **7** $-\dfrac{3}{2} - \dfrac{5x}{2} - \dfrac{11x^2}{4} - \dfrac{13x^3}{4}$ **9** $\dfrac{2}{3}$

10 (a) $-\dfrac{1}{6}$ (b) $\dfrac{1}{2}$ (c) 2 **11** $\dfrac{(n-r+2)x}{r-1}$; 1.426 **13** $\ln\cos x = -\dfrac{x^2}{2} - \dfrac{x^4}{12} - \ldots$

16 (a) $-\dfrac{1}{6}$ (b) $\dfrac{1}{2}$ **17** $1 - \dfrac{7x}{2} + 8x^2$ **19** $x^2 - x^3 + \dfrac{11x^4}{12}$; max. at $x = 0$

Test exercise 5 (page 176)

2 $y = a_0\left\{1 + \dfrac{5x^2}{2} + \dfrac{15x^4}{8} + \dfrac{5x^6}{16} + \ldots\right\} + a_1\left\{x + \dfrac{4x^3}{3} + \dfrac{8x^5}{15} + \ldots\right\}$

3 (a) $y = A\left\{1 - \dfrac{x}{1 \times 2} + \dfrac{x^2}{(1 \times 2)(2 \times 5)} - \dfrac{x^3}{(1 \times 2)(2 \times 5)(3 \times 8)} + \ldots\right\}$

$\quad + Bx^{\frac{1}{3}}\left\{1 - \dfrac{x}{1 \times 4} + \dfrac{x^2}{(1 \times 4)(2 \times 7)} - \dfrac{x^3}{(1 \times 4)(2 \times 7)(3 \times 10)} + \ldots\right\}$

(b) $y = a_0\left\{1 - \dfrac{x^4}{3 \times 4} + \dfrac{x^8}{(3 \times 4)(7 \times 8)} + \ldots\right\}$

$\quad\quad\quad + a_1\left\{x - \dfrac{x^5}{4 \times 5} + \dfrac{x^9}{(4 \times 5)(8 \times 9)} + \ldots\right\}$

(c) $y_A = A\left\{-\dfrac{1}{2} - \dfrac{x}{6} - \ldots\right\}$

$\quad y_B = B\left\{\ln x\left(-\dfrac{1}{2} - \dfrac{x}{6} - \ldots\right) + x^{-2}\left(1 - x + \dfrac{x^2}{4} + \ldots\right)\right\}$ **5** $\dfrac{1}{3}P_0(x) - \dfrac{4}{3}P_2(x)$

Further problems 5 (page 176)

1 $y_5 = 64e^{4x}\{16x^3 + 60x^2 + 60x + 15\}$

2 $y_n = (-1)^n e^{-x}\{x^3 - 3nx^2 + n(n-1)3x - n(n-1)(n-2)\}$, $n > 3$

3 $y_4 = 480x + 96$ **4** $y_6 = -\{(x^4 - 180x^2 + 360)\cos x + (24x^3 - 480x)\sin x\}$

5 $y_4 = -4e^{-x}\sin x$ **6** $y_3 = 2x(13 + 12\ln x)$ **8** $y_6 = -1018$

10 (a) $y_{2n} = \{x^2 + 2n(2n-1)\}\sinh x + 4nx\cosh x$

(b) $y_{2n} = \{x^3 + 6n(2n-1)x\}\cosh x + \{6nx^2 + 2n(2n-1)(2n-2)\}\sinh x$

11 $y_6 = 2^5 e^{2x}\{2x^3 + 24x^2 + 81x + 75\}$ **12** $y_3 = 2\sqrt{2}a^3 e^{-ax}\{\cos(ax + \pi/4)\}$

14 $y = y_0\left\{1 + \dfrac{9x^2}{2} + \dfrac{15x^4}{8} - \dfrac{7x^6}{16} + \dfrac{27x^8}{128} + \ldots\right\} + y_1\left\{x + \dfrac{4x^3}{3}\right\}$

15 $y = A(1 + x^2) + Be^{-x}$

16 $y = y_0 \left\{ 1 + \dfrac{3^2 \times x^2}{2!} + \dfrac{3^2 \times 5^2 \times x^4}{4!} + \dfrac{3^2 \times 5^2 \times 7^2 \times x^6}{6!} + \cdots \right\}$

$\qquad\qquad + y_1 \left\{ x + \dfrac{4^2 \times x^3}{3!} + \dfrac{4^2 \times 6^2 \times x^5}{5!} + \cdots \right\}$

17 $y = y_1 x + y_0 \left\{ 1 - x^2 - \dfrac{x^4}{3} - \dfrac{x^6}{5} - \dfrac{x^8}{7} - \cdots \right\}$

18 $y = y_0 \left\{ 1 - \dfrac{2x}{2^2} + \dfrac{2^2 \times x^4}{2^2 \times 4^2} - \dfrac{2^3 \times x^6}{2^2 \times 4^2 \times 6^2} + \cdots \right\}$

$\qquad\qquad + y_1 \left\{ x - \dfrac{2x^3}{3^2} + \dfrac{2^2 \times x^5}{3^2 \times 5^2} - \dfrac{2^3 \times x^7}{3^2 \times 5^2 \times 7^2} + \cdots \right\}$

19 $y = A \left\{ 1 + x + \dfrac{x^2}{2 \times 4} + \dfrac{x^3}{(2 \times 3)(4 \times 7)} + \dfrac{x^4}{(2 \times 3 \times 4)(4 \times 7 \times 10)} + \cdots \right\}$

$\qquad + Bx^{\frac{2}{3}} \left\{ 1 + \dfrac{x}{1 \times 5} + \dfrac{x^2}{(1 \times 2)(5 \times 8)} + \dfrac{x^3}{(1 \times 2 \times 3)(5 \times 8 \times 11)} + \cdots \right\}$

20 $y = a_0 \left\{ 1 - \dfrac{x^2}{2!} + \dfrac{x^4}{4!} + \cdots \right\} + a_1 \left\{ x - \dfrac{x^3}{3!} + \cdots \right\}$

21 $y = a_0 \left\{ 1 + \dfrac{x^3}{2 \times 3} + \dfrac{x^6}{(2 \times 3)(5 \times 6)} + \cdots \right\}$

$\qquad\qquad + a_1 \left\{ x + \dfrac{x^4}{3 \times 4} + \dfrac{x^7}{(3 \times 4)(6 \times 7)} + \cdots \right\}$

22 $y = A \left\{ 1 - \dfrac{x}{1 \times 4} + \dfrac{x^2}{(1 \times 2)(4 \times 7)} - \dfrac{x^3}{(1 \times 2 \times 3)(4 \times 7 \times 10)} + \cdots \right\}$

$\qquad + Bx^{-\frac{1}{3}} \left\{ 1 - \dfrac{x}{1 \times 2} + \dfrac{x^2}{(1 \times 2)(2 \times 5)} - \dfrac{x^3}{(1 \times 2 \times 3)(2 \times 5 \times 8)} + \cdots \right\}$

23 $y = a_1 x + a_0 \left\{ 1 - \dfrac{x^2}{2!} - \dfrac{x^4}{4!} - \dfrac{3x^6}{6!} - \dfrac{(3)(5)x^8}{8!} + \cdots \right\}$

24 $y = u + v$ where $u = A \left\{ \dfrac{-x^4}{4!\,3!} + \dfrac{x^5}{5!\,3!} - \cdots \right\}$

$\quad v = B \left\{ \ln x \left(\dfrac{-x^4}{4!\,3!} + \dfrac{x^5}{5!\,3!} - \cdots \right) + \left(1 + \dfrac{x}{1 \times 3} + \dfrac{x^2}{(1 \times 2)(2 \times 3)} + \cdots \right) \right\}$

25 $y = u + v$ where $u = A \left\{ 1 + \dfrac{3x}{1^2} + \dfrac{3^2 \times x^2}{1^2 \times 2^2} + \dfrac{3^3 \times x^3}{1^2 \times 2^2 \times 3^2} + \cdots \right\}$

$\quad v = B \left\{ \ln x \left(1 + \dfrac{3x}{1^2} + \dfrac{3^2 \times x^2}{1^2 \times 2^2} + \dfrac{3^3 \times x^3}{1^2 \times 2^2 \times 3^2} + \cdots \right) \right.$

$\qquad\quad \left. - \left(\dfrac{2 \times 3x}{1^2} + \dfrac{3 \times 3^2 \times x^2}{1^2 \times 2^2} + \dfrac{11 \times 3^3 \times x^3}{1^2 \times 2^2 \times 3^3} + \cdots \right) \right\}$

26 eigenfunctions: $y_n(x) = A_n \cos \sqrt{\lambda_n} x$; eigenvalues: $\lambda_n = \dfrac{(2n+1)^2 \pi^2}{4}$

27 $H_0 = 1$, $H_1 = 2x$, $H_2 = 4x^2 - 2$, $H_3 = 8x^3 - 12x$

28 $L_0 = 1$, $L_1 = 1 - x$, $L_2 = 2 - 4x + x^2$, $L_3 = 6 - 18x + 9x^2 - x^3$

Test exercise 6 (page 197)

1 (a) $F(s) = \dfrac{8}{s}$ (b) $F(s) = \dfrac{1}{s-5}$ (c) $F(s) = -\dfrac{4e^3}{s-2}$ **2** (a) $f(x) = -5xe^{2x}$

(b) $f(x) = e^3 x^2$ (c) $f(x) = \sin 3x$ (d) $f(x) = \dfrac{5}{\sqrt{3}} \sin \sqrt{3}x - 2\cos \sqrt{3}x$

3 $F(s) = \dfrac{2}{(s-3)^3}$ **4** (a) $f(x) = \dfrac{1}{4}(e^{-2x} + 2x - 1)$ (b) $f(x) = -\dfrac{1}{2}(e^x + e^{-x})$

(c) $f(x) = \dfrac{1}{2}x^2 e^{-2x}$ (d) $f(x) = 2 - e^{-3x/2} - e^{3x/2}$

Further problems 6 (page 198)

1 (a) $\dfrac{1}{s - k\ln a}$ (b) $\dfrac{k}{s^2 - k^2}$ (c) $\dfrac{s}{s^2 - k^2}$ (d) $\dfrac{k(1 - e^{-sa})}{s}$ **2** (a) $-\dfrac{2}{3}e^{4x/3}$

(b) $\dfrac{\sinh 2\sqrt{2}x}{2\sqrt{2}}$ (c) $3\cos 4x - \sin 4x$ (d) $f(x) = 3 + 4\cos 3x$ (e) $f(x) = x(e^x - e^{-x})$

(f) $f(x) = \dfrac{4}{5}\cos 2x + \dfrac{17}{5}\sin 2x - \dfrac{9}{5}e^x$ **3** (b) (i) $\dfrac{4s}{(s^2+4)^2}$ (ii) $\dfrac{2s(s^2-27)}{(s^2+9)^3}$

(c) $F^{(n)}(s) = (-1)^n L\{x^n f(x)\}$ **4** (a) $\dfrac{b}{(s-a)^2 + b^2}$ (b) $\dfrac{s-a}{(s-a)^2 + b^2}$

5 (a) $e^{3x} - e^{2x}$ (b) $\dfrac{1}{6} + \dfrac{1}{3}e^{3x} - \dfrac{1}{2}e^{2x}$ (c) $e^{3x} - (1+x)e^{2x}$ (d) $\dfrac{1}{20}e^{-3x} + \dfrac{8}{15}e^{-x/2} + \dfrac{17}{12}e^x$

(e) $\dfrac{1}{12}\left\{8e^{-x/2} + 7e^x - 15e^{-x} - 6xe^{-x}\right\}$ (f) $\sin 4x + \cos 4x$

(g) $-\dfrac{2}{5}e^{-x/2} - \dfrac{1}{5}\sin x + \dfrac{2}{5}\cos x$

Test exercise 7 (page 218)

1 (a)

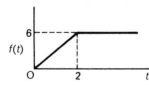

$F(s) = \dfrac{3}{s^2}\left\{1 - e^{-2s}\right\}$

(b)

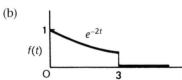

$F(s) = \dfrac{1}{s+2}\left\{1 - e^{-6}e^{-3s}\right\}$

(c)

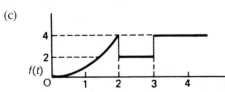

$F(s) = \dfrac{2}{s^3} - 2e^{-2s}\left\{\dfrac{1}{s^3} + \dfrac{2}{s^2} - \dfrac{1}{s}\right\}$

$+ \dfrac{2}{s}e^{-3s}$

(d)

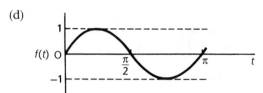

$$F(s) = \frac{2}{s^2+4}\{1 - e^{-\pi s}\}$$

2 $f(t) = 2 \cdot u(t) - 5 \cdot u(t-1) + 8 \cdot u(t-3)$

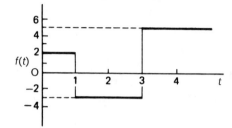

3 $f(t) = t \cdot u(t) + 3(t-2) \cdot u(t-2) - (t-3) \cdot u(t-3) - 3(t-5) \cdot u(t-5)$

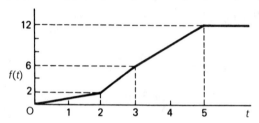

4 $f(t) = 2 \cdot u(t) - 2 \cdot u(t-1) + 2 \cdot u(t-3) - 2 \cdot u(t-4)$
$\qquad + 2 \cdot u(t-6) - 2 \cdot u(t-7) + \dots$

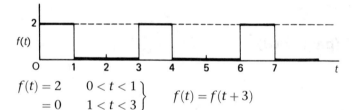

$$\left. \begin{array}{ll} f(t) = 2 & 0 < t < 1 \\ \quad\;\; = 0 & 1 < t < 3 \end{array} \right\} \qquad f(t) = f(t+3)$$

Further problems 7 (page 219)

1 $f(t) = 3 \cdot u(t) + 2(t-2) \cdot u(t-2) - 2(t-5) \cdot u(t-5)$

2 $f(t) = t \cdot u(t) - (t-1) \cdot u(t-1) + (t-2) \cdot u(t-2) - (t-3) \cdot u(t-3)$

3

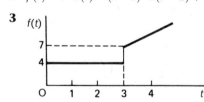

$$F(s) = \frac{4}{s} + \frac{3e^{-3s}}{s} + \frac{2e^{-3s}}{s^2}$$

4 (a) $f(t) = t^2 \cdot u(t) - (t^2 - 5t) \cdot u(t-3)$

  (b) $f(t) = \cos t \cdot u(t) + (\cos 2t - \cos t) \cdot u(t-\pi) + (\cos 3t - \cos 2t) \cdot u(t-2\pi)$

5 $F(s) = e^{-2s}\left\{\dfrac{1}{s^2} + \dfrac{3}{s}\right\} - e^{-3s}\left\{\dfrac{1}{s^2} + \dfrac{4}{s}\right\}$

6 (a) $f(t) = t^2 \cdot u(t) - t^2 \cdot u(t-2) + 4 \cdot u(t-2) - 4 \cdot u(t-5)$

   (b) $F(s) = \dfrac{2}{s^3} - \dfrac{2e^{-2s}}{s^3} - \dfrac{4e^{-2s}}{s^2} - \dfrac{4e^{-5s}}{s}$

Test exercise 8 (page 253)

1 $F(s) = \dfrac{2(1 - e^{-2s} - 2se^{-2s})}{s^2(1 - e^{-4s})}$ **2** (a) e^{-6} (b) 0   (c) 11

3 (a) $F(s) = 4e^{-3s}$ (b) $F(s) = e^{-2(3+s)}$

4

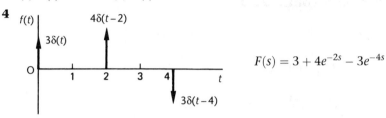

$F(s) = 3 + 4e^{-2s} - 3e^{-4s}$

5 $x = e^{-3t}\{4\sin t - \cos t\}$

6 $x = 3e^4 e^{-t} \cdot u(t-4) + e^{-2t}\{2 \cdot u(t) - 3e^8 \cdot u(t-4)\}$

7 (a) $f(t) = \sin t$, frequency 1 radian per unit of time, period 2π units of time

   (b) $f(t) = \dfrac{18}{\sqrt{53}} e^{-t/6} \sin\left(\dfrac{\sqrt{53}}{6}\right)t$, frequency $\dfrac{\sqrt{53}}{6}$ radian per unit of time,

   period $\dfrac{12\pi}{\sqrt{53}}$ units of time

8 Transient solution $\dfrac{e^{-t}}{19}\left(32\sqrt{2}\sin\sqrt{2}t - 40\cos\sqrt{2}t\right)$, steady-state solution $\dfrac{2}{19}e^{5t}$

Further problems 8 (page 253)

2 $L\{f(t)\} = \dfrac{a(1 + e^{-\pi s})}{(s^2 + 1)(1 - e^{-\pi s})}$ **3** (a) $F(s) = \dfrac{1}{s^2} - \dfrac{w}{s}\left\{\dfrac{e^{-ws}}{1 - e^{-ws}}\right\}$

   (b) $F(s) = \dfrac{1 - e^{2(1-s)\pi}}{(s-1)(1 - e^{-2\pi s})}$ (c) $F(s) = \dfrac{1 - e^{-s}(s+1)}{s^2(1 - e^{-2s})}$

   (d) $F(s) = \dfrac{1}{1 - e^{-3s}}\left\{\dfrac{2}{s^3} - \dfrac{2e^{-2s}}{s^3} - \dfrac{4e^{-2s}}{s^2} - \dfrac{4e^{-3s}}{s}\right\}$

4 $x = \dfrac{P}{M\omega}\sin\omega t$ **5** $i = \dfrac{E}{L}\cos\left(\dfrac{t}{\sqrt{LC}}\right)$

6 $x = 2e^{-2t}\{1 + 10e^8 \cdot u(t-4)\} - 2e^{-3t}\{1 + 10e^{12} \cdot u(t-4)\}$

7 (a) $f(t) = 4\sqrt{3}\sin\dfrac{t}{2\sqrt{3}} - \cos\dfrac{t}{2\sqrt{3}}$, frequency $\dfrac{1}{2\sqrt{3}}$ radian per unit of time,

   period $4\pi\sqrt{3}$ units of time   (b) $f(t) = 2\cos 2\sqrt{3}t - \dfrac{1}{2\sqrt{3}}\sin 2\sqrt{3}t$,

   frequency $2\sqrt{3}$ radian per unit of time, period $\pi\sqrt{3}$ units of time

8 (a) $f(t) = -4.48\sin 0.69t + 1.06\cos 0.69t$

   (b) $f(t) = \dfrac{\pi}{(3/2)^{\frac{1}{4}}}\sin[(1.5)^{\frac{1}{4}}t]$

9 Transient solution $e^{-3t/8}\left(\dfrac{421}{9\sqrt{23}}\sin\dfrac{\sqrt{23}}{8}t - \dfrac{1}{9}\cos\dfrac{\sqrt{23}}{8}t\right)$, steady-state solution $\dfrac{1}{9}e^t$

Test exercise 9 (page 280)

1 $\dfrac{z}{z+1}$ provided $|z| > 1$ **2** $-2\dfrac{z^3 - 4z^2 + (2a+1)z}{(z-1)^2(z-a)}$

3 (a) $\dfrac{z(3z-4)}{(z-1)^2}$, $|z| > 1$ (b) $\dfrac{25z}{z-5}$, $|z| > 5$

4 $\{2k + 3 - 2^{k+1}\}$ **5** $\{3u_k + 4k - 2^{k+1}\}$ **6** $\dfrac{z \sin T}{z^2 - 2z \cos T + 1}$

Further problems 9 (page 280)

1 $\dfrac{z}{z+a}$ provided $|z| > |a|$ **2** (a) $\left\{\dfrac{1}{12}u_k - \dfrac{3}{4}(-3)^k + \dfrac{2}{3}(-2)^k\right\}$

(b) $\left\{\dfrac{1}{4}u_k - \dfrac{k}{2} + \dfrac{3}{4}(1/3)^k\right\}$ (c) $\left\{\dfrac{2}{3}(3^k) + \dfrac{1}{3}(-3)^k - 2k\right\}$

3 $\left\{\dfrac{1}{2}(1+i)(-i)^{k-1} + \dfrac{1}{2}(1-i)(i)^{k-1}\right\}$ **4** (a) $\left\{u_k + \dfrac{3}{2}k(-2)^k\right\}$

(b) $\left\{\dfrac{1}{9}u_k - \dfrac{5}{6}k(-2)^k + \dfrac{8}{9}(-2)^k\right\}$ **5** (a) $\dfrac{z^2}{z^2 - 1}$ (b) $\dfrac{z}{z^2 - 1}$

(c) $\dfrac{z^7 + z^5 + z^4 + 1}{z^7}$ (d) $\dfrac{z^7 + z^6 + z^5 + z + 1}{z^7}$ (e) $\dfrac{z^7 + z^6 + z^5 + z + 1}{z^{10}}$

(f) $\dfrac{z^6 + z^5 + z + 1}{z^6}$ **6** (a) $\{x_k\} = \left\{\dfrac{1}{2}((-3)^k - 2(-2)^k + (-1)^k)\right\}$ for $k \geq 1$

(b) $\{x_k\} = \left\{\dfrac{1}{2}((-3^{k+1} - (-2)^{k+2} + (-1)^{k+1})\right\}$

(c) $\{x_k\} = \{10(3^k) - 7(2^k)\}$ (d) $\{x_k\} = \{6(2^k) - 3u_k\}$

9 3   **10** $-\dfrac{2}{7}$ **13** (a) $\dfrac{z \sinh T}{z^2 - 2z \cosh T + 1}$ (b) $\dfrac{z(z - \cosh aT)}{z^2 - 2z \cosh aT + 1}$

(c) $\dfrac{ze^{-aT}(ze^{aT} - \cosh bT)}{z^2 - 2ze^{-aT} \cosh bT + e^{-2aT}}$

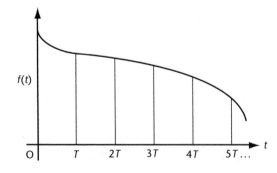

Test exercise 10 (page 345)

1 (a) solutions unique   (b) infinite number of solutions   **2** $x_1 = -4$, $x_2 = 2$, $x_3 = -3$ **3** $x_1 = -2$, $x_2 = 2$, $x_3 = 3$ **4** $x_1 = -3$, $x_2 = 4$, $x_3 = -2$

5 $x_1 = 1$, $x_2 = -2$, $x_3 = 2$ **6** $\lambda_1 = 1$, $\lambda_2 = -2$, $\lambda_3 = 3$. $x_1 = \begin{bmatrix} 1 \\ 0 \\ -1 \end{bmatrix}$;

$x_2 = \begin{bmatrix} 1 \\ -1 \\ 3 \end{bmatrix}$; $x_3 = \begin{bmatrix} 3 \\ 2 \\ -1 \end{bmatrix}$ **7** $\mathbf{M} = \begin{bmatrix} 1 & 1 \\ -2 & 1 \end{bmatrix}$; $\mathbf{M}^{-1} = \begin{bmatrix} 1/3 & -1/3 \\ 2/3 & 1/3 \end{bmatrix}$

$\mathbf{M}^{-1}\mathbf{AM} = \begin{bmatrix} -1 & 0 \\ 0 & 5 \end{bmatrix}$ **8** $f_1(x) = -\dfrac{10}{3}e^{6x} + \dfrac{1}{3}e^{3x}$; $f_2(x) = \dfrac{5}{3}e^{6x} + \dfrac{1}{3}e^{3x}$

9 $f_1(x) = \dfrac{1}{3}\cos\sqrt{5}x + \dfrac{4}{3\sqrt{5}}\sin\sqrt{5}x + \dfrac{2}{3}\cosh 2x + \dfrac{1}{3}\sinh 2x$

$f_2(x) = -\dfrac{1}{3}\cos\sqrt{5}x - \dfrac{4}{3\sqrt{5}}\sin\sqrt{5}x + \dfrac{1}{3}\cosh 2x + \dfrac{1}{6}\sinh 2x$

10 (a) $\begin{bmatrix} -8 \\ 1 \end{bmatrix}$ (b) $\begin{bmatrix} 7.196 \\ -0.464 \end{bmatrix}$

Further problems 10 (page 346)

1 $x_1 = 1$, $x_2 = -4$, $x_3 = 3$ **2** (a) $x_1 = 3$, $x_2 = 1$, $x_3 = -4$ (b) $x_1 = 4$, $x_2 = -2$, $x_3 = -1$ **3** (a) $x_1 = 4$, $x_2 = 2$, $x_3 = 5$, $x_4 = 3$ (b) $x_1 = 5$, $x_2 = -4$, $x_3 = 1$, $x_4 = 3$ (c) $x_1 = 3$, $x_2 = -2$, $x_3 = 0$, $x_4 = 5$

4 (a) $x_1 = -3$, $x_2 = 1$, $x_3 = 3$ (b) $x_1 = 5$, $x_2 = 2$, $x_3 = -1$ (c) $x_1 = 4$, $x_2 = 3$, $x_3 = -1$, $x_4 = -2$ **5** (a) $\lambda_1 = 2$, $\lambda_2 = 7$; $x_1 = \begin{bmatrix} 3 \\ -2 \end{bmatrix}$; $x_2 = \begin{bmatrix} 1 \\ 1 \end{bmatrix}$ (b) $\lambda_1 = 1$,

$\lambda_2 = -3$; $x_1 = \begin{bmatrix} 5 \\ 1 \end{bmatrix}$; $x_2 = \begin{bmatrix} 1 \\ 1 \end{bmatrix}$ (c) $\lambda_1 = -8$, $\lambda_2 = 4$; $x_1 = \begin{bmatrix} 5 \\ -2 \end{bmatrix}$; $x_2 = \begin{bmatrix} 1 \\ 2 \end{bmatrix}$

(d) $\lambda_1 = 4$, $\lambda_2 = -6$; $x_1 = \begin{bmatrix} 1 \\ 1 \end{bmatrix}$; $x_2 = \begin{bmatrix} 9 \\ -1 \end{bmatrix}$ (e) $\lambda_1 = 1$, $\lambda_2 = 3$; $\lambda_3 = 9$;

$x_1 = \begin{bmatrix} 7 \\ -1 \\ -5 \end{bmatrix}$; $x_2 = \begin{bmatrix} 7 \\ 1 \\ -7 \end{bmatrix}$; $x_3 = \begin{bmatrix} 1 \\ 1 \\ 5 \end{bmatrix}$ (f) $\lambda = 1, 2, 4$; $x = \begin{bmatrix} 0 \\ 1 \\ 6 \end{bmatrix}, \begin{bmatrix} 1 \\ 1 \\ 3 \end{bmatrix}, \begin{bmatrix} 3 \\ 1 \\ 3 \end{bmatrix}$

(g) $\lambda = -1, -3, 7$; $x = \begin{bmatrix} 6 \\ -5 \\ 2 \end{bmatrix}, \begin{bmatrix} 2 \\ -1 \\ 0 \end{bmatrix}, \begin{bmatrix} 6 \\ 27 \\ 10 \end{bmatrix}$ (h) $\lambda = -2, 4, 7$;

$x = \begin{bmatrix} 3 \\ -1 \\ 1 \end{bmatrix}, \begin{bmatrix} 3 \\ 4 \\ 5 \end{bmatrix}, \begin{bmatrix} 6 \\ 1 \\ -1 \end{bmatrix}$

6 (a) $f_1(x) = \dfrac{1}{4}(5e^x - e^{-3x})$; $f_2(x) = \dfrac{1}{4}(e^x - e^{-3x})$

(b) $f_1(x) = \dfrac{9}{5}(e^{-6x} - e^{4x})$; $f_2(x) = -\dfrac{1}{5}(e^{-6x} + 9e^{4x})$

(c) $f_1(x) = \dfrac{1}{2}(5e^{4x} - 3e^{2x})$; $f_2(x) = \dfrac{2}{3}e^x - \dfrac{3}{2}e^{2x} + \dfrac{5}{6}e^{4x}$;

$f_3(x) = 4e^x - \dfrac{9}{2}e^{2x} + \dfrac{5}{2}e^{4x}$ (d) $f_1(x) = 3e^{-2x} - e^{4x} + 2e^{7x}$;

$f_2(x) = -e^{-2x} - \dfrac{4}{3}e^{4x} + \dfrac{1}{3}e^{7x}$; $f_3(x) = e^{-2x} - \dfrac{5}{3}e^{4x} - \dfrac{1}{3}e^{7x}$ **7** $\lambda = 0, 7, 13$

8 $I_1 = 2$, $I_2 = -3$, $I_3 = 2$ **9** $k = 2$; $x_1 = -2$, $x_2 = \frac{1}{2}$, $x_3 = 1$

10 (a) $f_1(x) = \dfrac{3}{5}\cosh\sqrt{2}x + \dfrac{9}{5\sqrt{2}}\sinh\sqrt{2}x + \dfrac{2}{5}\cosh\sqrt{7}x + \dfrac{11}{5\sqrt{7}}\sinh\sqrt{7}x$;

$f_2(x) = -\dfrac{2}{5}\cosh\sqrt{2}x - \dfrac{6}{5\sqrt{2}}\sinh\sqrt{2}x + \dfrac{2}{5}\cosh\sqrt{7}x + \dfrac{11}{5\sqrt{7}}\sinh\sqrt{7}x$

(b) $f_1(x) = -\dfrac{5}{12}\cos 2\sqrt{2}x + \dfrac{5}{12\sqrt{2}}\sin 2\sqrt{2}x + \dfrac{5}{12}\cosh 2x + \dfrac{1}{12}\sinh 2x$;

$f_2(x) = \dfrac{1}{6}\cos 2\sqrt{2}x - \dfrac{1}{6\sqrt{2}}\sin 2\sqrt{2}x + \dfrac{5}{6}\cosh 2x + \dfrac{1}{6}\sinh 2x$

(c) $f_1(x) = -\dfrac{35}{16}\cosh x + \dfrac{7}{16}\sinh x + \dfrac{35}{12}\cosh\sqrt{3}x - \dfrac{7}{12\sqrt{3}}\sinh\sqrt{3}x$

$+ \dfrac{13}{48}\cosh 3x + \dfrac{7}{144}\sinh 3x$; $f_2(x) = \dfrac{5}{16}\cosh x - \dfrac{1}{16}\sinh x + \dfrac{5}{12}\cosh\sqrt{3}x$

$- \dfrac{1}{12\sqrt{3}}\sinh\sqrt{3}x + \dfrac{13}{48}\cosh 3x + \dfrac{7}{144}\sinh 3x$; $f_3(x) = \dfrac{25}{16}\cosh x - \dfrac{5}{16}\sinh x$

$- \dfrac{35}{12}\cosh\sqrt{3}x + \dfrac{7}{12\sqrt{3}}\sinh\sqrt{3}x + \dfrac{65}{48}\cosh 3x + \dfrac{35}{144}\sinh 3x$

(d) $f_1(x) = -\dfrac{9}{8}\cos x + \dfrac{9}{4}\sin x + \dfrac{19}{10}\cos\sqrt{3}x - \dfrac{12}{5\sqrt{3}}\sin\sqrt{3}x + \dfrac{9}{40}\cosh\sqrt{7}x$

$+ \dfrac{3}{20\sqrt{7}}\sinh\sqrt{7}x$; $f_2(x) = \dfrac{15}{16}\cos x - \dfrac{15}{8}\sin x - \dfrac{19}{20}\cos\sqrt{3}x + \dfrac{6}{5\sqrt{3}}\sin\sqrt{3}x$

$+ \dfrac{81}{80}\cosh\sqrt{7}x + \dfrac{27}{40\sqrt{7}}\sinh\sqrt{7}x$; $f_3(x) = -\dfrac{3}{8}\cos x + \dfrac{3}{4}\sin x + \dfrac{3}{8}\cosh\sqrt{7}x$

$+ \dfrac{1}{4\sqrt{7}}\sinh\sqrt{7}x$

Text exercise 11 (page 389)

1

x	y
0	1.0
0.1	1.1
0.2	1.211
0.3	1.3352
0.4	1.4753
0.5	1.6343

2

x	y
1	0
1.2	0.204
1.4	0.4211
1.6	0.6600
1.8	0.9264
2.0	1.2243

3

x	y
0	1.0
0.1	1.2052
0.2	1.4214
0.3	1.6499
0.4	1.8918
0.5	2.1487

4

x	y
2.0	3.0
2.1	3.005
2.2	3.0195
2.3	3.0427
2.4	3.0736
2.5	3.1117

5

x	y
1.0	0
1.1	0.1052
1.2	0.2215
1.3	0.3401
1.4	0.4717
1.5	0.6180

6

x	y
0.0	1.0000
0.1	1.0101
0.2	1.0202
0.3	1.0305
0.4	1.0408
0.5	1.0513
0.6	1.0619
0.7	1.0726
0.8	1.0834
0.9	1.0943
1.0	1.1053

Further problems 11 (page 390)

1

x	y
0	1.0
0.2	0.8
0.4	0.72
0.6	0.736
0.8	0.8288
1.0	0.9830

2

x	y
0	1.4
0.1	1.596
0.2	0.8707
0.3	2.2607
0.4	2.8318
0.5	3.7136

3

x	y
1.0	2.0
1.2	2.0333
1.4	2.1143
1.6	2.2250
1.8	2.3556
2.0	2.5000

4

x	y
0	0.5
0.1	0.543
0.2	0.5716
0.3	0.5863
0.4	0.5878
0.5	0.5768

5

x	y
0	1.0
0.1	1.1022
0.2	1.2085
0.3	1.3179
0.4	1.4296
0.5	1.5428

6

x	y
1.0	1.0
1.1	1.1871
1.2	1.3531
1.3	1.5033
1.4	1.6411
1.5	1.7688

7

x	y
0	0
0.1	0.1002
0.2	0.2015
0.3	0.3048
0.4	0.4110
0.5	0.5214

8

x	y
0	1.0
0.2	0.8562
0.4	0.8110
0.6	0.8465
0.8	0.9480
1.0	1.1037

9

x	y
0	1.0
0.1	0.9138
0.2	0.8512
0.3	0.8076
0.4	0.7798
0.5	0.7653

10

x	y
0	0.4
0.2	0.4259
0.4	0.4374
0.6	0.4319
0.8	0.4085
1.0	0.3689

11

x	y
1.0	2.0
1.2	2.4197
1.4	2.8776
1.6	3.3724
1.8	3.9027
2.0	4.4677

12

x	y
0	1.0
0.2	1.1997
0.4	1.3951
0.6	1.5778
0.8	1.7358
1.0	1.8540

13

x	y
0	1.0
0.2	1.1679
0.4	1.2902
0.6	1.3817
0.8	1.4497
1.0	1.4983

14

x	y
0	1.0
0.1	1.11
0.2	1.2422
0.3	1.4013
0.4	1.5937
0.5	1.8271

15

x	y
0	3.0
0.1	2.88
0.2	2.5224
0.3	1.9368
0.4	1.1424
0.5	0.1683

16

x	y
0	0
0.2	0.1987
0.4	0.3897
0.6	0.5665
0.8	0.7246
1.0	0.8624

17

x	y
0	1.0
0.2	1.1972
0.4	1.3771
0.6	1.5220
0.8	1.6161
1.0	1.6487

18

x	y
0	2.0
0.1	2.0845
0.2	2.1367
0.3	2.1554
0.4	2.1407
0.5	2.0943

19

x	y
0	1.0
0.2	1.0367
0.4	1.1373
0.6	1.2958
0.8	1.5145
1.0	1.8029

20

x	y
1.0	0
1.2	0.1833
1.4	0.3428
1.6	0.4875
1.8	0.6222
2.0	0.7500

21

x	y
1.0	2.0000
1.2	2.0333
1.4	2.1121
1.6	2.2219
1.8	2.3522
2.0	2.4965

22

x	y
0.0	1.0000
0.2	0.8600
0.4	0.8118
0.6	0.8452
0.8	0.9454
1.0	1.1002

23

x	y
1.0	2.0000
1.2	2.4191
1.4	2.8769
1.6	3.3715
1.8	3.9018
2.0	4.4666

Index